Finance Construction-2, *Tim Asikin, Steve Asikin, Indra Senihardja*

FINANCE

Construction-2

Corporate IFRS-GAAP (B/S-I/S) Engineering

Technologies No. 1,001-2,000 of 111,111 Laws

by

Tim Asikin, Steve Asikin,

Indra Senihardja

0
Corporate IFRS-GAAP (B/S-I/S), ISBN 13: **978-1542402415**, ISBN 10: **1542402417**

FINANCE CONSTRUCTION-2
Corporate IFRS-GAAP (B/S-I/S) Engineering
Technology No. 1,001-2,000 of 111,111 Laws
By Tim Asikin, Steve Asikin, Indra Senihardja

Creative Team:Steve Asikin
                 & BMC Research, (Non Profit), New York
An Amazon Paperback
First Published in United States 2017
By Amazon Createspace
This 1st edition published in 2017
AMZN, PO. Box 81226, Seattle
WA 98108-1226, United States

Copyright © 2017 by Steve Asikin,
gundul.ganesha@yahoo.com, indra.senihardja@yahoo.com
This book is sold subject to the condition
That it shall not, by the way of trade or otherwise,
Be lent, resold, hired out, or otherwise circulated
Without the publisher's prior consent in any form
Of binding and cover or other than that in which it
Is published and without a similar condition
Including this condition being imposed
On the subsequent purchaser.

ISBN-13: 978-1542402415
ISBN-10: 1542402417

Finance Construction-2, *Tim Asikin, Steve Asikin, Indra Senihardja*

PREFACE

This one is the only Real Finance Engineering Book, written by Engineering & Business Experts with strong Accounting Maths. It comes from a very long world wide experience, analyzing more than 100,000 Corporate Balancesheet and Income Statement reports, in conservative IFRS (International Finance Reporting Standard) and GAAP (Generally Accepted Accounting Principles) of 9,000 days works as Financing Banker, plus Business CEO expert & fresh Engineering views.

This is a manual for healthy Corporate Governance assurance dan practical business management. It is best for Government Authorities, Accounting Regulators, Business Owners & Small Public Investors (all majority & minority share holders as well).

This is not just a tool for making debt instrument unrelated to daily Corporate Finance. This only good for sincere Directors and Executive, and truly bad for the lazy and /or incapable one.

The nicest thing about it, is that only requires simple Junior Highschool Algebra. We just need to be consistent with its less number treatment of constellations. Its saddest part is that many professors and doctors are no longer able to understand Junior Highschool Algebra, and hiding in Accounting jargons. Please go back to its Introduction for glossary, at anywhere inside the book, so then we can still understand what are those mean with simple Algebraic Rule there. We try to avoid some sentences multiplied, added, eliminated divided, rooted or powered by another sentences, that make many traditional finance book successful (delivering few simple things, seen as difficult and scary), but less useful and seldom be accurate.

Finance Construction-2, *Tim Asikin, Steve Asikin, Indra Senihardja*

TABLE OF CONTENTS

Preface, p.002
Table of Contents, p.003
INTRODUCTION: The Math Fin Law Terminologies, p.004

Chapter-01: Growth of **S**= SALES, in Optimum Math Fin Law,
Chapter-02: Manageable **V**= VARIABLE COST, Optimization,
Chapter-03: **M**= MARGIN of CONTRIBUTION, Optimization,
Chapter-04: Manageable **F**= FIXED COST, Optimization,
Chapter-05: **O**= OPERATIONAL SURPLUS, Optimization,
Chapter-06: Manageable **I**= INTEREST COST, Optimization,
Chapter-07: **B**= BEFORE TAX INCOME, Optimization,
Chapter-08: Manageable **T**= TAX, Planning Optimization, p.6
Chapter-09: **A**= AFTER TAX INCOME, Optimization, p.22
Chapter-10: Manageable **D**= DIVIDEND, Optimization, p.255
Chapter-11: Manageable **Y**= YIELDING TAX, Optimization,
Chapter-12: **H**= HOME TAKEN DIVIDEND, Optimization,
Chapter-13: **R**= RETAINED EARNINGS, Optimization,
Chapter-14: **E**= EQUITY, or Capital Optimization,
Chapter-15: **L**= LIABILITIES, or Debt Optimization,
Chapter-16: **W**= WEALTH, or Total Assets Optimization,
Chapter-17: **P**= PROCURED TRADE INVENTORY, Optimization,
Chapter-18: **X**= XPRESS or Current Debt Optimization,
Chapter-19: **Q**= QUOTED LONGTERM, Debt Optimization,
Chapter-20: **G**= GOODS or Trade Payable Optimization,
Chapter-21: **Z**= ZERO TRADE or Current Debt Optimization,
Chapter-22: **J**= JOB or Trade Account Receivable Optimization,
Chapter-23: **C**= CURRENT ASSETS, Optimization,
Chapter-24: **K**=KIND of CASH, Optimization,
Chapter-25: **N**=NON CURRENT ASSETS, Optimization, p.006

Finance Construction-2, *Tim Asikin, Steve Asikin, Indra Senihardja*

INTRODUCTION:
The Math Fin Law Terminologies

Income Statement

S	V				
	M	F			
		O	I		
			B	T	
				A	D	Y
						H
						R

Balance Sheet

K	C	W	L	X	G
J					Z
P				Q	
		N		E	U
					R

    Please carefully look that all of 26 characters (**A** to **Z**) used once, except the **R**, that in Accountancy must be appeared in both diagrams at same identical value, with glossary (of well known abbreviations) in the next page. Only 2 (two) of them are historical as **U=E'** and **S'** means carried forward legal numbers from previous year.

    The 13 (thirteen) common measures (**c, d, f, g, i, j, l, p, q, s, t, v,** and **y**), are put up front as desired criteria along the way, so then later analyst could be surprised.

    In Balance Sheet, we found that **W= C+N= K+J+P+N= L+E= X+Q+U+R= G+Z+Q+U+R** as well, so then we put **W** in the middle of its T-Account balance.

    In Income Statement, same applied as **S= V+F+I +T+Y+H+R= V+M= V+F+ O= V+F+I+B= V+F+I+T+A= V+F+I+B= V+F+I+T+D+R**.

Finance Construction-2, *Tim Asikin, Steve Asikin, Indra Senihardja*

A= After Tax Income;
B= Before Tax Income,
C= Current Assets, c= C/X= Current Ratio,
D= Dividend Paid, d= D/A=Dividend Portion or Payout,
E= Equity or Capital, E'= Equity of Past Year,
F= Fixed Cost, f= F/S=Fixed Portion,
G= Goods or Trade Payables,
    g= 360G/V= Goods Payable Days,
H= Home Taken Dividend,
I= Interest Expense, i= I/S= Interest Portion,
J= Jobs or Account Receivable,
    j= 360J/S= Jobs or Trade Account ReceivableDays,
K= Kind of Cash,
L= Liabilities or Debt, l= L/E= Leverage or Gearing Ratio,
M= Margin of Contribution,
N= Non Current Assets,
O= Operational Surplus,
P= Procured Inventories,
    p= 360P/V= Procured Inventory Days,
Q= Quoted Longterm Liabilities,
    q= [C-P]/X= Quick or Acid Test Ratio,
R= Retained Earnings,
S= Sales or Revenues, S'= Sales of Past Year,
    s= [S/S']-1= Sales Growth,
T= Tax Paid, t= T/B= Tax Rate,
U= Utilized or Starting Capital,
V= Variable Cost, v= V/S= Variable Portion,
W= Wealth or Total Assets,
X= Xpress or Current Debt,
Y= Yielding Tax of Dividend,
    y= Y/D= Yielding Tax Rate of Dividend,
Z= Zero Trade Current Debt.

Finance Construction-2, *Tim Asikin, Steve Asikin, Indra Senihardja*

CHAPTER-08:
Manageable T= Tax Planning Optimization

Based on 12/31/15 Pfizer Annual Report
(http://www.nasdaq.com/symbol/pfe/financials?query=income-statement)

Continuing thw Laws No.1-1,000
at Finance Construction-1

Law-1001:
If ($= Sales or Revenues= 51,294 millions USD), (T= Tax Paid= 3,385 millions USD), ($'= Sales of Past Year= 48,851 millions USD), (f= F/S=Fixed Portion= 58.00%), (s= [S/S']-1= Sales Growth= 5.00%), (v= V/S= Variable Portion= 19.00%), and (t= T/B= Tax Rate= 30.00%), are known, then it's (i= Interest Portion planned) is:
i= {$'[1+s][1-v]}-T/t}/$-f
= {48,851[1+0.05][1-0.1900]-3,385
/0.3000}/ 51,294-0.5800
=1.00%

Law-1002:

If ($= Sales or Revenues= 51,294 millions USD), (i= I/S= Interest Portion= 1.00%), (i= I/S= Interest Portion= 1.00%), ($'= Sales of Past Year= 48,851 millions USD), (f= F/S=Fixed Portion= 58.00%), (s= [S/S']-1= Sales Growth= 5.00%), (v= V/S= Variable Portion= 19.00%), and (t= T/B= Tax Rate= 30.00%), are known, then it's (T= Tax Paid planned) is:

$$T= t\{\$'[1+s]-\$v[1+s]\}-\$f-\$i\}$$
$$= t\{\$'[1+s][1-v-i]-\$f\}$$
$$= 0.3000\{48,851[1+0.0500][1-0.1900-0.0100] -51,294*0.5800\}$$
$$=3,385 \text{ millions USD}$$

Law-1003:

If ($= Sales or Revenues= 51,294 millions USD),

(i= I/S= Interest Portion= 1.00%), (i= I/S= Interest Portion= 1.00%), (T= Tax Paid= 3,385 millions USD), (f= F/S=Fixed Portion= 58.00%), (s= [S/S']-1= Sales Growth= 5.00%), (v= V/S= Variable Portion= 19.00%), and (t= T/B= Tax Rate= 30.00%), are known, then it's ($'= Sales Past) must be:

$$\$'= \{\$f+T/t\}/\{[1+s][1-v-i]\}$$
$$=\{51,294*0.5800+3,385/0.3000\} /\{[1+0.0500][1-0.1900-0.0100]\}$$
$$=48,851 \text{ millions USD}$$

Law-1004:

If ($= Sales or Revenues= 51,294 millions USD), (i= I/S= Interest Portion= 1.00%), (i= I/S= Interest Portion= 1.00%), (T= Tax Paid= 3,385 millions USD), (f= F/S=Fixed Portion= 58.00%), ($'= Sales of Past Year= 48,851 millions USD), (v= V/S= Variable Portion= 19.00%), and (t= T/B= Tax Rate= 30.00%), are known, then it's (s= Sales Growth planned) is:

$$s = \{\$f + T/t\} / \{\$'[1-v-i]\} - 1$$
$$= \{51,294*0.5800 + 3,385/0.3000\} / \{48,851[1-0.1900-0.0100]\} - 1$$
$$= 5.00\%$$

Law-1005:

If ($= Sales or Revenues= 51,294 millions USD), (i= I/S= Interest Portion= 1.00%), (i= I/S= Interest Portion= 1.00%), (T= Tax Paid= 3,385 millions USD), (f= F/S=Fixed Portion= 58.00%), ($'= Sales of Past Year= 48,851 millions USD), (s= [S/S']-1= Sales Growth= 5.00%), and (t= T/B= Tax Rate= 30.00%), are known, then it's (v= Variable Portion planned) is:

$$v = 1 - \{\$f + T/t\} / \{\$'[1+s]\} - i$$
$$= 1 - \{51,294*0.5800 + 3,385/0.3000\} / \{48,851[1+0.0500]\} - 0.0100$$
$$= 19.00\%$$

Law-1006:

If ($= Sales or Revenues= 51,294 millions USD), (i= I/S= Interest Portion= 1.00%), (v= V/S= Variable Portion= 19.00%), (T= Tax Paid= 3,385 millions USD), (f= F/S=Fixed Portion= 58.00%), ($'= Sales of Past Year= 48,851 millions USD), (s= [S/S']-1= Sales Growth= 5.00%), and (t= T/B= Tax Rate= 30.00%), are known, then it's (i= Interest Portion planned) is:

$$i= 1-\{\$f+T/t\}/\{\$'[1+s]\}-v$$
$$=1-\{51,294*0.5800+3,385/0.3000\}/\{48,851[1+0.0500]\} -0.1900$$
$$=\underline{1.00\%}$$

Law-1007:

If (i= I/S= Interest Portion= 1.00%), (v= V/S= Variable Portion= 19.00%), (T= Tax Paid= 3,385 millions USD), (f= F/S=Fixed Portion= 58.00%), ($'= Sales of Past Year= 48,851 millions USD), (s= [S/S']-1= Sales Growth= 5.00%), and (t= T/B= Tax Rate= 30.00%), are known, then it's ($= Sales or Revenues planned) is:

$$\$= \{\$'[1+s][1-v-i]-T/t\}/f$$
$$=\{48,851[1+0.0500][1 -0.1900-0.0100] -3,385/0.3000\}/0.5800$$
$$=\underline{51,294 \text{ millions USD}}$$

Law-1008:

If (i= I/S= Interest Portion= 1.00%), (v= V/S= Variable Portion= 19.00%), (T= Tax Paid= 3,385 millions USD), (S= Sales or Revenues= 51,294 millions USD), (S'= Sales of Past Year= 48,851 millions USD), (s= [S/S']-1= Sales Growth= 5.00%), and (t= T/B= Tax Rate= 30.00%), are known, then it's (f= Fixed Portion planned) is:

$f = \{S'[1+s][1-v-i]-T/t\}/S$
$= \{48,851[1+0.0500][1-0.1900-0.0100]$
$\quad -3,385/0.3000\}/51,294$
$= 58.00\%$

Law-1009:

If (i= I/S= Interest Portion= 1.00%), (v= V/S= Variable Portion= 19.00%), (T= Tax Paid= 3,385 millions USD), (S= Sales or Revenues= 51,294 millions USD), (S'= Sales of Past Year= 48,851 millions USD), (s= [S/S']-1= Sales Growth= 5.00%), and (f= F/S=Fixed Portion= 58.00%), are known, then it's (t= Tax Rate planned) is:

$t = T/\{S'[1+s]-S'v[1+s]-Sf-S'i[1+s]\}$
$\quad = T/\{S'[1+s][1-v-i]-Sf\}$
$\quad = 3,385\{48,851/[1+0.0500][1-0.1900$
$\quad\quad -0.0100]-51,294*0.5800\}$
$\quad = 30.00\%$

Law-1010:

If (I= Interest Expense= 513 millions USD), (v= V/S= Variable Portion= 19.00%), (t= T/B= Tax Rate= 30.00%), (S= Sales or Revenues= 51,294 millions USD), (S'= Sales of Past Year= 48,851 millions USD), (s= [S/S']-1= Sales Growth= 5.00%), and (f= F/S=Fixed Portion= 58.00%), are known, then it's (T= Tax Paid planned) is:

$T = t\{S'[1+s] - S'v[1+s] - S'f[1+s] - I\}$
$= t\{S'[1+s][1-v-f] - I\}$
$= 0.3000\{48,851[1+0.0500][1-0.1900$
$\phantom{T=0.3000\{}-0.5800\,] - 513\}$
$= \underline{3,385}$ millions USD

Law-1011:

If (I= Interest Expense= 513 millions USD), (v= V/S= Variable Portion= 19.00%), (t= T/B= Tax Rate= 30.00%), (S= Sales or Revenues= 51,294 millions USD), (T= Tax Paid= 3,385 millions USD), (s= [S/S']-1= Sales Growth= 5.00%), and (f= F/S=Fixed Portion= 58.00%), are known, then it's (S'= Sales Past) must be:

$S' = \{I + T/t\} / [1+s][1-v-f]$
$= \{513 + 3,385/0.3000\} / [1+0.0500]$
$\phantom{S'=\{513}[1-0.1900-0.5800\,]$
$= \underline{48,851}$ millions USD

Law-1012:

If (I= Interest Expense= 513 millions USD), (v= V/S= Variable Portion= 19.00%), (t= T/B= Tax Rate= 30.00%), (S= Sales or Revenues= 51,294 millions USD), (T= Tax Paid= 3,385 millions USD), (S'= Sales of Past Year= 48,851 millions USD), and (f= F/S= Fixed Portion= 58.00%), are known, then it's (s= Sales Growth planned) is:

$$s = \{I+T/t\}/\{S'[1-v-f]\}-1$$
$$= \{513+3,385/0.3000\}/\{48,851[1-0.1900-0.5800]\}-1$$
$$= \underline{5.00\%}$$

Law-1013:

If (I= Interest Expense= 513 millions USD), (s= [S/S']-1= Sales Growth= 5.00%), (t= T/B= Tax Rate= 30.00%), (S= Sales or Revenues= 51,294 millions USD), (T= Tax Paid= 3,385 millions USD), (S'= Sales of Past Year= 48,851 millions USD), and (f= F/S= Fixed Portion= 58.00%), are known, then it's (v= Variable Portion planned) is:

$$v = 1-f-\{I+T/t\}/\{S'[1+s]\}$$
$$= 1-0.5800-\{513+3,385/0.3000\}/\{48,851[1+0.0500]\}$$
$$= \underline{19.00\%}$$

Law-1014:

If ($I=$ Interest Expense= 513 millions USD), ($s=$ [S/S']-1= Sales Growth= 5.00%), ($t=$ T/B= Tax Rate= 30.00%), ($\$=$ Sales or Revenues= 51,294 millions USD), ($T=$ Tax Paid= 3,385 millions USD), ($\$'=$ Sales of Past Year= 48,851 millions USD), and ($v=$ V/S= Variable Portion= 19.00%), are known, then it's ($f=$ Fixed Portion planned) is:

$f = 1 - v - \{I + T/t\}/\{\$'[1+s]\}$
 = 1 - 0.1900 - {513 + 3,385/0.3000}
   /{48,851[1+0.0500]}
 = 58.00%

Law-1015:

If ($f=$ F/S=Fixed Portion= 58.00%), ($s=$ [S/S']-1= Sales Growth= 5.00%), ($t=$ T/B= Tax Rate= 30.00%), ($\$=$ Sales or Revenues= 51,294 millions USD), ($T=$ Tax Paid= 3,385 millions USD), ($\$'=$ Sales of Past Year= 48,851 millions USD), and ($v=$ V/S= Variable Portion= 19.00%), are known, then it's ($I=$ Interest Expense planned) is:

$I = \$'[1+s][1-v-f] - T/t$
 = 48,851[1+0.0500][1- 0.19000.5800]
   -3,385/0.3000
 = 513 millions USD

Law-1016:

If (**f**= F/S=Fixed Portion= 58.00%), (**s**= [S/S']-1= Sales Growth= 5.00%),c(**I**= Interest Expense= 513 millions USD), (**$**= Sales or Revenues= 51,294 millions USD), (**T**= Tax Paid= 3,385 millions USD), (**$'**= Sales of Past Year= 48,851 millions USD), and (**v**= V/S= Variable Portion= 19.00%), are known, then it's (**t**= Tax Rate planned) is:

$$t = T/\{\$'[1+s]-\$'v[1+s]-\$'f[1+s]-I\}$$
$$= T/\{\$'[1+s][1-v-f]-I\}$$
$$= 3,385/\{48,851[1+0.0500][1-0.1900-0.5800]-513\}$$
$$= 30.00\%$$

Law-1017:

If (**f**= F/S=Fixed Portion= 58.00%), (**s**= [S/S']-1= Sales Growth= 5.00%), (**i**= I/S= Interest Portion= 1.00%), (**$**= Sales or Revenues= 51,294 millions USD), (**t**= T/B= Tax Rate= 30.00%), (**$'**= Sales of Past Year= 48,851 millions USD), and (**v**= V/S= Variable Portion= 19.00%), are known, then it's (**T**= Tax Paid planned) is:

$$T = t\{\$'[1+s]-\$'v[1+s]-\$'f[1+s]-\$i\}$$
$$= t\{\$'[1+s][1-v-f]-\$i\}$$
$$= 0.3000\{48,851[1+0.0500][1-0.1900-0.5800]-51,294*0.0100\}$$
$$= 3,385 \text{ millions USD}$$

Law-1018:

If (f= F/S=Fixed Portion= 58.00%), (s= [S/S']-1= Sales Growth= 5.00%), (i= I/S= Interest Portion= 1.00%), (S= Sales or Revenues= 51,294 millions USD), (t= T/B= Tax Rate= 30.00%), (T= Tax Paid= 3,385 millions USD), and (v= V/S= Variable Portion= 19.00%), are known, then it's (S'= Sales Past) must be:

S' = {$Si+T/t$}/{[1+s][1-v-f}-1
= {51,294*0.0100+3,385/0.3000}/{[1 +0.0500][1- 0.1900-0.5800]}
= 48,851 millions USD

Law-1019:

If (f= F/S=Fixxed Portion planned= 58.00%), (S'= Sales of Past Year= 48,851 millions USD), (i= I/S= Interest Portion= 1.00%), (S= Sales or Revenues= 51,294 millions USD), (t= T/B= Tax Rate= 30.00%), (T= Tax Paid= 3,385 millions USD), and (v= V/S= Variable Portion= 19.00%), are known, then it's (s= Sales Growth planned) is:

s= {$Si+T/t$}/{S'[1-v-f}-1
= {51,294*0.0100+3,385/0.3000} /{48,851[1- 0.1900-0.5800]}-1
= 5.00%

Law-1020:
If (f= F/S=Fixed Portion= 58.00%), ($\$'$= Sales of Past Year= 48,851 millions USD), (i= I/S= Interest Portion= 1.00%), ($\$$= Sales or Revenues= 51,294 millions USD), (t= T/B= Tax Rate= 30.00%), (T= Tax Paid= 3,385 millions USD), and are known, then it's (v= Variable Portion planned) is:
v= 1-f-{$\$i$+$T/t$}/{$\$'$[1+$s$]}
   = 1-0.5800-{51,294*0.0100+3,385
            /0.3000}/{48,851[1+0.05]}
   = 19.00%

Law-1021:
If (v= V/S= Variable Portion= 19.00%), ($\$'$= Sales of Past Year= 48,851 millions USD), (i= I/S= Interest Portion= 1.00%), ($\$$= Sales or Revenues= 51,294 millions USD), (t= T/B= Tax Rate= 30.00%), (T= Tax Paid= 3,385 millions USD), and (s= [S/S']-1= Sales Growth= 5.00%), are known, then it's (f= Fixed Portion planned) is:
f= 1-v-{$\$i$+$T/t$}/{$\$'$[1+$s$]}
   = 1-0.1900-{51,294*0.0100+3,385
            /0.3000}/{48,851[1+0.05]]}
   = 58.00%

Law-1022:

If (v= V/S= Variable Portion= 19.00%), ($\$'$= Sales of Past Year= 48,851 millions USD), (i= I/S= Interest Portion= 1.00%), (f= F/S=Fixed Portion= 58.00%), (t= T/B= Tax Rate= 30.00%), (T= Tax Paid= 3,385 millions USD), and (s= [S/S']-1= Sales Growth= 5.00%), are known, then it's ($\$$= Sales or Revenues planned) is:

$$\$ = \{\$'[1+s][1-v-f]-T/t\}/i$$
$$= \{48,851[1+0.05]][1-0.1900-0.5800] -3,385/0.3000\}/0.0100$$
$$= 51,294 \text{ millions USD}$$

Law-1023:

If (v= V/S= Variable Portion= 19.00%), ($\$'$= Sales of Past Year= 48,851 millions USD), ($\$$= Sales or Revenues= 51,294 millions USD), (f= F/S=Fixed Portion= 58.00%), (t= T/B= Tax Rate= 30.00%), (T= Tax Paid= 3,385 millions USD), and (s= [S/S']-1= Sales Growth= 5.00%), are known, then it's (i= Interest Portion planned) is:

$$i = \{\$'[1+s][1-v-f]-T/t\}/\$$$
$$= 48,851[1+0.0500][1-0.1900-0.5800] -3,385/0.3000\}/51,294$$
$$= 1.00\%$$

Law-1024:

If (v= V/S= Variable Portion= 19.00%), ($\$'$= Sales of Past Year= 48,851 millions USD), (i= I/S= Interest Portion= 1.00%), (f= F/S=Fixed Portion= 58.00%), (f= F/S=Fixed Portion= 58.00%), (T= Tax Paid= 3,385 millions USD), and (s= [S/S']-1= Sales Growth= 5.00%), are known, then it's (t= Tax Rate planned) is:

t = $T/\{\$'[1+s]-\$'v[1+s]-\$'f[1+s]-\$i\}$
　　= $T/\{\$'[1+s][1-v-f]-\$i\}$
　　= $3,385/\{48,851[1+0.05][1-0.1900$
　　　　$-0.5800]-51,294*0.0100\}$
　　= 30.00%

Law-1025:

If (v= V/S= Variable Portion= 19.00%), ($\$'$= Sales of Past Year= 48,851 millions USD), ($\$$= Sales or Revenues= 51,294 millions USD), (i= I/S= Interest Portion= 1.00%), (f= F/S=Fixed Portion= 58.00%), (t= T/B= Tax Rate= 30.00%), and (s= [S/S']-1= Sales Growth= 5.00%), are known, then it's (T= Tax Paid planned) is:

T = $t\{\$'[1+s]-\$'v[1+s]-\$'f[1+s]-\$'i[1+s]\}$
　　= $\$'t[1+s][1-v-f-i]$
　　= $48,851*0.3000[1+0.0500][1-0.1900$
　　　　$-0.5800-0.0100]\}$
　　= 3,385 millions USD

Law-1026:
If (v= V/S= Variable Portion= 19.00%), (T= Tax Paid= 3,385 millions USD), ($= Sales or Revenues= 51,294 millions USD), (i= I/S= Interest Portion= 1.00%), (f= F/S=Fixed Portion= 58.00%), (t= T/B= Tax Rate= 30.00%), and (s= [S/S']-1= Sales Growth= 5.00%), are known, then it's ($'= Sales Past) must be:

$$\$' = T/t[1+s][1-v-f-i]$$
$$= 3,385/0.3000[1+0.0500][1-0.1900-0.5800-0.0100]$$
$$= 48,851 \text{ millions USD}$$

Law-1027:
If (v= V/S= Variable Portion= 19.00%), (T= Tax Paid= 3,385 millions USD), ($= Sales or Revenues= 51,294 millions USD), (i= I/S= Interest Portion= 1.00%), (f= F/S=Fixed Portion= 58.00%), (t= T/B= Tax Rate= 30.00%), and ($'= Sales of Past Year= 48,851 millions USD), are known, then it's (s= Sales Growth planned) is:

$$s = T/\{\$'t[1-v-f-i]\}-1$$
$$= 3,385/\{48,851*0.3000[1-0.1900-0.5800-0.0100]\}-1$$
$$= 5.00\%$$

Law-1028:
If (s= [S/S']-1= Sales Growth= 5.00%), (T= Tax Paid= 3,385 millions USD), ($\$$= Sales or Revenues= 51,294 millions USD), (i= I/S= Interest Portion= 1.00%), (f= F/S=Fixed Portion= 58.00%), (t= T/B= Tax Rate= 30.00%), and ($\$'$= Sales of Past Year= 48,851 millions USD), are known, then it's (v= Variable Portion planned) is:

v= 1-f-i-T/{$\$'t$[1+$s$]}
 = 1-0.5800-0.0100-3,385/{48,851
  *0.3000[1+0.0500]}
 = 19.00%

Law-1029:
If (s= [S/S']-1= Sales Growth= 5.00%), (T= Tax Paid= 3,385 millions USD), ($\$$= Sales or Revenues= 51,294 millions USD), (i= I/S= Interest Portion= 1.00%), (v= V/S= Variable Portion= 19.00%), (t= T/B= Tax Rate= 30.00%), and ($\$'$= Sales of Past Year= 48,851 millions USD), are known, then it's (f= Fixed Portion planned) is:

f= 1-v-i-T/{$\$'t$[1+$s$]}
 = 1-0.1900-0.0100-3,385/{48,851
  *0.3000[1+0.0500]}
 = 58.00%

Law-1030:

If (s= [S/S']-1= Sales Growth= 5.00%), (**T**= Tax Paid= 3,385 millions USD), ($= Sales or Revenues= 51,294 millions USD), (**f**= F/S=Fixed Portion= 58.00%), (**v**= V/S= Variable Portion= 19.00%), (**t**= T/B= Tax Rate= 30.00%), and ($'= Sales of Past Year= 48,851 millions USD), are known, then it's (**i**= Interest Portion planned) is:

$$i = 1\text{-}f\text{-}v\text{-}T/\{\$'t[1+s]\}$$
$$= 1\text{-}0.5800\text{-}0.1900\text{-}3,385/\{48,851 *0.3000[1+0.0500]\}$$
$$= \underline{1.00\%}$$

Law-1031:

If (s= [S/S']-1= Sales Growth= 5.00%), (**T**= Tax Paid= 3,385 millions USD), ($= Sales or Revenues= 51,294 millions USD), (**f**= F/S=Fixed Portion= 58.00%), (**v**= V/S= Variable Portion= 19.00%), (**i**= I/S= Interest Portion= 1.00%), and ($'= Sales of Past Year= 48,851 millions USD), are known, then it's (**t**= Tax Rate planned) is:

$$t = T/\{\$'[1+s]\text{-}\$'v[1+s]\text{-}\$'f[1+s]\text{-}\$'i[1+s]\}$$
$$= T/\{\$'[1+s][1\text{-}v\text{-}f\text{-}i]\}$$
$$= 3,385/\{48,851[1+0.0500][1\text{-}0.1900\text{-}0.5800\text{-}0.0100]\}$$
$$= \underline{30.00\%}$$

CHAPTER-09:
Manageable A= After Tax Income Optimization

Based on 12/31/15 Pfizer Annual Report
(http://www.nasdaq.com/symbol/pfe/financials?query=income-statement)

Law-1032:
 If (**B**= Before Tax Income= 11,285 millions USD), and (**T**= Tax Paid= 3,385 millions USD), are known, then it's (**A**= After Tax Income planned) is:
 A= **B**-**T**= 11,285-3,385
 = 7,899 millions USD

Law-1033:
 If (**A**= After Tax Income= 7,899 millions USD), and (**T**= Tax Paid= 3,385 millions USD), are known, then it's (**B**= Before Tax Income planned) is:
 B= **A**+**T**= 7,899+3,385
 = 11,285 millions USD

Law-1034:
 If (**A**= After Tax Income= 7,899 millions USD), and (**B**= Before Tax Income= 11,285 millions USD), are known, then it's (**T**= Tax Paid planned) is:
 T= **B**-**A**= 11,285-7,899
 = 3,385 millions USD

Law-1035:
If (t= T/B= Tax Rate= 30.00%), and (B= Before Tax Income= 11,285 millions USD), are known, then it's (A= After Tax Income planned) is:
$$A = B - Bt = B[1-t]$$
$$= 11,285[1-0.3000]$$
$$= \underline{7,899} \text{ millions USD}$$

Law-1036:
If (t= T/B= Tax Rate= 30.00%), and (A= After Tax Income= 7,899 millions USD), are known, then it's (B= Before Tax Income planned) is:
$$B = A/[1-t] = 7,899/[1-0.3000]$$
$$= \underline{11,285} \text{ millions USD}$$

Law-1037:
If (B= Before Tax Income= 11,285 millions USD), and (A= After Tax Income= 7,899 millions USD), are known, then it's (t= Tax Rate planned) is:
$$t = 1 - A/B = 1 - [7,899/11,285]$$
$$= \underline{30.00\%}$$

Law-1038:
If (I= Interest Expense= 513 millions USD), (O= Operational Surplus= 11,798 millions USD), and (t= T/B= Tax Rate= 30.00%), are known, then it's (A= After Tax Income planned) is:
$$A = [1-t][O-I] = [1-0.3000][11,798-513]$$
$$= \underline{7,899} \text{ millions USD}$$

Law-1039:

If (**I**= Interest Expense= 513 millions USD), (**O**= Operational Surplus= 11,798 millions USD), and (**A**= After Tax Income= 7,899 millions USD), are known, then it's (**t**= Tax Rate planned) is:

$$t = 1 - A/[O-I]$$
$$= 1 - 7{,}899/[11{,}798 - 513]$$
$$= 30.00\%$$

Law-1040:

If (**I**= Interest Expense= 513 millions USD), (**A**= After Tax Income= 7,899 millions USD), and (**t**= T/B= Tax Rate= 30.00%), are known, then it's (**O**= Operational Surplus planned) is:

$$O = I + A/[1-t]$$
$$= 513 + 7{,}899/[1 - 0.3000]$$
$$= 11{,}798 \text{ millions USD}$$

Law-1041:

If (**O**= Operational Surplus= 11,798 millions USD), (**A**= After Tax Income= 7,899 millions USD), and (**t**= T/B= Tax Rate= 30.00%), are known, then it's (**I**= Interest Expense planned) is:

$$I = O - A/[1-t]$$
$$= 11{,}798 - 7{,}899/[1 - 0.3000]$$
$$= 513 \text{ millions USD}$$

Law-1042:
   If (O= Operational Surplus= 11,798 millions USD), (i= I/S= Interest Portion= 1.00%), ($\$$= Sales or Revenues= 51,294 millions USD), and (t= T/B= Tax Rate= 30.00%), are known, then it's (A= After Tax Income planned) is:
   $A = [1-t][O-\$i]$
   $= [1-0.3000][11,798-51,294*0.0100]$
   $= 7,899$ millions USD

Law-1043:
   If (O= Operational Surplus= 11,798 millions USD), (i= I/S= Interest Portion= 1.00%), ($\$$= Sales or Revenues= 51,294 millions USD), and (A= After Tax Income= 7,899 millions USD), are known, then it's (t= Tax Rate planned) is:
   $t = 1-A/[O-\$i]$
   $= 1-7,899/[11,798-51,294*0.0100]$
   $= 30.00\%$

Law-1044:
   If (t= T/B= Tax Rate= 30.00%), (i= I/S= Interest Portion= 1.00%), ($\$$= Sales or Revenues= 51,294 millions USD), and (A= After Tax Income= 7,899 millions USD), are known, then it's (O= Operational Surplus planned) is:
   $O = \$i + A/[1-t]$
   $= 51,294*0.0100 + 7,899/[1-0.3000]$
   $= 11,798$ millions USD

Law-1045:
If (t= T/B= Tax Rate= 30.00%), (i= I/S= Interest Portion= 1.00%), (O= Operational Surplus= 11,798 millions USD), and (A= After Tax Income= 7,899 millions USD), are known, then it's (S= Sales or Revenues planned) is:
$$S = \{O-A/[1-t]\}/i$$
$$= \{11,798 - 7,899/[1-0.3000]\}/0.0100$$
$$= \underline{51,294} \text{ millions USD}$$

Law-1046:
If (t= T/B= Tax Rate= 30.00%), (S= Sales or Revenues= 51,294 millions USD), (O= Operational Surplus= 11,798 millions USD), and (A= After Tax Income= 7,899 millions USD), are known, then it's (i= Interest Portion planned) is:
$$i = \{O-A/[1-t]\}/S$$
$$= \{11,798 - 7,899/[1-0.3000]\}/51,294$$
$$= \underline{1.00}\%$$

Law-1047:
If (t= T/B= Tax Rate= 30.00%), (s= [S/S']-1= Sales Growth= 5.00%), (S'= Sales of Past Year= 48,851 millions USD), (O= Operational Surplus= 11,798 millions USD), and (i= I/S= Interest Portion= 1.00%), are known, then it's (A= After Income planned) is:
$$A = [1-t]\{O-S'i[1+s]\}$$
$$= [1-0.3000]\{11,798 - 48,851$$
$$*0.0100[1+0.0500]\}$$
$$= \underline{7,899} \text{ millions USD}$$

Law-1048:

If (t = T/B = Tax Rate = 30.00%), (s = [S/S']-1 = Sales Growth = 5.00%), ($\$'$ = Sales of Past Year = 48,851 millions USD), (A = After Tax Income = 7,899 millions USD), and (i = I/S = Interest Portion = 1.00%), are known, then it's (O = Operational Surplus planned) is:

$O = \$'i[1+s] + \{A/[1-t]\}$
$= 48{,}851*0.0100[1+0.0500] + \{7{,}899 / [1-0.3000]\}$
$= \underline{11{,}798}$ millions USD

Law-1049:

If (t = T/B = Tax Rate = 30.00%), (s = [S/S']-1 = Sales Growth = 5.00%), (O = Operational Surplus = 11,798 millions USD), (A = After Tax Income = 7,899 millions USD), and (i = I/S = Interest Portion = 1.00%), are known, then it's ($\$'$ = Sales Past) is:

$\$' = \{O - A/[1-t]\} / \{i[1+s]\}$
$= \{11{,}798 - 7{,}899/[1-0.3000]\} / \{0.0100/[1+0.0500]\}$
$= \underline{48{,}851}$ millions USD

Law-1050:

If (t= T/B= Tax Rate= 30.00%), (s= [S/S']-1= Sales Growth= 5.00%), (O= Operational Surplus= 11,798 millions USD), (A= After Tax Income= 7,899 millions USD), and (S'= Sales of Past Year= 48,851 millions USD), are known, then it's (i= Interest Portion planned) is:

$i = \{O-A/[1-t]\}/\{S'[1+s]\}$
$= \{11,798 - 7,899/[1-0.3000]\}/\{48,851[1+0.0500]\}$
$= 1.00\%$

Law-1051:

If (t= T/B= Tax Rate= 30.00%), (O= Operational Surplus= 11,798 millions USD), (A= After Tax Income= 7,899 millions USD), (i= I/S= Interest Portion= 1.00%), and (S'= Sales of Past Year= 48,851 millions USD), are known, then it's (s= Sales Growth planned) is:

$s = \{O-A/[1-t]\}/[S'i]-1$
$= \{11,798 - 7,899/[1-0.3000]\}/[48,851*0.0100]-1$
$= 5.00\%$

Law-1052:

If (s= [S/S']-1= Sales Growth= 5.00%), (**O**= Operational Surplus= 11,798 millions USD), (**A**= After Tax Income= 7,899 millions USD), (**i**= I/S= Interest Portion= 1.00%), and (**S'**= Sales of Past Year= 48,851 millions USD), are known, then it's (**t**= Tax Rate planned) is:

$$t = 1-\mathbf{A}/\{\mathbf{O}-\mathbf{S'i}[1+s]\}$$
$$= 1-7,899/\{11,798-48,851*0.0100$$
$$[1+0.0500]\}$$
$$= \underline{30.00\%}$$

Law-1053:

If (**M**= Margin of Contribution= 41,548 millions USD), (**t**= T/B= Tax Rate= 30.00%), (**I**= Interest Expense= 513 millions USD), and (**F**= Fixed Cost= 29,740 millions USD), are known, then it's (**A**= After Tax Income planned) is:

$$\mathbf{A} = [1-\mathbf{t}][\mathbf{M}-\mathbf{F}-\mathbf{I}]$$
$$= [1-0.3000][41,548-29,740-513]$$
$$= \underline{7,899} \text{ millions USD}$$

Law-1054:
If (A= After Tax Income= 7,899 millions USD), (t= T/B= Tax Rate= 30.00%), (I= Interest Expense= 513 millions USD), and (F= Fixed Cost= 29,740 millions USD), are known, then it's (M= Margin of Contribution planned) is:
$$M = F + I + A/[1-t]$$
$$= 29,740 + 513 + 7,899/[1-0.3000]$$
$$= 41,548 \text{ millions USD}$$

Law-1055:
If (A= After Tax Income= 7,899 millions USD), (t= T/B= Tax Rate= 30.00%), (I= Interest Expense= 513 millions USD), and (M= Margin of Contribution= 41,548 millions USD), are known, then it's (F= Fixed Cost planned) is:
$$F = M - I - A/[1-t]$$
$$= 41,548 - 513 - 7,899/[1-0.3000]$$
$$= 29,740 \text{ millions USD}$$

Law-1056:
If (A= After Tax Income= 7,899 millions USD), (t= T/B= Tax Rate= 30.00%), (F= Fixed Cost= 29,740 millions USD), and (M= Margin of Contribution= 41,548 millions USD), are known, then it's (I= Interest Expense planned) is:
$$I = M - F - A/[1-t]$$
$$= 41,548 - 29,740 - 7,899/[1-0.3000]$$
$$= 513 \text{ millions USD}$$

Law-1057:

If (A= After Tax Income= $\underline{7,899}$ millions USD), (I= Interest Expense= $\underline{513}$ millions USD), (F= Fixed Cost= $\underline{29,740}$ millions USD), and (M= Margin of Contribution= $\underline{41,548}$ millions USD), are known, then it's (I= Interest Expense planned) is:

$t = 1 - A/[M-F-I]$
$= 1 - 7,899/[41,548 - 29,740 - 513]$
$= \underline{30.00\%}$

Law-1058:

If (t= T/B= Tax Rate= $\underline{30.00\%}$), (i= I/S= Interest Portion= $\underline{1.00\%}$), (F= Fixed Cost= $\underline{29,740}$ millions USD), (S= Sales or Revenues= $\underline{51,294}$ millions USD), and (M= Margin of Contribution= $\underline{41,548}$ millions USD), are known, then it's (A= After Tax Income planned) is:

$A = [1-t][M-F-Si]$
$= [1 - 0.3000][41,548 - 29,740 - 51,294*0.0100]$
$= \underline{7,899}$ millions USD

Law-1059:
   If (t= T/B= Tax Rate= 30.00%), (i= I/S= Interest Portion= 1.00%), (F= Fixed Cost= 29,740 millions USD), (S= Sales or Revenues= 51,294 millions USD), and (A= After Tax Income= 7,899 millions USD), are known, then it's (M= Margin of Contribution planned) is:
   $M = F + Si + A/[1-t]$
   $= 29,740 + 51,294*0.0100 + 7,899$
   $/[1-0.3000]$
   = 41,548 millions USD

Law-1060:
   If (t= T/B= Tax Rate= 30.00%), (i= I/S= Interest Portion= 1.00%), (M= Margin of Contribution= 41,548 millions USD), (S= Sales or Revenues= 51,294 millions USD), and (A= After Tax Income= 7,899 millions USD), are known, then it's (F= Fixed Cost planned) is:
   $F = M - Si - A/[1-t]$
   $= 41,548 - 51,294*0.0100 - 7,899$
   $/[1-0.3000]$
   = 29,740 millions USD

Law-1061:

If (t= T/B= Tax Rate= 30.00%), (i= I/S= Interest Portion= 1.00%), (M= Margin of Contribution= 41,548 millions USD), (F= Fixed Cost= 29,740 millions USD), and (A= After Tax Income= 7,899 millions USD), are known, then it's (S= Sales or Revenues planned) is:

$$S = \{M-F-A/[1-t]\}/i$$
$$= \{41,548-29,740 -7,899/[1-0.3000]\}/0.0100$$
$$= 51,294 \text{ millions USD}$$

Law-1062:

If (t= T/B= Tax Rate= 30.00%), (S= Sales or Revenues= 51,294 millions USD), (M= Margin of Contribution= 41,548 millions USD), (F= Fixed Cost= 29,740 millions USD), and (A= After Tax Income= 7,899 millions USD), are known, then it's (i= Interest Portion planned) is:

$$i = \{M-F-A/[1-t]\}/S$$
$$= \{41,548-29,740 -7,899/[1-0.3000]\}/51,294$$
$$= 1.00\%$$

Law-1063:
If (i= I/S= Interest Portion= 1.00%), (S= Sales or Revenues= 51,294 millions USD), (M= Margin of Contribution= 41,548 millions USD), (F= Fixed Cost= 29,740 millions USD), and (A= After Tax Income= 7,899 millions USD), are known, then it's (t= Tax Rate planned) is:

$t = 1 - A/[M-F-Si]$
$= 1 - 7,899/[41,548 - 29,740$
$\quad -51,294*0.0100]$
$= 30.00\%$

Law-1064:
If (i= I/S= Interest Portion= 1.00%), (S= Sales or Revenues= 51,294 millions USD), (M= Margin of Contribution= 41,548 millions USD), (F= Fixed Cost= 29,740 millions USD), and (t= T/B= Tax Rate= 30.00%), are known, then it's (A= After Tax Income planned) is:

$A = [1-t]\{M-F-S'i[1+s]\}$
$= [1-0.3000]\{41,548 - 29,740 - 48,851$
$\quad *0.0100[1+0.0500]$
$= 7,899$ millions USD

Law-1065:
If (**i**= I/S= Interest Portion= 1.00%), (**S'**= Sales of Past Year= 48,851 millions USD), (**A**= After Tax Income= 7,899 millions USD), (**F**= Fixed Cost= 29,740 millions USD), and (**t**= T/B= Tax Rate= 30.00%), are known, then it's (**M**= Margin of Contribution planned) is:

M= **F**+**S'i**[1+**s**]+**A**/[1-**t**]
= 29,740 + 48,851*0.0100[1+0.0500]
+7,899/[1-0.3000]
= 41,548 millions USD

Law-1066:
If (**i**= I/S= Interest Portion= 1.00%), (**S'**= Sales of Past Year= 48,851 millions USD), (**A**= After Tax Income= 7,899 millions USD), (**M**= Margin of Contribution= 41,548 millions USD), and (**t**= T/B= Tax Rate= 30.00%), are known, then it's (**F**= Fixed Cost planned) is:

F= **M**-**S'i**[1+**s**]-**A**/[1-**t**]
= 41,548-48,851*0.0100[1+0.0500]
-7,899/[1-0.3000]
= 29,740 millions USD

Law-1067:

If (**i**= I/S= Interest Portion= 1.00%), (**F**= Fixed Cost= 29,740 millions USD), (**A**= After Tax Income= 7,899 millions USD), (**s**= [S/S']-1= Sales Growth= 5.00%), (**M**= Margin of Contribution= 41,548 millions USD), and (**t**= T/B= Tax Rate= 30.00%), are known, then it's (**S'**= Sales Past) must be:

$S' = \{M\text{-}F\text{-}A/[1\text{-}t]\}/\{i[1+s]\}$
$= \{41,548\text{-}29,740 - 7,899/[1\text{-}0.3000]\}$
$\quad /\{0.0100[1+0.0500]\}$
$= 48,851$ millions USD

Law-1068:

If (**S'**= Sales of Past Year= 48,851 millions USD), (**F**= Fixed Cost= 29,740 millions USD), (**A**= After Tax Income= 7,899 millions USD), (**s**= [S/S']-1= Sales Growth= 5.00%), (**M**= Margin of Contribution= 41,548 millions USD), and (**t**= T/B= Tax Rate= 30.00%), are known, then it's (**i**= Interest Portion planned) is:

$i = \{M\text{-}F\text{-}A/[1\text{-}t]\}/\{S'[1+s]\}$
$= \{41,548\text{-}29,740\text{-} 7,899/[1\text{-}0.3000]\}$
$\quad /\{48,851[1+0.0500]\}$
$= 1.00\%$

Law-1069:

If (S'= Sales of Past Year= 48,851 millions USD), (**F**= Fixed Cost= 29,740 millions USD), (**A**= After Tax Income= 7,899 millions USD), (**i**= I/S= Interest Portion= 1.00%), (**M**= Margin of Contribution= 41,548 millions USD), and (**t**= T/B= Tax Rate= 30.00%), are known, then it's (**s**= Sales Growth planned) is:

s= {**M-F-A**/[1-**t**]}/[**S'i**]-1
 = {41,548-29,740- 7,899/[1-0.3000]}
   /[48,851*0.0100]-1
 = 5.00%

Law-1070:

If (S'= Sales of Past Year= 48,851 millions USD), (**F**= Fixed Cost= 29,740 millions USD), (**A**= After Tax Income= 7,899 millions USD), (**i**= I/S= Interest Portion= 1.00%), (**M**= Margin of Contribution= 41,548 millions USD), and (**s**= [S/S']-1= Sales Growth= 5.00%), are known, then it's (**t**= Tax Rate planned) is:

t= 1-**A**/{**M-F-S'i**[1+**s**]}
 = 1-7,899/{ 41,548-29,740-48,851
   *0.0100 [1+0.0500]}
 = 30.00%

Law-1071:

If ($= Sales or Revenues= 51,294 millions USD), (f= F/S=Fixed Portion= 58.00%), (t= T/B= Tax Rate= 30.00%), (M= Margin of Contribution= 41,548 millions USD), and (I= Interest Expense= 513 millions USD), are known, then it's (A= After Tax Income planned) is:

$$A = [1-t][M-\$f-I]$$
$$= [1-0.3000]\{[41,548-51,294 *0.5800-513]$$
$$= \underline{7,899} \text{ millions USD}$$

Law-1072:

If ($= Sales or Revenues= 51,294 millions USD), (f= F/S=Fixed Portion= 58.00%), (t= T/B= Tax Rate= 30.00%), (A= After Tax Income= 7,899 millions USD), and (I= Interest Expense= 513 millions USD), are known, then it's (M= Margin of Contribution planned) is:

$$M = \$f+I+A/[1-t]$$
$$= 51,294*0.5800+513+7,899/[1-0.3000]$$
$$= \underline{41,548} \text{ millions USD}$$

Law-1073:

If (**M**= Margin of Contribution= 41,548 millions USD), (**f**= F/S=Fixed Portion= 58.00%), (**t**= T/B= Tax Rate= 30.00%), (**A**= After Tax Income= 7,899 millions USD), and (**I**= Interest Expense= 513 millions USD), are known, then it's (**S**= Sales or Revenues planned) is:

$$S = \{M-I-A/[1-t]\}/f$$
$$= 41,548 - 513 - 7,899/[1-0.3000]\}$$
$$/0.5800$$
$$= 51,294 \text{ millions USD}$$

Law-1074:

If (**M**= Margin of Contribution= 41,548 millions USD), (**S**= Sales or Revenues= 51,294 millions USD), (**t**= T/B= Tax Rate= 30.00%), (**A**= After Tax Income= 7,899 millions USD), and (**I**= Interest Expense= 513 millions USD), are known, then it's (**f**= Fixed Portion planned) is:

$$f = \{M-I-A/[1-t]\}/S$$
$$= 41,548 - 513 - 7,899/[1-0.3000]\}$$
$$/51,294$$
$$= 58.00\%$$

Law-1075:
 If (**M**= Margin of Contribution= 41,548 millions USD), (**S**= Sales or Revenues= 51,294 millions USD), (**t**= T/B= Tax Rate= 30.00%), (**A**= After Tax Income= 7,899 millions USD), and (**f**= F/S=Fixed Portion= 58.00%), are known, then it's (**I**= Interest Expense planned) is:
 $I = M-Sf-A/[1-t]$
 $= 41,548 - 51,294*0.5800 - 7,899/[1-0.3000]\}$
 $= 513$ millions USD

Law-1076:
 If (**M**= Margin of Contribution= 41,548 millions USD), (**S**= Sales or Revenues= 51,294 millions USD), (**I**= Interest Expense= 513 millions USD), (**A**= After Tax Income= 7,899 millions USD), and (**f**= F/S= Fixed Portion= 58.00%), are known, then it's (**t**= Tax Rate planned) is:
 $t = 1-A/[M-Sf-I]$
 $= 1 - 7,899/[41,548 - 51,294*0.5800 - 513]$
 $= 30.00\%$

Law-1077:

If (**M**= Margin of Contribution= 41,548 millions USD), (**S**= Sales or Revenues= 51,294 millions USD), (**i**= I/S= Interest Portion= 1.00%), (**t**= T/B= Tax Rate= 30.00%), and (**f**= F/S=Fixed Portion= 58.00%), are known, then it's (**A**= After Tax Income planned) is:

$$A = [1-t][M-Sf-Si] = [1-t]\{M-S[f+i]\}$$
$$= [1-0.3000]\{41,548 - 51,294 [0.5800+0.0100]\}$$
$$= 7,899 \text{ millions USD}$$

Law-1078:

If (**A**= After Tax Income= 7,899 millions USD), (**S**= Sales or Revenues= 51,294 millions USD), (**i**= I/S= Interest Portion= 1.00%), (**t**= T/B= Tax Rate= 30.00%), and (**f**= F/S=Fixed Portion= 58.00%), are known, then it's (**M**= Margin of Contribution planned) is:

$$M = S[f+i]] + A/[1-t]$$
$$= 51,294[0.5800+0.0100] + 7,899 /[1-0.3000]$$
$$= 41,548 \text{ millions USD}$$

Law-1079:

If (**A**= After Tax Income= 7,899 millions USD), (**M**= Margin of Contribution= 41,548 millions USD), (**i**= I/S= Interest Portion= 1.00%), (**t**= T/B= Tax Rate= 30.00%), and (**f**= F/S=Fixed Portion= 58.00%), are known, then it's (**$**= Sales or Revenues planned) is:

$$\$ = \{M-A/[1-t]\}/[f+i]$$
$$= \{41,548-7,899/[1-0.3000]\}$$
$$/[0.5800+0.0100]$$
$$= 51,294 \text{ millions USD}$$

Law-1080:

If (**A**= After Tax Income= 7,899 millions USD), (**M**= Margin of Contribution= 41,548 millions USD), (**i**= I/S= Interest Portion= 1.00%), (**t**= T/B= Tax Rate= 30.00%), and (**$**= Sales or Revenues= 51,294 millions USD), are known, then it's (**f**= Fix Rate planned) is:

$$f = \{M-A/[1-t]\}/\$-i$$
$$= \{41,548-7,899/[1-0.3000]\}$$
$$/51,294 - 0.0100$$
$$= 58.00\%$$

Finance Construction-2, *Tim Asikin, Steve Asikin, Indra Senihardja*

Law-1081:
If (**A**= After Tax Income= 7,899 millions USD), (**M**= Margin of Contribution= 41,548 millions USD), (**f**= F/S=Fixed Portion= 58.00%), (**t**= T/B= Tax Rate= 30.00%), and (**$**= Sales or Revenues= 51,294 millions USD), are known, then it's (**i**= Interest Portion planned) is:

$$i = \{M-A/[1-t]\}/\$-f$$
$$= \{41{,}548 - 7{,}899/[1 - 0.3000]\}$$
$$/51{,}294 - 0.5800$$
$$= \underline{1.00\%}$$

Law-1082:
If (**A**= After Tax Income= 7,899 millions USD), (**M**= Margin of Contribution= 41,548 millions USD), (**f**= F/S=Fixed Portion= 58.00%), (**i**= I/S= Interest Portion= 1.00%), and (**$**= Sales or Revenues= 51,294 millions USD), are known, then it's (**t**= Tax Rate planned) is:

$$t = 1 - A/\{M - \$[f+i]\}$$
$$= 1 - 7{,}899/\{41{,}548 - 51{,}294$$
$$[0.5800 - 0.0100]\}$$
$$= \underline{30.00\%}$$

Law-1083:

If (t= T/B= Tax Rate= 30.00%), (M= Margin of Contribution= 41,548 millions USD), (f= F/S=Fixed Portion= 58.00%), (i= I/S= Interest Portion= 1.00%), (S'= Sales of Past Year= 48,851 millions USD), (s= [S/S']-1= Sales Growth= 5.00%), and (S= Sales or Revenues= 51,294 millions USD), are known, then it's (A= After Tax Income planned) is:

$$A = [1-t]\{M-Sf-S'i[1+s]\}$$
$$= [1-0.3000]\{41,548 - 51,294 * 0.5800 - 48,851 * 0.0100[1+0.0500]\}$$
$$= \underline{7,899} \text{ millions USD}$$

Law-1084:

If (t= T/B= Tax Rate= 30.00%), (A= After Tax Income= 7,899 millions USD), (f= F/S=Fixed Portion= 58.00%), (i= I/S= Interest Portion= 1.00%), (S'= Sales of Past Year= 48,851 millions USD), (s= [S/S']-1= Sales Growth= 5.00%), and (S= Sales or Revenues= 51,294 millions USD), are known, then it's (M= Margin of Contribution planned) is:

$$M = Sf + S'i[1+s] + A/[1-t]$$
$$= 51,294 * 0.5800 + 48,851 * 0.0100[1+0.0500] + 7,899/[1-0.3000]$$
$$= \underline{41,548} \text{ millions USD}$$

Law-1085:

If (**t**= T/B= Tax Rate= 30.00%), (**A**= After Tax Income= 7,899 millions USD), (**f**= F/S=Fixed Portion= 58.00%), (**i**= I/S= Interest Portion= 1.00%), (**S'**= Sales of Past Year= 48,851 millions USD), (**s**= [S/S']-1= Sales Growth= 5.00%), and (**M**= Margin of Contribution= 41,548 millions USD), are known, then it's (**S**= Sales or Revenues planned) is:

$S = \{M - S'i[1+s] - A/[1-t]\}/f$
$= \{41,548 - 48,851 * 0.0100[1+0.0500]$
$\quad -7,899/[1-0.3000]/0.5800$
$= \underline{51,294}$ millions USD

Law-1086:

If (**t**= T/B= Tax Rate= 30.00%), (**A**= After Tax Income= 7,899 millions USD), (**S**= Sales or Revenues= 51,294 millions USD), (**i**= I/S= Interest Portion= 1.00%), (**S'**= Sales of Past Year= 48,851 millions USD), (**s**= [S/S']-1= Sales Growth= 5.00%), and (**M**= Margin of Contribution= 41,548 millions USD), are known, then it's (**f**= Fixed Portion planned) is:

$f = \{M - S'i[1+s] - A/[1-t]\}/S$
$= \{41,548 - 48,851 * 0.0100[1+0.0500]$
$\quad -7,899/[1-0.3000]\}/51,294$
$= \underline{58.00\%}$

Law-1087:
If (t= T/B= Tax Rate= 30.00%), (A= After Tax Income= 7,899 millions USD), (S= Sales or Revenues= 51,294 millions USD), (i= I/S= Interest Portion= 1.00%), (f= F/S=Fixed Portion= 58.00%), (s= [S/S']-1= Sales Growth= 5.00%), and (M= Margin of Contribution= 41,548 millions USD), are known, then it's (S'= Sales Past) must be:

S' = {M-Sf-A/[1-t]}/{i[1+s]}
   = {41,548 - 51,294 * 0.5800
      -7,899/ [1-0.3000]}
      /{0.0100[1+0.0500]}
   = 48,851 millions USD

Law-1088:
If (t= T/B= Tax Rate= 30.00%), (A= After Tax Income= 7,899 millions USD), (S= Sales or Revenues= 51,294 millions USD), (S'= Sales of Past Year= 48,851 millions USD), (f= F/S=Fixed Portion= 58.00%), (s= [S/S']-1= Sales Growth= 5.00%), and (M= Margin of Contribution= 41,548 millions USD), are known, then it's (i= Interest Portion planned) is:

i = {M-Sf-A/[1-t]}/{S'[1+s]}
   = {41,548 -51,294 * 0.5800 -7,899
      / [1-0.3000]}
      /{48,851[1+0.0500]}
   = 1.00%

Law-1089:

If (t= T/B= Tax Rate= 30.00%), (A= After Tax Income= 7,899 millions USD), (S= Sales or Revenues= 51,294 millions USD), (S'= Sales of Past Year= 48,851 millions USD), (f= F/S=Fixed Portion= 58.00%), (i= I/S= Interest Portion= 1.00%), and (M= Margin of Contribution= 41,548 millions USD), are known, then it's (s= Sales Growth planned) is:

s = {M-$S f$-A/[1-t]}/[$S'i$]-1
 = {41,548 - 51,294 * 0.5800 -7,899
   / [1-0.3000]}/[48,851*0.0100]-1
 = 5.00%

Law-1090:

If (s= [S/S']-1= Sales Growth= 5.00%), (A= After Tax Income= 7,899 millions USD), (S= Sales or Revenues= 51,294 millions USD), (S'= Sales of Past Year= 48,851 millions USD), (f= F/S=Fixed Portion= 58.00%), (i= I/S= Interest Portion= 1.00%), and (M= Margin of Contribution= 41,548 millions USD), are known, then it's (t= Tax Rate planned) is:

t = 1-A/{M-Sf-$S'i$[1+s]}
 = 1-7,899/{41,548 -51,294 * 0.5800
   - 48,851*0.0100[1+0.0500]}
 = 30.00%

Law-1091:
If (s= [S/S']-1= Sales Growth= 5.00%), (t= T/B= Tax Rate= 30.00%), (S= Sales or Revenues= 51,294 millions USD), (S'= Sales of Past Year= 48,851 millions USD), (f= F/S=Fixed Portion= 58.00%), (i= I/S= Interest Portion= 1.00%), and (M= Margin of Contribution= 41,548 millions USD), are known, then it's (A= After Tax Income planned) is:
A= [1-t]/{M-I-$S'f$[1+s]}
 = [1-0.3000] {41,548 -513- 48,851
  *0.5800[1+0.0500]}
 = 7,899 millions USD

Law-1092:
If (s= [S/S']-1= Sales Growth= 5.00%), (t= T/B= Tax Rate= 30.00%), (S= Sales or Revenues= 51,294 millions USD), (S'= Sales of Past Year= 48,851 millions USD), (f= F/S=Fixed Portion= 58.00%), (i= I/S= Interest Portion= 1.00%), and (A= After Tax Income= 7,899 millions USD), are known, then it's (M= Margin of Contribution planned) is:
M= {$S'f$[1+s]+I+A/[1-t]}
 = {48,851 * 0.5800[1+0.0500]+513
  +7,899 /[1-0.3000]}
 = 41,548 millions USD

Law-1093:
If (s= [S/S']-1= Sales Growth= 5.00%), (t= T/B= Tax Rate= 30.00%), (S= Sales or Revenues= 51,294 millions USD), (M= Margin of Contribution= 41,548 millions USD), (f= F/S=Fixed Portion= 58.00%), (I= Interest Expense= 513 millions USD), and (A= After Tax Income= 7,899 millions USD), are known, then it's (S'= Sales Past) must be:

$$S' = \{M-I-A/[1-t]\}/\{f[1+s]\}$$
$$= \{41,548-513-7,899/[1-0.3000]\}/\{0.5800[1+0.0500]\}$$
$$= 48,851 \text{ millions USD}$$

Law-1094:
If (s= [S/S']-1= Sales Growth= 5.00%), (t= T/B= Tax Rate= 30.00%), (S= Sales or Revenues= 51,294 millions USD), (M= Margin of Contribution= 41,548 millions USD), (S'= Sales of Past Year= 48,851 millions USD), (I= Interest Expense= 513 millions USD), and (A= After Tax Income= 7,899 millions USD), are known, then it's (f= Fixed Portion planned) is:

$$f = \{M-I-A/[1-t]\}/\{S'[1+s]\}$$
$$= \{41,548-513-7,899/[1-0.3000]\}/\{48,851[1+0.0500]\}$$
$$= 58.00\%$$

Law-1095:
   If (**f**= F/S=Fixed Portion= 58.00%), (**t**= T/B= Tax Rate= 30.00%), (**$**= Sales or Revenues= 51,294 millions USD), (**M**= Margin of Contribution= 41,548 millions USD), (**$'**= Sales of Past Year= 48,851 millions USD), (**I**= Interest Expense= 513 millions USD), and (**A**= After Tax Income= 7,899 millions USD), are known, then it's (**s**= Sales Growth planned) is:

$$s = \{M-I-A/[1-t]\}/[\$'f]-1$$
$$= \{41{,}548-513-7{,}899/[1-0.3000]\}/[48{,}851*0.5800]-1$$
$$= 5.00\%$$

Law-1096:
   If (**f**= F/S=Fixed Portion= 58.00%), (**t**= T/B= Tax Rate= 30.00%), (**$**= Sales or Revenues= 51,294 millions USD), (**M**= Margin of Contribution= 41,548 millions USD), (**$'**= Sales of Past Year= 48,851 millions USD), (**s**= [S/S']-1= Sales Growth= 5.00%), and (**A**= After Tax Income= 7,899 millions USD), are known, then it's (**I**= Interest Expense planned) is:

$$I = M-\$'f[1+s]-A/[1-t]$$
$$= \{41{,}548-48{,}851*0.5800[1+0.0500]-7{,}899/[1-0.3000]$$
$$= 513 \text{ millions USD}$$

Law-1097:

If (f= F/S=Fixed Portion= 58.00%), (I= Interest Expense= 513 millions USD), (S= Sales or Revenues= 51,294 millions USD), (M= Margin of Contribution= 41,548 millions USD), (S'= Sales of Past Year= 48,851 millions USD), (s= [S/S']-1= Sales Growth= 5.00%), and (A= After Tax Income= 7,899 millions USD), are known, then it's (t= Tax Rate planned) is:

t=1-A/{M-$S'f$[1+s]−I}
 = 1-7,899/{41,548-48,851*0.5800
  [1+0.0500]-513}
 = 30.00%

Law-1098:

If (f= F/S=Fixed Portion= 58.00%), (i= I/S= Interest Portion= 1.00%), (S= Sales or Revenues= 51,294 millions USD), (M= Margin of Contribution= 41,548 millions USD), (S'= Sales of Past Year= 48,851 millions USD), (s= [S/S']-1= Sales Growth= 5.00%), and (t= T/B= Tax Rate= 30.00%), are known, then it's (A= After Tax Income planned) is:

A= [1-t]{M-$S'f$[1+s]-Si}
 = [1-0.3000][41,548-48,851*0.5800
  [1+0.0500]-51,294*0.0100}
 = 7,899 millions USD

Law-1099:

If (f= F/S=Fixed Portion= 58.00%), (i= I/S= Interest Portion= 1.00%), ($\$$= Sales or Revenues= 51,294 millions USD), (A= After Tax Income= 7,899 millions USD), ($\$'$= Sales of Past Year= 48,851 millions USD), (s= [S/S']-1= Sales Growth= 5.00%), and (t= T/B= Tax Rate= 30.00%), are known, then it's (M= Margin of Contribution planned) is:

$M = \$'f[1+s] + \$i + A/[1-t]$
$= 48,851*0.5800[1+0.0500] + 51,294$
$\qquad *0.0100 + 7,899/[1-0.3000]$
$= \underline{41,548}$ millions USD

Law-1100:

If (f= F/S=Fixed Portion= 58.00%), (i= I/S= Interest Portion= 1.00%), ($\$$= Sales or Revenues= 51,294 millions USD), (A= After Tax Income= 7,899 millions USD), (M= Margin of Contribution= 41,548 millions USD), (s= [S/S']-1= Sales Growth= 5.00%), and (t= T/B= Tax Rate= 30.00%), are known, then it's ($\$'$= Sales Past) must be:

$\$' = \{M - \$i - A/[1-t]\} / \{f[1+s]\}$
$= \{41,548 - 51,294*0.0100 - 7,899$
$\qquad /[1-0.3000]\}$
$\qquad /\{0.5800[1+0.0500]\}$
$= \underline{48,851}$ millions USD

Law-1101:

If ($'= Sales of Past Year= 48,851 millions USD), (i= I/S= Interest Portion= 1.00%), ($= Sales or Revenues= 51,294 millions USD), (A= After Tax Income= 7,899 millions USD), (M= Margin of Contribution= 41,548 millions USD), (s= [S/S']-1= Sales Growth= 5.00%), and (t= T/B= Tax Rate= 30.00%), are known, then it's (f= Fixed Portion planned) is:

$$f = \{M-Si-A/[1-t]\}/\{\$'[1+s]\}$$
$$= \{41,548-51,294*0.0100-7,899$$
$$/[1-0.3000]\}$$
$$/\{48,851[1+0.0500]\}$$
$$= 58.00\%$$

Law-1102:

If ($'= Sales of Past Year= 48,851 millions USD), (i= I/S= Interest Portion= 1.00%), ($= Sales or Revenues= 51,294 millions USD), (A= After Tax Income= 7,899 millions USD), (M= Margin of Contribution= 41,548 millions USD), (f= F/S=Fixed Portion= 58.00%), and (t= T/B= Tax Rate= 30.00%), are known, then it's (s= Sales Growth planned) is:

$$s = \{M-Si-A/[1-t]\}/[\$'f]-1$$
$$= \{41,548-51,294*0.0100-7,899$$
$$/[1-0.3000]\}/[48,851*0.5800]-1$$
$$= 5.00\%$$

Law-1103:
If ($'$ = Sales of Past Year= 48,851 millions USD), (i= I/S= Interest Portion= 1.00%), (s= [S/S']-1= Sales Growth= 5.00%), (A= After Tax Income= 7,899 millions USD), (M= Margin of Contribution= 41,548 millions USD), (f= F/S=Fixed Portion= 58.00%), and (t= T/B= Tax Rate= 30.00%), are known, then it's ($\$$= Sales or Revenues planned) is:

$\$ = \{M-\$'f[1+s]-A/[1-t]\}/i$
$= \{41,548-48,851*0.5800[1+0.0500]$
$\qquad -7,899/[1-0.3000]\}/0.0100$
$= 51,294$ millions USD

Law-1104:
If ($'$ = Sales of Past Year= 48,851 millions USD), ($\$$= Sales or Revenues= 51,294 millions USD), (s= [S/S']-1= Sales Growth= 5.00%), (A= After Tax Income= 7,899 millions USD), (M= Margin of Contribution= 41,548 millions USD), (f= F/S=Fixed Portion= 58.00%), and (t= T/B= Tax Rate= 30.00%), are known, then it's (i= Interest Portion planned) is:

$i = \{M-\$'f[1+s]-A/[1-t]\}/\$$
$= \{41,548-48,851*0.5800[1+0.0500]$
$\qquad -7,899/[1-0.3000]\}/51,294$
$= 1.00\%$

Law-1105:

If ($\$'$ = Sales of Past Year= 48,851 millions USD), (s= [S/S']-1= Sales Growth= 5.00%), (A= After Tax Income= 7,899 millions USD), (M= Margin of Contribution= 41,548 millions USD), (f= F/S=Fixed Portion= 58.00%), and (i= I/S= Interest Portion= 1.00%), are known, then it's (t= Tax Rate planned) is:

t= 1-A/{M-$\$'f$[1+$s$]-$\$i$}
 = 1-7,899/{41,548-48,851*0.5800
   [1+0.0500]- 51,294*0.0100}
 = 30.00%

Law-1106:

If ($\$'$ = Sales of Past Year= 48,851 millions USD), (s= [S/S']-1= Sales Growth= 5.00%), (t= T/B= Tax Rate= 30.00%), (M= Margin of Contribution= 41,548 millions USD), (f= F/S=Fixed Portion= 58.00%), and (i= I/S= Interest Portion= 1.00%), are known, then it's (A= After Tax Income planned) is:

A= [1-t]{M-$\$'f$[1+$s$]-$\$'i$[1+$s$]}
 = [1-t]{M-$\$'$[1+$s$][$f$+$i$]}
 = [1-0.3000]{41,548-48,851[1+0.0500]
   [0.5800+0.0100]}
 = 7,899 millions USD

Law-1107:

If (S'= Sales of Past Year= 48,851 millions USD), (s= [S/S']-1= Sales Growth= 5.00%), (t= T/B= Tax Rate= 30.00%), (A= After Tax Income= 7,899 millions USD), (f= F/S=Fixed Portion= 58.00%), and (i= I/S= Interest Portion= 1.00%), are known, then it's (M= Margin of Contribution planned) is:

$M = S'[1+s][f+i] + A/[1-t]$
$= 48,851\ [1+0.0500][0.5800+0.0100]$
$\quad +7,899/[1-0.3000]$
$= 41,548$ millions USD

Law-1108:

If (M= Margin of Contribution= 41,548 millions USD), (s= [S/S']-1= Sales Growth= 5.00%), (t= T/B= Tax Rate= 30.00%), (A= After Tax Income= 7,899 millions USD), (f= F/S=Fixed Portion= 58.00%), and (i= I/S= Interest Portion= 1.00%), are known, then it's (S'= Sales Past) must be:

$S' = \{M - A/[1-t]\}/\{[1+s][f+i]\}$
$= \{41584 - 7,899/[1-0.3000]\}/\{[1+0.0500]$
$\quad [0.5800+0.0100]\}$
$= 48,851$ millions USD

Law-1109:

If (**M**= Margin of Contribution= 41,548 millions USD), (**S'**= Sales of Past Year= 48,851 millions USD), (**t**= T/B= Tax Rate= 30.00%), (**A**= After Tax Income= 7,899 millions USD), (**f**= F/S=Fixed Portion= 58.00%), and (**i**= I/S= Interest Portion= 1.00%), are known, then it's (**s**= Sales Growth planned) is:

$$s = \{M-A/[1-t]\}/\{S'[f+i]\}-1$$
$$= \{41584-7{,}899/[1-0.3000]\}/\{48{,}851 [0.5800+0.0100]\}-1$$
$$= 5.00\%$$

Law-1110:

If (**M**= Margin of Contribution= 41,548 millions USD), (**S'**= Sales of Past Year= 48,851 millions USD), (**t**= T/B= Tax Rate= 30.00%), (**A**= After Tax Income= 7,899 millions USD), (**s**= [S/S']-1= Sales Growth= 5.00%), and (**i**= I/S= Interest Portion= 1.00%), are known, then it's (**f**= Fixed Portion planned) is:

$$f = \{M-A/[1-t]\}/\{S'[1+s]\}-i$$
$$= \{41548-7{,}899/[1-0.3000]\}/\{48{,}851 [1+0.0500]\}-0.0100$$
$$= 58.00\%$$

Law-1111:

If (**M**= Margin of Contribution= 41,548 millions USD), (**S'**= Sales of Past Year= 48,851 millions USD), (**t**= T/B= Tax Rate= 30.00%), (**A**= After Tax Income= 7,899 millions USD), (**s**= [S/S']-1= Sales Growth= 5.00%), and (**f**= F/S=Fixed Portion= 58.00%), are known, then it's (**i**= Interest Portion planned) is:

$$\begin{aligned}
\mathbf{i} &= \{\mathbf{M}\text{-}\mathbf{A}/[1\text{-}\mathbf{t}]\}/\{\mathbf{S'}[1\text{+}\mathbf{s}]\}\text{-}\mathbf{f} \\
&= \{41548\text{-}7{,}899/[1\text{-}0.3000]\}/\{48{,}851 \\
&\qquad [1\text{+}0.0500]\}\text{-}0.5800 \\
&= 1.00\%
\end{aligned}$$

Law-1112:

If (**M**= Margin of Contribution= 41,548 millions USD), (**S'**= Sales of Past Year= 48,851 millions USD), (**i**= I/S= Interest Portion= 1.00%), (**A**= After Tax Income= 7,899 millions USD), (**s**= [S/S']-1= Sales Growth= 5.00%), and (**f**= F/S=Fixed Portion= 58.00%), are known, then it's (**t**= Tax Rate planned) is:

$$\begin{aligned}
\mathbf{t} &= 1\text{-}\mathbf{A}/\{\mathbf{M}\text{-}\mathbf{S'}\mathbf{f}[1\text{+}\mathbf{s}]\text{-}\mathbf{S'}\mathbf{i}[1\text{+}\mathbf{s}]\} \\
&= 1\text{-}\mathbf{A}/\{\mathbf{M}\text{-}\mathbf{S'}[1\text{+}\mathbf{s}][\mathbf{f}\text{+}\mathbf{i}]\} \\
&= 1\text{-}7{,}899/\{41{,}548\text{-}48{,}851[\,1\text{+}0.0500] \\
&\qquad [0.5800\text{+}0.0100]\} \\
&= 30.00\%
\end{aligned}$$

Law-1113:

If (S= Sales or Revenues= 51,294 millions USD), (V= Variable Cost= 9,746 millions USD), (I= Interest Expense= 513 millions USD), (t= T/B= Tax Rate= 30.00%), and (F= Fixed Cost= 29,750 millions USD), are known, then it's (A= After Tax Income planned) is:

$$A = [1-t][S-V-F-I]$$
$$= [1-0.3000][51,294-9,746-29,750-513]$$
$$= \underline{7,899} \text{ millions USD}$$

Law-1114:

If (A= After Tax Income= 7,899 millions USD), (V= Variable Cost= 9,746 millions USD), (I= Interest Expense= 513 millions USD), (t= T/B= Tax Rate= 30.00%), and (F= Fixed Cost= 29,750 millions USD), are known, then it's (S= Sales or Revenues planned) is:

$$S = V+F+I+A/[1-t]$$
$$= 9,746+29,750+513+7,899/[1-0.3000]$$
$$= \underline{51,294} \text{ millions USD}$$

Law-1115:

If (**A**= After Tax Income= 7,899 millions USD), (**$**= Sales or Revenues= 51,294 millions USD), (**I**= Interest Expense= 513 millions USD), (**t**= T/B= Tax Rate= 30.00%), and (**F**= Fixed Cost= 29,750 millions USD), are known, then it's (**V**= Variable Cost planned) is:

$$V = \$ - F - I - A/[1-t]$$
$$= 51,294 - 29,750 - 513 - 7,899/[1 - 0.3000]$$
$$= 9,746 \text{ millions USD}$$

Law-1116:

If (**A**= After Tax Income= 7,899 millions USD), (**$**= Sales or Revenues= 51,294 millions USD), (**I**= Interest Expense= 513 millions USD), (**t**= T/B= Tax Rate= 30.00%), and (**V**= Variable Cost= 9,746 millions USD), are known, then it's (**F**= Fixed Cost planned) is:

$$F = \$ - V - I - A/[1-t]$$
$$= 51,294 - 9,746 - 513 - 7,899/[1 - 0.3000]$$
$$= 29,750 \text{ millions USD}$$

Law-1117:
If (A= After Tax Income= 7,899 millions USD), (S= Sales or Revenues= 51,294 millions USD), (F= Fixed Cost= 29,750 millions USD), (t= T/B= Tax Rate= 30.00%), and (V= Variable Cost= 9,746 millions USD), are known, then it's (I= Interest Expense planned) is:

$I = S-V-F-A/[1-t]$
 = 51,294-9,746-29,750-7,899/[1-0.3000]
 = 513 millions USD

Law-1118:
If (A= After Tax Income= 7,899 millions USD), (S= Sales or Revenues= 51,294 millions USD), (F= Fixed Cost= 29,750 millions USD), (I= Interest Expense= 513 millions USD), and (V= Variable Cost= 9,746 millions USD), are known, then it's (t= Tax Rate planned) is:

$t = 1-A/[S-V-F-I]$
 = 1-7,899/[51,294-9,746-29,750-513]
 = 30.00%

Law-1119:
If (t= T/B= Tax Rate= 30.00%), (S= Sales or Revenues= 51,294 millions USD), (F= Fixed Cost= 29,750 millions USD), (i= I/S= Interest Portion= 1.00%), and (V= Variable Cost= 9,746 millions USD), are known, then it's (A= After Tax Income planned) is:
A= [1-t][S-V-F-Si]= [1-t]{S[1-i]-V-F}
  = [1-0.3000] [51,294[1-0.0100]
      -9,746-29,750]
  = 7,899 millions USD

Law-1120:
If (t= T/B= Tax Rate= 30.00%), (A= After Tax Income= 7,899 millions USD), (F= Fixed Cost= 29,750 millions USD), (i= I/S= Interest Portion= 1.00%), and (V= Variable Cost= 9,746 millions USD), are known, then it's (S= Sales or Revenues planned) is:
S= {V+F+A/[1-t]}/[1-i]
  = {9,746+29,750+7,899/[1-0.3000]}
      /[1-0.0100]
  = 51,294 millions USD

Law-1121:

If (t= T/B= Tax Rate= 30.00%), (A= After Tax Income= 7,899 millions USD), (F= Fixed Cost= 29,750 millions USD), (S= Sales or Revenues= 51,294 millions USD), and (V= Variable Cost= 9,746 millions USD), are known, then it's (i= Interest Portion planned) is:

$i = 1 - \{V+F+A/[1-t]\}/S$
$= 1 - \{9,746+29,750+7,899/[1-0.3000]\}/51,294$
$= \underline{1.00\%}$

Law-1122:

If (t= T/B= Tax Rate= 30.00%), (A= After Tax Income= 7,899 millions USD), (F= Fixed Cost= 29,750 millions USD), (S= Sales or Revenues= 51,294 millions USD), and (i= I/S= Interest Portion= 1.00%), are known, then it's (V= Variable Cost planned) is:

$V = S[1-i] - F - A/[1-t]$
$= 51,294[1-0.0100] - 29,750 - 7,899/[1-0.3000]$
$= \underline{9,746}$ millions USD

Law-1123:
If (t= T/B= Tax Rate= 30.00%), (A= After Tax Income= 7,899 millions USD), (V= Variable Cost= 9,746 millions USD), (S= Sales or Revenues= 51,294 millions USD), and (i= I/S= Interest Portion= 1.00%), are known, then it's (F= Fixed Cost planned) is:
$$F = S[1-i] - V - A/[1-t]$$
$$= 51,294[1-0.0100] - 9,746 - 7,899 / [1-0.3000]$$
$$= 29,750 \text{ millions USD}$$

Law-1124:
If (F= Fixed Cost= 29,750 millions USD), (A= After Tax Income= 7,899 millions USD), (V= Variable Cost= 9,746 millions USD), (S= Sales or Revenues= 51,294 millions USD), and (i= I/S= Interest Portion= 1.00%), are known, then it's (t= Tax Rate planned) is:
$$t = 1 - A/[S-V-F-Si] = 1 - A/\{S[1-i] - V - F\}$$
$$= 1 - 7,899/\{51,294[1-0.0100] - 9,746 - 29,750\}$$
$$= 30.00\%$$

Law-1125:
If (**F**= Fixed Cost= 29,750 millions USD), (**t**= T/B= Tax Rate= 30.00%), (**V**= Variable Cost= 9,746 millions USD), (**S'**= Sales of Past Year= 48,851 millions USD), (**s**= [S/S']-1= Sales Growth= 5.00%), (**S**= Sales or Revenues= 51,294 millions USD), and (**i**= I/S= Interest Portion= 1.00%), are known, then it's (**A**= After Tax Income planned) is:

A= [1-**t**]{**S-V-F-S'i**[1+**s**]}
= [1-0.3000]{51,294 -9,746-29,750
-48,851*0.0100[1+0.0500]}
= 7,899 millions USD

Law-1126:
If (**F**= Fixed Cost= 29,750 millions USD), (**t**= T/B= Tax Rate= 30.00%), (**V**= Variable Cost= 9,746 millions USD), (**A**= After Tax Income= 7,899 millions USD and (**S'**= Sales of Past Year= 48,851 millions USD), (**s**= [S/S']-1= Sales Growth= 5.00%), (**i**= I/S= Interest Portion= 1.00%), are known, then it's (**S**= Sales or Revenues planned) is:

S= **S'i**[1+**s**]+**V**+**F**+**A**/[1-**t**]
= 48,851*0.0100[1+0.0500]+9,746
+29,750+7,899/[1-0.3000]
= 51,294 millions USD

Law-1127:
If (**F**= Fixed Cost= 29,750 millions USD), (**t**= T/B= Tax Rate= 30.00%), (**$**= Sales or Revenues= 51,294 millions USD), (**$'**= Sales of Past Year= 48,851 millions USD), (**s**= [S/S']-1= Sales Growth= 5.00%), (**A**= After Tax Income= 7,899 millions USD) and (**i**= I/S= Interest Portion= 1.00%), are known, then it's (**V**= Variable Cost planned) is:

$$V = \$'i[1+s] - \$ - F - A/[1-t]$$
$$= 48{,}851 * 0.0100[1+0.0500] - 51{,}294$$
$$-29{,}750 - 7{,}899/[1-0.3000]$$
$$= 9{,}746 \text{ millions USD}$$

Law-1128:
If (**V**= Variable Cost= 9,746 millions USD), (**t**= T/B= Tax Rate= 30.00%), (**$**= Sales or Revenues= 51,294 millions USD), (**$'**= Sales of Past Year= 48,851 millions USD), (**s**= [S/S']-1= Sales Growth= 5.00%), (**A**= After Tax Income= 7,899 millions USD) and (**i**= I/S= Interest Portion= 1.00%), are known, then it's (**V**= Variable Cost planned) is:

$$F = \$ - \$'i[1+s] - V - A/[1-t]$$
$$= 51{,}294 - 48{,}851 * 0.0100[1+0.0500]$$
$$-9{,}746 - 7{,}899/[1-0.3000]$$
$$= 29{,}750 \text{ millions USD}$$

Law-1129:
If (**V**= Variable Cost= 9,746 millions USD), (**t**= T/B= Tax Rate= 30.00%), (**s**= [S/S']-1= Sales Growth= 5.00%), (**F**= Fixed Cost= 29,750 millions USD), (**S**= Sales or Revenues= 51,294 millions USD), (**A**= After Tax Income= 7,899 millions USD and (**i**= I/S= Interest Portion= 1.00%), are known, then it's (**S'**= Sales Past) must be:

$$S' = \{S-V-F-A/[1-t]\}/\{i[1+s]\}$$
$$= \{51{,}294 - 9{,}746 - 29{,}750 - 7{,}899 /[1-0.3000]\} /\{0.0100[1+0.0500]\}$$
$$= 48{,}851 \text{ millions USD}$$

Law-1130:
If (**V**= Variable Cost= 9,746 millions USD), (**t**= T/B= Tax Rate= 30.00%), (**s**= [S/S']-1= Sales Growth= 5.00%), (**F**= Fixed Cost= 29,750 millions USD), (**S**= Sales or Revenues= 51,294 millions USD), (**A**= After Tax Income= 7,899 millions USD and (**S'**= Sales of Past Year= 48,851 millions USD), are known, then it's (**i**= Interest Portion planned) is:

$$i = \{S-V-F-A/[1-t]\}/\{S'[1+s]\}$$
$$= \{51{,}294 - 9{,}746 - 29{,}750 - 7{,}899 /[1-0.3000]\}/ \{48{,}851 [1+0.0500]\}$$
$$= 1.00\%$$

Law-1131:
 If (**V**= Variable Cost= 9,746 millions USD), (**t**= T/B= Tax Rate= 30.00%), (**$'**= Sales of Past Year= 48,851 millions USD), (**F**= Fixed Cost= 29,750 millions USD), (**$**= Sales or Revenues= 51,294 millions USD), (**A**= After Tax Income= 7,899 millions USD and (**i**= I/S= Interest Portion= 1.00%), are known, then it's (**s**= Sales Growth planned) is:
$$s = \{\$-V-F-A/[1-t]\}/[\$'i]-1$$
$$= \{51,294-9,746-29,750-7,899 /[1-0.3000]\}/[48,851*0.0100]-1$$
$$= 5.00\%$$

Law-1132:
 If (**V**= Variable Cost= 9,746 millions USD), (**s**= [S/S']-1= Sales Growth= 5.00%), (**$'**= Sales of Past Year= 48,851 millions USD), (**F**= Fixed Cost= 29,750 millions USD), (**$**= Sales or Revenues= 51,294 millions USD), (**A**= After Tax Income= 7,899 millions USD and (**i**= I/S= Interest Portion= 1.00%), are known, then it's (**t**= Tax Rate planned) is:
$$t = 1-A/\{\$-V-F-\$'i[1+s]\}$$
$$= 1-7,899/\{51,294-9,746-29,750 -48,851*0.0100[1+0.0500]\}$$
$$= 30.00\%$$

Law-1133:

If (V= Variable Cost= 9,746 millions USD), (s= [S/S']-1= Sales Growth= 5.00%), (f= F/S=Fixed Portion= 58.00%), (S= Sales or Revenues= 51,294 millions USD), (t= T/B= Tax Rate= 30.00%), and (I= Interest Expense= 513 millions USD), are known, then it's (A= After Tax Income planned) is:

$$A = [1-t][S-V-Sf-I] = [1-t]\{S[1-f]-V-I\}$$
$$= [1-0.3000]\{51,294[1-0.5800] -9,746-513\}$$
$$= \underline{7,899} \text{ millions USD}$$

Law-1134:

If (V= Variable Cost= 9,746 millions USD), (s= [S/S']-1= Sales Growth= 5.00%), (f= F/S=Fixed Portion= 58.00%), (A= After Tax Income= 7,899 millions USD), (t= T/B= Tax Rate= 30.00%), and (I= Interest Expense= 513 millions USD), are known, then it's (S= Sales or Revenues planned) is:

$$S = \{V+I+A/[1-t]\}/[1-f]$$
$$= \{9,746+513+7,899/[1-0.3000]\} [1-0.5800]$$
$$= \underline{51,294} \text{ millions USD}$$

Law-1135:
If (**V**= Variable Cost= 9,746 millions USD), (**s**= [S/S']-1= Sales Growth= 5.00%), (**S**= Sales or Revenues= 51,294 millions USD), (**A**= After Tax Income= 7,899 millions USD), (**t**= T/B= Tax Rate= 30.00%), and (**I**= Interest Expense= 513 millions USD), are known, then it's (**f**= Fixed Portion planned) is:

$$f = 1 - \{V+I+A/[1-t]\}/S$$
$$= 1 - \{9,746+513+7,899/[1-0.3000]\}/51,294$$
$$= 58.00\%$$

Law-1136:
If (**f**= F/S=Fixed Portion= 58.00%), (**s**= [S/S']-1= Sales Growth= 5.00%), (**S**= Sales or Revenues= 51,294 millions USD), (**A**= After Tax Income= 7,899 millions USD), (**t**= T/B= Tax Rate= 30.00%), and (**I**= Interest Expense= 513 millions USD), are known, then it's (**f**= Fixed Portion planned) is:

$$V = S[1-f] - I - A/[1-t]$$
$$= 51,294[1-0.5800] - 513 - 7,899/[1-0.3000]$$
$$= 9,746 \text{ millions USD}$$

Law-1137:
If ($f=$ F/S=Fixed Portion= 58.00%), ($s=$ [S/S']-1= Sales Growth= 5.00%), ($\$=$ Sales or Revenues= 51,294 millions USD), ($A=$ After Tax Income= 7,899 millions USD), ($t=$ T/B= Tax Rate= 30.00%), and ($V=$ Variable Cost= 9,746 millions USD), are known, then it's ($I=$ Intense Expense planned) is:
$I = \$[1-f]-V-A/[1-t]$
$= 51,294[1-0.5800]-9,746-7,899$
$/[1-0.3000]$
$= 513$ millions USD

Law-1138:
If ($f=$ F/S=Fixed Portion= 58.00%), ($s=$ [S/S']-1= Sales Growth= 5.00%), ($\$=$ Sales or Revenues= 51,294 millions USD), ($A=$ After Tax Income= 7,899 millions USD), ($I=$ Interest Expense= 513 millions USD), and ($V=$ Variable Cost= 9,746 millions USD), are known, then it's ($t=$ Tax Rate planned) is:
$t = 1-A/[\$-V-\$f-I] = 1-A/\{\$[1-f]-V-I\}$
$= 1-7,899/\{51,294[1-0.5800]-9,746-513\}$
$= 30.00\%$

Law-1139:

If (**f**= F/S=Fixed Portion= 58.00%), (**s**= [S/S']-1= Sales Growth= 5.00%), (**S**= Sales or Revenues= 51,294 millions USD), (**t**= T/B= Tax Rate= 30.00%), (**i**= I/S= Interest Portion= 1.00%), and (**V**= Variable Cost= 9,746 millions USD), are known, then it's (**A**= After Tax Income planned) is:

$$A = [1-t][S-V-Sf-Si] = [1-t]\{S[1-f-i]-V\}$$
$$= [1-0.3000]\{51,294[1-0.5800-0.0100]$$
$$-9,746\}$$
$$= 7,899 \text{ millions USD}$$

Law-1140:

If (**f**= F/S=Fixed Portion= 58.00%), (**s**= [S/S']-1= Sales Growth= 5.00%), (**A**= After Tax Income= 7,899 millions USD), (**t**= T/B= Tax Rate= 30.00%), (**i**= I/S= Interest Portion= 1.00%), and (**V**= Variable Cost= 9,746 millions USD), are known, then it's (**S**= Sales or Revenues planned) is:

$$S = \{V+A/[1-t]\}/[1-f-i]$$
$$= \{9,746+7,899/[1-0.3000]\}$$
$$/[1-0.5800-0.0100]$$
$$= 51,294 \text{ millions USD}$$

Law-1141:
If ($= Sales or Revenues= 51,294 millions USD), (s= [S/S']-1= Sales Growth= 5.00%), (A= After Tax Income= 7,899 millions USD), (t= T/B= Tax Rate= 30.00%), (i= I/S= Interest Portion= 1.00%), and (V= Variable Cost= 9,746 millions USD), are known, then it's (f= Fixed Portion planned) is:

$$f = 1-i-\{V+A/[1-t]\}/\$$$
$$= 1-0.0100-\{9,746+7,899/[1-0.3000]\}/51,294$$
$$= \underline{58.00\%}$$

Law-1142:
If ($= Sales or Revenues= 51,294 millions USD), (s= [S/S']-1= Sales Growth= 5.00%), (A= After Tax Income= 7,899 millions USD), (t= T/B= Tax Rate= 30.00%), (f= F/S=Fixed Portion= 58.00%), and (V= Variable Cost= 9,746 millions USD), are known, then it's (i= Interest Portion planned) is:

$$i = 1-f-\{V+A/[1-t]\}/\$$$
$$= 1-0.5800-\{9,746+7,899/[1-0.3000]\}/51,294$$
$$= \underline{1.00\%}$$

Law-1143:

If ($= Sales or Revenues= 51,294 millions USD), (s= [S/S']-1= Sales Growth= 5.00%), (A= After Tax Income= 7,899 millions USD), (t= T/B= Tax Rate= 30.00%), (f= F/S=Fixed Portion= 58.00%), and (i= I/S= Interest Portion= 1.00%), are known, then it's (V= Variable Cost planned) is:

$$V = \$[1-f-i] - A/[1-t]$$
$$= 51,294[1-0.5800-0.0100] - 7,899 / [1-0.3000]$$
$$= 9,746 \text{ millions USD}$$

Law-1144:

If ($= Sales or Revenues= 51,294 millions USD), (s= [S/S']-1= Sales Growth= 5.00%), (A= After Tax Income= 7,899 millions USD), (V= Variable Cost= 9,746 millions USD), (f= F/S=Fixed Portion= 58.00%), and (i= I/S= Interest Portion= 1.00%), are known, then it's (t= Tax Rate planned) is:

$$t = 1 - A/[\$-V-\$f-\$i] = 1 - A/\{\$[1-f-i]-V\}$$
$$= 1 - 7,899/\{51,294[1-0.5800-0.0100] - 9,746\}$$
$$= 30.00\%$$

Law-1145:
If ($'$= Sales of Past Year= 48,851 millions USD), ($$= Sales or Revenues= 51,294 millions USD), (s= [S/S']-1= Sales Growth= 5.00%), (t= T/B= Tax Rate= 30.00%), (V= Variable Cost= 9,746 millions USD), (f= F/S=Fixed Portion= 58.00%), and (i= I/S= Interest Portion= 1.00%), are known, then it's (A= After Tax Income planned) is:

A= [1-t][$$-V-$$$f$-$$'i[1+s]]
   = [1-t]{$$[1-f]-V-$$'i[1+s]}
   = [1-0.3000]{51,294[1-0.5800]-9,746
      -48,851*0.0100[1+0.0500}
   = 7,899 millions USD

Law-1146:
If ($'$= Sales of Past Year= 48,851 millions USD), (A= After Tax Income= 7,899 millions USD), (s= [S/S']-1= Sales Growth= 5.00%), (t= T/B= Tax Rate= 30.00%), (V= Variable Cost= 9,746 millions USD), (f= F/S=Fixed Portion= 58.00%), and (i= I/S= Interest Portion= 1.00%), are known, then it's ($$= Sales or Revenues planned) is:

$$= {V+$$'i[1+s]+A/[1-t]}/[1-f]}
   = 9,746+48,851*0.0100[1+0.0500}
      +7,899/[1-0.3000]}/[1-0.5800]
   = 51,294 millions USD

Law-1147:
If ($= Sales of Past Year= 48,851 millions USD), (A= After Tax Income= 7,899 millions USD), (s= [S/S']-1= Sales Growth= 5.00%), (t= T/B= Tax Rate= 30.00%), ($= Sales or Revenues= 51,294 millions USD), (f= F/S=Fixed Portion= 58.00%), and (i= I/S= Interest Portion= 1.00%), are known, then it's (V= Variable Cost planned) is:

$V = \$[1-f] - A/[1-t] - \$'i[1+s]$
$ = 51,294[1-0.5800] - 7,899/[1-0.3000]$
$ -48,851*0.0100[1+0.0500]$
$ = 9,746$ millions USD

Law-1148:
If (V= Variable Cost= 9,746 millions USD), (A= After Tax Income= 7,899 millions USD), (s= [S/S']-1= Sales Growth= 5.00%), (t= T/B= Tax Rate= 30.00%), ($= Sales or Revenues= 51,294 millions USD), (f= F/S=Fixed Portion= 58.00%), and (i= I/S= Interest Portion= 1.00%), are known, then it's ($'= Sales Past) must be:

$\$' = \$[1-f] - V - A/[1-t]/\{i[1+s]\}$
$ = 51,294[1-0.5800] - 9,746 - 7,899$
$ /[1-0.3000]/\{0.0100[1+0.0500]\}$
$ = 48,851$ millions USD

Law-1149:
If (V= Variable Cost= 9,746 millions USD), (A= After Tax Income= 7,899 millions USD), (s= [S/S']-1= Sales Growth= 5.00%), (t= T/B= Tax Rate= 30.00%), (S= Sales or Revenues= 51,294 millions USD), (S'= Sales of Past Year= 48,851 millions USD), and (i= I/S= Interest Portion= 1.00%), are known, then it's (f= Fixed Portion planned) is:

$f = 1 - \{V + S'i[1+s] + A/[1-t]\}/S$
$= 1 - \{9{,}746 + 48{,}851 * 0.0100[1+0.0500]$
$\quad + 7{,}899/[1-0.3000]\}/51{,}294$
$= 58.00\%$

Law-1150:
If (V= Variable Cost= 9,746 millions USD), (A= After Tax Income= 7,899 millions USD), (f= F/S=Fixed Portion= 58.00%), (t= T/B= Tax Rate= 30.00%), (S= Sales or Revenues= 51,294 millions USD), (S'= Sales of Past Year= 48,851 millions USD), and (i= I/S= Interest Portion= 1.00%), are known, then it's (s= Sales Growth planned) is:

$s = \{S[1-f] - V - A/[1-t]\}[S'i] - 1$
$= \{51{,}294[1-0.58] - 9{,}746 - 7{,}899$
$\quad /[1-0.3000]\}$
$\quad /[48{,}851 * 0.0100] - 1$
$= 5.00\%$

Law-1151:
If (**V**= Variable Cost= 9,746 millions USD), (**A**= After Tax Income= 7,899 millions USD), (**f**= F/S=Fixed Portion= 58.00%), (**t**= T/B= Tax Rate= 30.00%), (**$**= Sales or Revenues= 51,294 millions USD), (**$'**= Sales of Past Year= 48,851 millions USD), and (**s**= [S/S']-1= Sales Growth= 5.00%), are known, then it's (**i**= Interest Portion planned) is:

$$i = \{\$[1-f]-V-A/[1-t]\}[\$'i]-1$$
$$= \{51{,}294[1-0.58]-9{,}746-7{,}899$$
$$\qquad /[1-0.3000]\}/\{48{,}851[1+0.0500]\}$$
$$= 1.00\%$$

Law-1152:
If (**V**= Variable Cost= 9,746 millions USD), (**A**= After Tax Income= 7,899 millions USD), (**f**= F/S=Fixed Portion= 58.00%), (**i**= I/S= Interest Portion= 1.00%), (**$**= Sales or Revenues= 51,294 millions USD), (**$'**= Sales of Past Year= 48,851 millions USD), and (**s**= [S/S']-1= Sales Growth= 5.00%), are known, then it's (**t**= Tax Rate planned) is:

$$t = 1-A/\{\$-V-\$f-\$'i[1+s]\}$$
$$= 1-A/\{\$[1-f]-V-\$'i[1+s]\}$$
$$= 1-7{,}899/\{51{,}294[1-0.5800]-9{,}746$$
$$\qquad -48{,}851*0.0100[1+0.0500]\}$$
$$= 30.00\%$$

Law-1153:

If (V= Variable Cost= 9,746 millions USD), (t= T/B= Tax Rate= 30.00%), (f= F/S=Fixed Portion= 58.00%), (I= Interest Expense= 513 millions USD), (S= Sales or Revenues= 51,294 millions USD), (S'= Sales of Past Year= 48,851 millions USD), and (s= [S/S']-1= Sales Growth= 5.00%), are known, then it's (A= After Tax Income planned) is:

A= [1-t]{S-V-I-$S'f$[1+s]}
 = [1-0.3000]{51,294- 9,746 -513
   -48,851*0.5800[1+0.0500]
 = 7,899 millions USD

Law-1154:

If (V= Variable Cost= 9,746 millions USD), (t= T/B= Tax Rate= 30.00%), (f= F/S=Fixed Portion= 58.00%), (I= Interest Expense= 513 millions USD), (A= After Tax Income= 7,899 millions USD), (S'= Sales of Past Year= 48,851 millions USD), and (s= [S/S']-1= Sales Growth= 5.00%), are known, then it's (S= Sales or Revenues planned) is:

S= V+I+$S'f$[1+s]+A/[1-t]
 = 9,746 +513+48,851*0.5800[1+0.0500]
   +7,899/[1-0.3000]
 = 51,294 millions USD

Law-1155:
If (t= T/B= Tax Rate= 30.00%), (f= F/S=Fixed Portion= 58.00%), (S= Sales or Revenues= 51,294 millions USD), (I= Interest Expense= 513 millions USD), (A= After Tax Income= 7,899 millions USD), (S'= Sales of Past Year= 48,851 millions USD), and (s= [S/S']-1= Sales Growth= 5.00%), are known, then it's (V= Variable Cost planned) is:

V= S-I-$S'f$[1+s]-A/[1-t]
   = 51,294 -513-48,851*0.5800[1+0.0500]
       -7,899/[1-0.3000]
   = 9,746 millions USD

Law-1156:
If (t= T/B= Tax Rate= 30.00%), (f= F/S=Fixed Portion= 58.00%), (S= Sales or Revenues= 51,294 millions USD), (I= Interest Expense= 513 millions USD), (A= After Tax Income= 7,899 millions USD), (V= Variable Cost= 9,746 millions USD), and (s= [S/S']-1= Sales Growth= 5.00%), are known, then it's (S'= Sales Past) must be:

S'= {S-V-I-A/[1-t]}/{f[1+s]}
   = {51,294 -9,746-513-7,899/[1-0.3000]}
          /{0.5800[1+0.0500]}
   = 48,851 millions USD

Law-1157:
If (t= T/B= Tax Rate= 30.00%), (S'= Sales of Past Year= 48,851 millions USD), (S= Sales or Revenues= 51,294 millions USD), (I= Interest Expense= 513 millions USD), (A= After Tax Income= 7,899 millions USD), (V= Variable Cost= 9,746 millions USD), and (s= [S/S']-1= Sales Growth= 5.00%), are known, then it's (f= Fixed Portion planned)is:

f = {S-V-I-A/[1-t]}/{S'[1+s]}
= {51,294 -9,746-513-7,899/[1-0.3000]}
   /{48,851[1+0.0500]}
= 58.00%

Law-1158:
If (t= T/B= Tax Rate= 30.00%), (S'= Sales of Past Year= 48,851 millions USD), (S= Sales or Revenues= 51,294 millions USD), (I= Interest Expense= 513 millions USD), (A= After Tax Income= 7,899 millions USD), (V= Variable Cost= 9,746 millions USD), and (f= F/S=Fixed Portion= 58.00%), are known, then it's (s= Sales Growth planned) is:

s= {S-V-I-A/[1-t]}/[$S'f$]-1
= {51,294 -9,746-513-7,899/[1-0.3000]}
   /[48,851*0.5800]- 1
= 5.00%

Law-1159:
If (**t**= T/B= Tax Rate= 30.00%), (**$'**= Sales of Past Year= 48,851 millions USD), (**$**= Sales or Revenues= 51,294 millions USD), (**s**= [S/S']-1= Sales Growth= 5.00%), (**A**= After Tax Income= 7,899 millions USD), (**V**= Variable Cost= 9,746 millions USD), and (**f**= F/S=Fixed Portion= 58.00%), are known, then it's (**I**= Interest Expense planned) is:

I= **$**-**V**-**$'f**[1+**s**]-**A**/[1-**t**]
 = 51,294 -9,746-48,851*0.5800
    [1+0.0500]-7,899/[1-0.3000]}
 = 513 millions USD

Law-1160:
If (**I**= Interest Expense= 513 millions USD), (**$'**= Sales of Past Year= 48,851 millions USD), (**$**= Sales or Revenues= 51,294 millions USD), (**s**= [S/S']-1= Sales Growth= 5.00%), (**A**= After Tax Income= 7,899 millions USD), (**V**= Variable Cost= 9,746 millions USD), and (**f**= F/S=Fixed Portion= 58.00%), are known, then it's (**t**= Tax Rate planned) is:

t= 1-**A**/{**$**-**V**-**I**-**$'f**[1+**s**]}
 = 1-7,899/{ 51,294 -9,746-513-48,851
    *0.5800[1+0.0500]}
 = 30.00%

Law-1161:

If (i= I/S= Interest Portion= 1.00%), ($\$'$= Sales of Past Year= 48,851 millions USD), ($\$$= Sales or Revenues= 51,294 millions USD), (s= [S/S']-1= Sales Growth= 5.00%), (t= T/B= Tax Rate= 30.00%), (V= Variable Cost= 9,746 millions USD), and (f= F/S=Fixed Portion= 58.00%), are known, then it's (A= After Tax Income planned) is:

A= [1-t]{$\$$-V-$\$'f$[1+$s$]-$\$i$}
   = [1-t]{$\$$[1-i]-V-$\$'f$[1+$s$]}
   = [1-0.3000]/{ 51,294[1-0.0100] -9,746
      -48,851*0.5800[1+0.0500]}
   = 7,899 millions USD

Law-1162:

If (i= I/S= Interest Portion= 1.00%), ($\$'$= Sales of Past Year= 48,851 millions USD), (A= After Tax Income= 7,899 millions USD), (s= [S/S']-1= Sales Growth= 5.00%), (t= T/B= Tax Rate= 30.00%), (V= Variable Cost= 9,746 millions USD), and (f= F/S=Fixed Portion= 58.00%), are known, then it's ($\$$= Sales or Revenues planned) is:

$\$$= {V+$\$'f$[1+$s$]+$A$/[1-$t$]/[1-$i$]
   = {9,746+48,851*0.5800[1+0.0500]
      +7,899/[1-0.3000]}/[1-0.0100]
   = 51,294 millions USD

Law-1163:

If ($ = Sales or Revenues= 51,294 millions USD), ($' = Sales of Past Year= 48,851 millions USD), (**A**= After Tax Income= 7,899 millions USD), (**s**= [S/S']-1= Sales Growth= 5.00%), (**t**= T/B= Tax Rate= 30.00%), (**V**= Variable Cost= 9,746 millions USD), and (**f**= F/S=Fixed Portion= 58.00%), are known, then it's (**i**= Interest Portion planned) is:

$$i = 1 - \{V + S'f[1+s] + A/[1-t]\}/S$$
$$= 1 - \{9,746 + 48,851*0.5800[1+0.0500] + 7,899/[1-0.3000]\}/51,294$$
$$= 1.00\%$$

Law-1164:

If ($ = Sales or Revenues= 51,294 millions USD), ($' = Sales of Past Year= 48,851 millions USD), (**A**= After Tax Income= 7,899 millions USD), (**s**= [S/S']-1= Sales Growth= 5.00%), (**t**= T/B= Tax Rate= 30.00%), (**i**= I/S= Interest Portion= 1.00%), and (**f**= F/S=Fixed Portion= 58.00%), are known, then it's (**V**= Variable Cost planned) is:

$$V = S[1-i] - S'f[1+s] - A/[1-t]$$
$$= 51,294[1-0.0100] - 48,851*0.5800 [1+0.0500] - 7,899/[1-0.3000]$$
$$= 9,746 \text{ millions USD}$$

Law-1165:

If ($ = Sales or Revenues = 51,294 millions USD), (V = Variable Cost = 9,746 millions USD), (A = After Tax Income = 7,899 millions USD), (s = [S/S']-1 = Sales Growth = 5.00%), (t = T/B = Tax Rate = 30.00%), (i = I/S = Interest Portion = 1.00%), and (f = F/S = Fixed Portion = 58.00%), are known, then it's ($' = Sales Past) must be:

$$\$' = \{\$[1-i]-V-A/[1-t]\}/\{f[1+s]\}$$
$$= 51,294[1-0.0100]-9,746-7,899$$
$$/[1-0.3000]/\{0.5800[1+0.0500]\}$$
$$= 48,851 \text{ millions USD}$$

Law-1166:

If ($ = Sales or Revenues = 51,294 millions USD), (V = Variable Cost = 9,746 millions USD), (A = After Tax Income = 7,899 millions USD), (s = [S/S']-1 = Sales Growth = 5.00%), (t = T/B = Tax Rate = 30.00%), (i = I/S = Interest Portion = 1.00%), and ($' = Sales of Past Year = 48,851 millions USD), are known, then it's (f = Fixed Portion planned) is:

$$f = \{\$[1-i]-V-A/[1-t]\}/\{\$'[1+s]\}$$
$$= 51,294[1-0.0100]-9,746-7,899$$
$$/[1-0.3000]/\{48,851[1+0.0500]\}$$
$$= 58.00\%$$

Law-1167:
If ($= Sales or Revenues= 51,294 millions USD), (V= Variable Cost= 9,746 millions USD), (A= After Tax Income= 7,899 millions USD), (f= F/S=Fixed Portion= 58.00%), (t= T/B= Tax Rate= 30.00%), (i= I/S= Interest Portion= 1.00%), and ($'= Sales of Past Year= 48,851 millions USD), are known, then it's (s= Sales Growth planned) is:

s= {$[1-i]-V-A/[1-t]}/[$'f]-1
 = 51,294[1-0.0100]-9,746 -7,899
       /[1-0.3000]/[48,851*0.5800]-1
 = 5.00%

Law-1168:
If ($= Sales or Revenues= 51,294 millions USD), (V= Variable Cost= 9,746 millions USD), (A= After Tax Income= 7,899 millions USD), (f= F/S=Fixed Portion= 58.00%), (s= [S/S']-1= Sales Growth= 5.00%), (i= I/S= Interest Portion= 1.00%), and ($'= Sales of Past Year= 48,851 millions USD), are known, then it's (t= Tax Rate planned) is:

t= 1-A/{$-V-$'f[1+s]-$i}
 = 1-A/{$[1-i]-V-$'f[1+s]
 = 1-7,899/{51,294[1-0.0100]-9,746
       -48,851*0.5800[1+0.0500]}
 = 30.00%

Law-1169:

If ($= Sales or Revenues= 51,294 millions USD), (V= Variable Cost= 9,746 millions USD), (t= T/B= Tax Rate= 30.00%), (f= F/S=Fixed Portion= 58.00%), (s= [S/S']-1= Sales Growth= 5.00%), (i= I/S= Interest Portion= 1.00%), and ($'= Sales of Past Year= 48,851 millions USD), are known, then it's (A= After Tax Income planned) is:

$$A = [1-t]\{\$-V-\$'f[1+s]-\$'i[1+s]\}$$
$$= [1-t]\{\$-V-\$'[1+s][f+i]\}$$
$$= [1-0.3000]\{51,294 - 9,746 - 48,851$$
$$[1+0.0500][0.5800+0.0100]\}$$
$$= \underline{7,899} \text{ millions USD}$$

Law-1170:

If (A= After Tax Income= 7,899 millions USD), (V= Variable Cost= 9,746 millions USD), (t= T/B= Tax Rate= 30.00%), (f= F/S=Fixed Portion= 58.00%), (s= [S/S']-1= Sales Growth= 5.00%), (i= I/S= Interest Portion= 1.00%), and ($'= Sales of Past Year= 48,851 millions USD), are known, then it's ($= Sales or Revenues planned) is:

$$\$ = \$'[1+s][f+i]+V+A/[1-t]$$
$$= 48,851 [1+0.0500][0.5800+0.0100]$$
$$+9,746+7,899/[1-0.3000]$$
$$= \underline{51,294} \text{ millions USD}$$

Law-1171:
If (**A**= After Tax Income= 7,899 millions USD), (**$**= Sales or Revenues= 51,294 millions USD), (**t**= T/B= Tax Rate= 30.00%), (**f**= F/S=Fixed Portion= 58.00%), (**s**= [S/S']-1= Sales Growth= 5.00%), (**i**= I/S= Interest Portion= 1.00%), and (**$'**= Sales of Past Year= 48,851 millions USD), are known, then it's (**V**= Variable Cost planned) is:

$$V = \$ - \$'[1+s][f+i] - A/[1-t]$$
$$= 51{,}294 - 48{,}851 [1+0.0500] [0.5800 + 0.0100] - 7{,}899/[1-0.3000]$$
$$= 9{,}746 \text{ millions USD}$$

Law-1172:
If (**A**= After Tax Income= 7,899 millions USD), (**$**= Sales or Revenues= 51,294 millions USD), (**t**= T/B= Tax Rate= 30.00%), (**f**= F/S=Fixed Portion= 58.00%), (**s**= [S/S']-1= Sales Growth= 5.00%), (**i**= I/S= Interest Portion= 1.00%), and (**V**= Variable Cost= 9,746 millions USD), are known, then it's (**$'**= Sales Past) must be:

$$\$' = \{\$ - V - A/[1-t]\} / \{[1+s][f+i]\}$$
$$= \{51{,}294 - 9{,}746 - 7{,}899/[1-0.3000]\} / \{[1+0.0500] [0.5800+0.0100]\}$$
$$= 48{,}851 \text{ millions USD}$$

Law-1173:

If (A= After Tax Income= 7,899 millions USD), (S= Sales or Revenues= 51,294 millions USD), (t= T/B= Tax Rate= 30.00%), (f= F/S=Fixed Portion= 58.00%), (S'= Sales of Past Year= 48,851 millions USD), (i= I/S= Interest Portion= 1.00%), and (V= Variable Cost= 9,746 millions USD), are known, then it's (s= Sales Growth planned) is:

$s = \{S-V-A/[1-t]\}/\{S'[f+i]\}-1$
$= \{51,294 - 9,746 - 7,899/[1-0.3000]\}$
$\quad /\{48,851[0.5800+0.0100]\}-1$
$= 5.00\%$

Law-1174:

If (A= After Tax Income= 7,899 millions USD), (S= Sales or Revenues= 51,294 millions USD), (t= T/B= Tax Rate= 30.00%), (s= [S/S']-1= Sales Growth= 5.00%), (S'= Sales of Past Year= 48,851 millions USD), (i= I/S= Interest Portion= 1.00%), and (V= Variable Cost= 9,746 millions USD), are known, then it's (f= Fixed Portion planned) is:

$f = \{S-V-A/[1-t]\}/\{S'[1+s]\}-i$
$= \{51,294 - 9,746 - 7,899/[1-0.3000]\}$
$\quad /\{48,851[1+0.0500]\}-0.0100$
$= 58.00\%$

Law-1175:
If (A= After Tax Income= 7,899 millions USD), (S= Sales or Revenues= 51,294 millions USD), (t= T/B= Tax Rate= 30.00%), (s= [S/S']-1= Sales Growth= 5.00%), (S'= Sales of Past Year= 48,851 millions USD), (f= F/S=Fixed Portion= 58.00%), and (V= Variable Cost= 9,746 millions USD), are known, then it's (i= Interest Portion planned) is:

$i = \{S-V-A/[1-t]\}/\{S'[1+s]\}-f$
$= \{51,294- 9,746-7,899/[1-0.3000]\}$
$\quad /\{48,851[1+0.0500]\}-0.5800$
$= \underline{1.00\%}$

Law-1176:
If (A= After Tax Income= 7,899 millions USD), (S= Sales or Revenues= 51,294 millions USD), (i= I/S= Interest Portion= 1.00%), (s= [S/S']-1= Sales Growth= 5.00%), (S'= Sales of Past Year= 48,851 millions USD), (f= F/S=Fixed Portion= 58.00%), and (V= Variable Cost= 9,746 millions USD), are known, then it's (t= Tax Rate planned) is:

$t = 1-A/\{S- V-S'f[1+s]\}-S'i[1+s]\}$
$= 1-A/\{S-V-S'[1+s][f+i]\}$
$= 1-7,899/\{51,294-9746- 48,851$
$\quad [1+0.0500][0.5800+0.0100]\}$
$= \underline{30.00\%}$

Law-1177:
If (t= T/B= Tax Rate= 30.00%), (S= Sales or Revenues= 51,294 millions USD), (I= Interest Expense= 513 millions USD), (F= Fixed Cost= 29,750 millions USD), and (v= V/S= Variable Portion= 19.00%), are known, then it's (A= After Tax Income planned) is:

A= [1-t][S-Sv-F-I]= [1-t]/{S[1-v]-F-I}
= [1-0.3000]/51,294[1-0.1900]
-29,750-513}
= 7,899 millions USD

Law-1178:
If (t= T/B= Tax Rate= 30.00%), (A= After Tax Income= 7,899 millions USD), (I= Interest Expense= 513 millions USD), (F= Fixed Cost= 29,750 millions USD), and (v= V/S= Variable Portion= 19.00%), are known, then it's (S= Sales or Revenues planned) is:

S= {F+I+A/[1-t]}/[1-v]
={29,750+513+7,899/[1-0.3000]}
/[1-0.1900]
= 51,294 millions USD

Law-1179:

If (t= T/B= Tax Rate= 30.00%), (A= After Tax Income= 7,899 millions USD), (I= Interest Expense= 513 millions USD), (F= Fixed Cost= 29,750 millions USD), and (S= Sales or Revenues= 51,294 millions USD), are known, then it's (v= Variable Portion planned) is:

$$v = 1 - \{F+I+A/[1-t]\}/S$$
$$= 1 - \{29,750+513+7,899/[1-0.3000]\}/51,294$$
$$= 19.00\%$$

Law-1180:

If (t= T/B= Tax Rate= 30.00%), (A= After Tax Income= 7,899 millions USD), (I= Interest Expense= 513 millions USD), (v= V/S= Variable Portion= 19.00%), and (S= Sales or Revenues= 51,294 millions USD), are known, then it's (F= Fixed Cost planned) is:

$$F = S[1-v] - I - A/[1-t]$$
$$= 51,294[1-0.1900] - 513 - 7,899/[1-0.3000]\}$$
$$= 29,750 \text{ millions USD}$$

Law-1181:
 If (t= T/B= Tax Rate= 30.00%), (A= After Tax Income= 7,899 millions USD), (F= Fixed Cost= 29,750 millions USD), (v= V/S= Variable Portion= 19.00%), and ($\$$= Sales or Revenues= 51,294 millions USD), are known, then it's (I= Interest Expense planned) is:
$$I= \$[1-v]-F-A/[1-t]$$
$$= 51,294[1-0.1900]-29,750-7,899$$
$$/[1-0.3000]\}$$
$$= 513 \text{ millions USD}$$

Law-1182:
 If (I= Interest Expense= 513 millions USD), (A= After Tax Income= 7,899 millions USD), (F= Fixed Cost= 29,750 millions USD), (v= V/S= Variable Portion= 19.00%), and ($\$$= Sales or Revenues= 51,294 millions USD), are known, then it's (t= Tax Rate planned) is:
$$t= 1-A/[\$-\$v-F-I]= 1-A/\{\$[1-v]-F-I\}$$
$$= 1-7,899/\{51,294[1-0.1900]-29,750$$
$$-513\}$$
$$= 30.00\%$$

Law-1183:

If (i= I/S= Interest Portion= 1.00%), (t= T/B= Tax Rate= 30.00%), (F= Fixed Cost= 29,750 millions USD), (v= V/S= Variable Portion= 19.00%), and (S= Sales or Revenues= 51,294 millions USD), are known, then it's (A= After Tax Income planned) is:

$A = [1-t][S-Sv-F-Si] = [1-t]\{S[1-v-i]-F\}$
$= [1-0.3000]\{51,294[1-0.1900-0.0100]$
$-29,750\}$
$= 7,899$ millions USD

Law-1184:

If (i= I/S= Interest Portion= 1.00%), (t= T/B= Tax Rate= 30.00%), (F= Fixed Cost= 29,750 millions USD), (v= V/S= Variable Portion= 19.00%), and (A= After Tax Income= 7,899 millions USD), are known, then it's (S= Sales or Revenues planned) is:

$S = \{F+A/[1-t]\}/[1-v-i]$
$= \{29,750+7,899/[1-0.3000]\}$
$/ [1-0.1900-0.0100]$
$= 51,294$ millions USD

Law-1185:
If (i= I/S= Interest Portion= 1.00%), (t= T/B= Tax Rate= 30.00%), (F= Fixed Cost= 29,750 millions USD), ($\$$= Sales or Revenues= 51,294 millions USD), and (A= After Tax Income= 7,899 millions USD), are known, then it's (v= Variable Portion planned) is:
v= 1-i-{F+A/[1-t]}/$\$$
= 1-0.0100-{29,750+7,899/[1-0.3000]}/51,294
= 19.00%

Law-1186:
If (v= V/S= Variable Portion= 19.00%), (t= T/B= Tax Rate= 30.00%), (F= Fixed Cost= 29,750 millions USD), ($\$$= Sales or Revenues= 51,294 millions USD), and (A= After Tax Income= 7,899 millions USD), are known, then it's (i= Interest Portion planned) is:
i= 1-v-{F+A/[1-t]}/$\$$
= 1-0.1900-{29,750+7,899/[1-0.3000]}/51,294
= 1.00%

Law-1187:
If (v= V/S= Variable Portion= 19.00%), (t= T/B= Tax Rate= 30.00%), (i= I/S= Interest Portion= 1.00%), ($\$$= Sales or Revenues= 51,294 millions USD), and (A= After Tax Income= 7,899 millions USD), are known, then it's (F= Fixed Cost planned) is:

$F = \$[1-v-i] - A/[1-t]\}$
= 51,294[1-0.1900-0.0100]-7,899
/[1-0.3000]}
= 29,750 millions USD

Law-1188:
If (v= V/S= Variable Portion= 19.00%), (F= Fixed Cost= 29,750 millions USD), (i= I/S= Interest Portion= 1.00%), ($\$$= Sales or Revenues= 51,294 millions USD), and (A= After Tax Income= 7,899 millions USD), are known, then it's (t= Tax Rate planned) is:

$t = 1 - A/[\$ - \$v - F - \$i] = 1 - A/\{\$[1-v-i] - F\}$
= 1-7,899/{51,294[1-0.1900-0.0100]
-29,750]}
= 30.00%

Law-1189:

If (v= V/S= Variable Portion= 19.00%), (F= Fixed Cost= 29,750 millions USD), (i= I/S= Interest Portion= 1.00%), ($\$'$= Sales of Past Year= 48,851 millions USD), ($\$$= Sales or Revenues= 51,294 millions USD), and (t= T/B= Tax Rate= 30.00%), are known, then it's (A= After Tax Income planned) is:

A= [1-t]{$\$$-$\$v$-$F$-$\$'i$[1+$s$]}]
 = [1-t]{$\$$[1-v]-F-$\$'i$[1+$s$]}
 = [1-0.3000]{51,294[1-0.1900]-29,750
   -48,851*0.0100[1+0.0500]}
 = 7,899 millions USD

Law-1190:

If (v= V/S= Variable Portion= 19.00%), (F= Fixed Cost= 29,750 millions USD), (i= I/S= Interest Portion= 1.00%), ($\$'$= Sales of Past Year= 48,851 millions USD), (A= After Tax Income= 7,899 millions USD), and (t= T/B= Tax Rate= 30.00%), are known, then it's ($\$$= Sales or Revenues planned) is:

$\$$= {F+$\$'i$[1+$s$]+$A$/[1-$t$]}/[1-$v$]
 = {29,750+48,851*0.0100[1+0.0500]
   +7,899/ [1-0.3000]}/[1-0.1900]
 = 51,294 millions USD

Law-1191:
If ($= Sales or Revenues= 51,294 millions USD), ($'= Sales of Past Year= 48,851 millions USD), (F= Fixed Cost= 29,750 millions USD), (i= I/S= Interest Portion= 1.00%), (A= After Tax Income= 7,899 millions USD), and (t= T/B= Tax Rate= 30.00%), are known, then it's (v= Variable Portion planned) is:

$$v= 1-\{F+\$'i[1+s]+A/[1-t]\}/\$$$
$$= 1-\{29,750+48,851*0.0100$$
$$[1+0.0500]+7,899/[1-0.3000]\}$$
$$/51,294$$
$$= 19.00\%$$

Law-1192:
If ($= Sales or Revenues= 51,294 millions USD), ($'= Sales of Past Year= 48,851 millions USD), (v= V/S= Variable Portion= 19.00%), (i= I/S= Interest Portion= 1.00%), (A= After Tax Income= 7,899 millions USD), and (t= T/B= Tax Rate= 30.00%), are known, then it's (F= Fixed Cost planned) is:

$$F= \$[1-v]-\$'i[1+s]-A/[1-t]$$
$$= 51,294 [1-0.1900] -48,851*0.0100$$
$$[1+0.0500]-7,899/[1-0.3000]$$
$$= 29,750 \text{ millions USD}$$

Law-1193:

If (**F**= Fixed Cost= 29,750 millions USD), (**$**= Sales or Revenues= 51,294 millions USD), (**v**= V/S= Variable Portion= 19.00%), (**i**= I/S= Interest Portion= 1.00%), (**A**= After Tax Income= 7,899 millions USD), and (**s**= [S/S']-1= Sales Growth= 5.00%), (**t**= T/B= Tax Rate= 30.00%), are known, then it's (**$'**= Sales Past) must be:

$$\$' = \{\$[1-v]-F-A/[1-t]\}/\{i[1+s]\}$$
$$= \{51,294 [1-0.1900] -29,750 -7,899$$
$$/[1-0.3000]\}$$
$$/\{0.0100[1+0.0500]\}$$
$$= 48,851 \text{ millions USD}$$

Law-1194:

If (**F**= Fixed Cost= 29,750 millions USD), (**$**= Sales or Revenues= 51,294 millions USD), (**v**= V/S= Variable Portion= 19.00%), (**$'**= Sales of Past Year= 48,851 millions USD), (**A**= After Tax Income= 7,899 millions USD), and (**s**= [S/S']-1= Sales Growth= 5.00%), (**t**= T/B= Tax Rate= 30.00%), are known, then it's (**i**= Interest Portion planned) is:

$$i = \{\$[1-v]-F-A/[1-t]\}/\{\$'[1+s]\}$$
$$= \{51,294 [1-0.1900] -29,750 -7,899$$
$$/[1-0.3000]\}$$
$$/\{48,851[1+0.0500]\}$$
$$= 1.00\%$$

Law-1195 :
If (**F**= Fixed Cost= 29,750 millions USD), (**$**= Sales or Revenues= 51,294 millions USD), (**υ**= V/S= Variable Portion= 19.00%), (**$'**= Sales of Past Year= 48,851 millions USD), (**A**= After Tax Income= 7,899 millions USD), and (**i**= I/S= Interest Portion= 1.00%), (**t**= T/B= Tax Rate= 30.00%), are known, then it's (**s**= Sales growth planned) is:

$s = \{\$[1-υ]-F-A/[1-t]\}/\{[\$'i]\}-1$
$= \{51,294 [1-0.1900] -29750 -7,899$
$/[1-0.3000]\}/[48,851*0.0100]-1$
$= 5.00\%$

Law-1196 :
If (**F**= Fixed Cost= 29,750 millions USD), (**$**= Sales or Revenues= 51,294 millions USD), (**υ**= V/S= Variable Portion= 19.00%), (**$'**= Sales of Past Year= 48,851 millions USD), (**A**= After Tax Income= 7,899 millions USD), and (**i**= I/S= Interest Portion= 1.00%), (**s**= [S/S']-1= Sales Growth= 5.00%), are known, then it's (**t**= Tax Rate planned) is:

$t = 1-A/\{\$-\$υ-F-\$'i[1+s]\}$
$= 1-A/\{\$[1-υ]-F-\$'i[1+s]\}$
$= 1-7,899/\{51,294 [1-0.1900] -29,750$
$-48,851*0.0100[1+0.0500]\}$
$= 30.00 \%$

Law-1197:
If (**f**= F/S=Fixed Portion= 58.00%), (**$**= Sales or Revenues= 51,294 millions USD), (**v**= V/S= Variable Portion= 19.00%), (**t**= T/B= Tax Rate= 30.00%), and (**I**= Interest Expense= 513 millions USD), are known, then it's (**A**= After Tax Income planned) is:

$$A = [1-t]\{S - Sv - Sf - I\} = [1-t]\{S[1-v-f] - I\}$$
$$= [1 - 0.3000]\{51,294 [1 - 0.1900 - 0.5800] - 513\}$$
$$= 7,899 \text{ millions USD}$$

Law-1198:
If (**f**= F/S=Fixed Portion= 58.00%), (**A**= After Tax Income= 7,899 millions USD), (**v**= V/S= Variable Portion= 19.00%), (**t**= T/B= Tax Rate= 30.00%), and (**I**= Interest Expense= 513 millions USD), are known, then it's (**$**= Sales or Revenues planned) is:

$$S = \{I + A/[1-t]\}/[1-v-f]$$
$$= \{513 + 7,899/[1-0.3000]\} / [1 - 0.1900 - 0.5800]$$
$$= 51,294 \text{ millions USD}$$

Law-1199 :

If (f= F/S=Fixed Portion= 58.00%), (A= After Tax Income= 7,899 millions USD), (S= Sales or Revenues= 51,294 millions USD), (t= T/B= Tax Rate= 30.00%), and (I= Interest Expense= 513 millions USD), are known, then it's (v= Variable Portion planned) is:

$$v = 1-f-\{I+A/[1-t]\}/S$$
$$= 1-0.5800-\{513+7,899/[1-0.3000]\}/51,294$$
$$= 19.00\%$$

Law-1200 :

If (v= V/S= Variable Portion= 19.00%), (A= After Tax Income= 7,899 millions USD), (S= Sales or Revenues= 51,294 millions USD), (t= T/B= Tax Rate= 30.00%), and (I= Interest Expense= 513 millions USD), are known, then it's (f= Fixed Portion planned) is:

$$f = 1-v-\{I+A/[1-t]\}/S$$
$$= 1-0.1900-\{513+7,899/[1-0.3000]\}/51,294$$
$$= 58.00\%$$

Law-1201 :
If (v= V/S= Variable Portion= 19.00%), (**A**= After Tax Income= 7,899 millions USD), ($= Sales or Revenues= 51,294 millions USD), (**t**= T/B= Tax Rate= 30.00%), and (**f**= F/S=Fixed Portion= 58.00%), are known, then it's (**I**= Interest Expense planned) is:

$$I = \$[1-v-f] - A/[1-t]$$
$$= 51,294[1-0.1900-0.5800] - 7,899$$
$$/[1-0.3000]$$
$$= \underline{513} \text{ millions USD}$$

Law-1202 :
If (v= V/S= Variable Portion= 19.00%), (**A**= After Tax Income= 7,899 millions USD), ($= Sales or Revenues= 51,294 millions USD), (**I**= Interest Expense= 513 millions USD), and (**f**= F/S=Fixed Portion= 58.00%), are known, then it's (**t**= Tax Rate planned) is:

$$t = 1 - A/[\$ - \$v - \$f - I] = 1 - A/\{\$[1-v-f] - I\}$$
$$= 1 - 7,899/\{51,294[1-0.1900-0.5800] - 513\}$$
$$= \underline{30.00\%}$$

Law-1203 :

If (v= V/S= Variable Portion= 19.00%), (t= T/B= Tax Rate= 30.00%), ($\$$= Sales or Revenues= 51,294 millions USD), (i= I/S= Interest Portion= 1.00%), and (f= F/S=Fixed Portion= 58.00%), are known, then it's (A= After Tax Income planned) is:

A= [1-t][$\$$-$\$v$-$\$f$-$\$i$]= {$\$$[1-t][1-v-f-i]}
  = [1-0.3000]{51,294[[1-0.1900-0.5800
     -0.0100]}
  = 7,899 millions USD

Law-1204 :

If (i= I/S= Interest Portion= 1.00%), (A= After Tax Income= 7,899 millions USD), (t= T/B= Tax Rate= 30.00%), (v= V/S= Variable Portion= 19.00%), and (f= F/S=Fixed Portion= 58.00%), are known, then it's ($\$$= Sales or Revenues planned) is:

$\$$= A/{[1-t][1-v-f-i]
   = 7,899/[1-0.3000][1-0.1900-0.5800
       -0.0100]
   = 51,294 millions USD

Law-1205:

If (i= I/S= Interest Portion= 1.00%), (A= After Tax Income= 7,899 millions USD), (t= T/B= Tax Rate= 30.00%), ($\$$= Sales or Revenues= 51,294 millions USD), and (f= F/S=Fixed Portion= 58.00%), are known, then it's (v= Variable Portion planned) is:

v = 1-f-i-A/{$\$$[1-t]}
  = 1-0.5800-0.0100-7,899
    /{51,294[1-0.3000]}
  = 19.00%

Law-1206:

If (i= I/S= Interest Portion= 1.00%), (A= After Tax Income= 7,899 millions USD), (t= T/B= Tax Rate= 30.00%), ($\$$= Sales or Revenues= 51,294 millions USD), and (v= V/S= Variable Portion= 19.00%), are known, then it's (f= Fixed Portion planned) is:

f = 1-v-i-A/{$\$$[1-t]}
  =1-0.1900-0.0100-7,899
    /{51,294 [1-0.3000]}
  = 58.00%

Law-1207 :
If (f= F/S=Fixed Portion= 58.00%), (A= After Tax Income= 7,899 millions USD), (t= T/B= Tax Rate= 30.00%), ($\$$= Sales or Revenues= 51,294 millions USD), and (v= V/S= Variable Portion= 19.00%), are known, then it's (i= Interest Portion planned) is:

$i = 1-v-f-A/\{\$[1-t]\}$
$= 1-0.1900-0.5800-7,899$
$\qquad /\{51,294[1-0.3000]\}$
$= 1.00\%$

Law-1208 :
If (f= F/S=Fixed Portion= 58.00%), (A= After Tax Income= 7,899 millions USD), (i= I/S= Interest Portion= 1.00%), ($\$$= Sales or Revenues= 51,294 millions USD), and (v= V/S= Variable Portion= 19.00%), are known, then it's (t =Tax Rate planned) is:

$t = 1-A/[\$-\$v-\$f-\$i] = 1-A\{\$[1-v-f-i]$
$= 1-7,899\{51,294[1-0.1900-0.5800$
$\qquad -0.0100]\}$
$= 30.00\%$

Law-1209:

If (**f**= F/S=Fixed Portion= 58.00%), (**t**= T/B= Tax Rate= 30.00%), (**i**= I/S= Interest Portion= 1.00%), (**$'**= Sales of Past Year= 48,851 millions USD), (**s**= [S/S']-1= Sales Growth= 5.00%), (**$**= Sales or Revenues= 51,294 millions USD), and (**v**= V/S= Variable Portion= 19.00%), are known, then it's (**A**= After Tax Income planned) is:

$$\begin{aligned}\mathbf{A} &= [1\text{-}\mathbf{t}]\{\mathbf{\$}\text{-}\mathbf{\$v}\text{-}\mathbf{\$f}\text{-}\mathbf{\$'i}[1\text{+}\mathbf{s}]\} \\ &= [1\text{-}\mathbf{t}]\{\mathbf{\$}[1\text{-}\mathbf{v}\text{-}\mathbf{f}]\text{-}\mathbf{\$'i}[1\text{+}\mathbf{s}]\} \\ &= [1\text{-}0.3000]\{51,294[1\text{-}0.1900\text{-}0.5800] \\ &\quad\quad -48,851*0.0100[1+0.0500]\} \\ &= \underline{7,899} \text{ millions USD}\end{aligned}$$

Law-1210:

If (**f**= F/S=Fixed Portion= 58.00%), (**t**= T/B= Tax Rate= 30.00%), (**i**= I/S= Interest Portion= 1.00%), (**$'**= Sales of Past Year= 48,851 millions USD), (**s**= [S/S']-1= Sales Growth= 5.00%), (**A**= After Tax Income= 7,899 millions USD), and (**v**= V/S= Variable Portion= 19.00%), are known, then it's (**$**= Sales or Revenues planned) is:

$$\begin{aligned}\mathbf{\$} &= \{\mathbf{\$'i}[1\text{+}\mathbf{s}]+\mathbf{A}/[1\text{-}\mathbf{t}]\}/[1\text{-}\mathbf{v}\text{-}\mathbf{f}] \\ &= \{48,851*0.0100[1+0.0500]+7,899 \\ &\quad\quad /[1\text{-}0.3000]\}/[1\text{-}0.1900\text{-}0.5800] \\ &= \underline{51,294} \text{ millions USD}\end{aligned}$$

Law-1211 :

If (f= F/S=Fixed Portion= 58.00%), (t= T/B= Tax Rate= 30.00%), (i= I/S= Interest Portion= 1.00%), (S'= Sales of Past Year= 48,851 millions USD), (s= [S/S']-1= Sales Growth= 5.00%), (A= After Tax Income= 7,899 millions USD), and (S= Sales or Revenues= 51,294 millions USD), are known, then it's (v= Variable Portion planned) is:

$$v = 1-f-\{S'i[1+s]+A/[1-t]\}/S$$
$$= 1-0.5800-\{48,851*0.0100[1+0.0500]+7,899/[1-0.3000]\}/51,294$$
$$= 19.00\%$$

Law-1212 :

If (v= V/S= Variable Portion= 19.00%), (t= T/B= Tax Rate= 30.00%), (i= I/S= Interest Portion= 1.00%), (S'= Sales of Past Year= 48,851 millions USD), (s= [S/S']-1= Sales Growth= 5.00%), (A= After Tax Income= 7,899 millions USD), and (S= Sales or Revenues= 51,294 millions USD), are known, then it's (f = Fixed Portion planned) is:

$$f = 1-v-\{S'i[1+s]+A/[1-t]\}/S$$
$$= 1-0.1900-\{48,851*0.0100[1+0.0500]+7,899/[1-0.3000]\}/51,294$$
$$= 58.00\%$$

Law-1213 :

If (v= V/S= Variable Portion= 19.00%), (t= T/B= Tax Rate= 30.00%), (i= I/S= Interest Portion= 1.00%), (f= F/S=Fixed Portion= 58.00%), (s= [S/S']-1= Sales Growth= 5.00%), (A= After Tax Income= 7,899 millions USD), and (S= Sales or Revenues= 51,294 millions USD), are known, then it's (S'= Sales Past) must be:

S' = {S[1-v-f]-A/[1-t]} {i[1+s]}
 = {51,294[1-0.1900-0.5800]-7,899
      /[1-0.3000]}
      /{0.0100[1+0.0500]}
 = 48,851 millions USD

Law-1214 :

If (v= V/S= Variable Portion= 19.00%), (t= T/B= Tax Rate= 30.00%), (S'= Sales of Past Year= 48,851 millions USD), (f= F/S=Fixed Portion= 58.00%), (s= [S/S']-1= Sales Growth= 5.00%), (A= After Tax Income= 7,899 millions USD), and (S= Sales or Revenues= 51,294 millions USD), are known, then it's (i= Interest Portion planned) is:

i = {S[1-v-f]-A/[1-t]} {S'[1+s]}
 = {51,294[1-0.1900-0.5800]-7,899
      /[1-0.3000]}
      /{48,851[1+0.0500]}
 = 1.00%

Law-1215:

If (v= V/S= Variable Portion= 19.00%), (t= T/B= Tax Rate= 30.00%), ($\$'$= Sales of Past Year= 48,851 millions USD), (f= F/S=Fixed Portion= 58.00%), (i= I/S= Interest Portion= 1.00%), (A= After Tax Income= 7,899 millions USD), and ($\$$= Sales or Revenues= 51,294 millions USD), are known, then it's (s= Sales Growth planned) is:

$$s = \{\$[1-v-f]-A/[1-t]\}[\$'i]-1$$
$$= \{51,294[1-0.1900-0.5800]-7,899 /[1-0.3000]\}/[48,851*0.0100]-1$$
$$= 5.00\%$$

Law-1216:

If (v= V/S= Variable Portion= 19.00%), (s= [S/S']-1= Sales Growth= 5.00%), ($\$'$= Sales of Past Year= 48,851 millions USD), (f= F/S=Fixed Portion= 58.00%), (i= I/S= Interest Portion= 1.00%), (A= After Tax Income= 7,899 millions USD), and ($\$$= Sales or Revenues= 51,294 millions USD), are known, then it's (t= Tax Rate planned) is:

$$t = 1-A/\{\$-\$v-\$f-\$'i[1+s]\}$$
$$= 1-A/\{\$[1-v-f]-\$'i[1+s]\}$$
$$=1-7,899/\{51,294[1-0.1900-0.5800] -48,851*0.0100[1+0.0500]\}$$
$$= 30.00\%$$

Law-1217:

If (v= V/S= Variable Portion= 19.00%), (s= [S/S']-1= Sales Growth= 5.00%), (S'= Sales of Past Year= 48,851 millions USD), (f= F/S=Fixed Portion= 58.00%), (I= Interest Expense= 513 millions USD), (t= T/B= Tax Rate= 30.00%), and (S= Sales or Revenues= 51,294 millions USD), are known, then it's (A= After Tax Income planned) is:

A = [1-t][S-Sv-$S'f$[1+s]-I]
    = [1-t]{S[1-v]-$S'f$[1+s]-I}
    = [1-0.3000]{51,294[1-0.1900]
        -48,851*0.5800[1+0.0500]-513}
    = 7,899 millions USD

Law-1218:

If (v= V/S= Variable Portion= 19.00%), (s= [S/S']-1= Sales Growth= 5.00%), (S'= Sales of Past Year= 48,851 millions USD), (f= F/S=Fixed Portion= 58.00%), (I= Interest Expense= 513 millions USD), (t= T/B= Tax Rate= 30.00%), and (A= After Tax Income= 7,899 millions USD), are known, then it's (S= Sales or Revenues planned) is:

S = {$S'f$[1+s]+I+A/[1-t]}/[1-v]
    = {48,851*0.5800[1+0.0500]+513
        +7,899/ [1-0.3000]}/[1-0.1900]

    = 51,294 millions USD

Law-1219:

If ($= Sales or Revenues= 51,294 millions USD), (s= [S/S']-1= Sales Growth= 5.00%), ($'= Sales of Past Year= 48,851 millions USD), (f= F/S=Fixed Portion= 58.00%), (I= Interest Expense= 513 millions USD), (t= T/B= Tax Rate= 30.00%), and (A= After Tax Income= 7,899 millions USD), are known, then it's (v= Variable Portion planned) is:

$$v = 1 - \{\$'f[1+s] + I + A/[1-t]\}/\$$$
$$= \{48,851*0.5800[1+0.0500] + 513 + 7,899/[1-0.3000]\}/51,294$$
$$= 19.00\%$$

Law-1220:

If ($= Sales or Revenues= 51,294 millions USD), (s= [S/S']-1= Sales Growth= 5.00%), (v= V/S= Variable Portion= 19.00%), (f= F/S=Fixed Portion= 58.00%), (I= Interest Expense= 513 millions USD), (t= T/B= Tax Rate= 30.00%), and (A= After Tax Income= 7,899 millions USD), are known, then it's ($'= Sales Past) must be:

$$\$' = \{\$[1-v] - I - A/[1-t]\}\{f[1+s]\}$$
$$= 51,294[1-0.1900] - 513 - 7,899 /[1-0.3000]\}/0.5800[1+0.0500]\}$$
$$= 48,851 \text{ millions USD}$$

Law-1221:

If ($= Sales or Revenues= 51,294 millions USD), (s= [S/S']-1= Sales Growth= 5.00%), (v= V/S= Variable Portion= 19.00%), ($'= Sales of Past Year= 48,851 millions USD), (I= Interest Expense= 513 millions USD), (t= T/B= Tax Rate= 30.00%), and (A= After Tax Income= 7,899 millions USD), are known, then it's (f= Fixed Portion planned) is:

$$f = \{\$[1-v]-I-A/[1-t]\}\{\$'[1+s]\}$$
$$= 51,294[1-0.1900]-513-7,899$$
$$/[1-0.3000]\}/48,851[1+0.0500]\}$$
$$= 58.00\%$$

Law-1222:

If ($= Sales or Revenues= 51,294 millions USD), (f= F/S=Fixed Portion= 58.00%), (v= V/S= Variable Portion= 19.00%), ($'= Sales of Past Year= 48,851 millions USD), (I= Interest Expense= 513 millions USD), (t= T/B= Tax Rate= 30.00%), and (A= After Tax Income= 7,899 millions USD), are known, then it's (s= Sales Growth planned) is:

$$s = \{\$[1-v]-I-A/[1-t]\}[\$'f]-1$$
$$= \{51,294[1-0.1900]-513-7,899$$
$$/[1-0.3000]\}/[48,851*0.5800]-1$$
$$= 5.00\%$$

Law-1223:

If ($= Sales or Revenues= 51,294 millions USD), (f= F/S=Fixed Portion= 58.00%), (v= V/S= Variable Portion= 19.00%), ($'= Sales of Past Year= 48,851 millions USD), (s= [S/S']-1= Sales Growth= 5.00%), (t= T/B= Tax Rate= 30.00%), and (A= After Tax Income= 7,899 millions USD), are known, then it's (I= Interest Expense planned) is:

$$I = \$[1-v]-\$'f][1+s]-A/[1-t]$$
$$= \{51,294[1-0.1900]- 48,851*0.5800\}$$
$$[1+0.0500]-7,899/[1-0.3000]$$
$$= 513 \text{ millions USD}$$

Law-1224:

If ($= Sales or Revenues= 51,294 millions USD), (f= F/S=Fixed Portion= 58.00%), (v= V/S= Variable Portion= 19.00%), ($'= Sales of Past Year= 48,851 millions USD), (s= [S/S']-1= Sales Growth= 5.00%), (I= Interest Expense= 513 millions USD), and (A= After Tax Income= 7,899 millions USD), are known, then it's (t= Tax Rate planned) is:

$$t = 1-A/\{\$-\$v-\$'f[1+s]-I\}$$
$$= 1-A/\{\$[1-v] -\$'f][1+s]-I\}$$
$$= 1-7,899/\{51,294[1-0.1900]- 48,851$$
$$*0.5800][1+0.0500]-513\}$$
$$= 30.00\%$$

Law-1225:
If ($ = Sales or Revenues= 51,294 millions USD), (f= F/S=Fixed Portion= 58.00%), (v= V/S= Variable Portion= 19.00%), ($'= Sales of Past Year= 48,851 millions USD), (s= [S/S']-1= Sales Growth= 5.00%), (i= I/S= Interest Portion= 1.00%), and (t= T/B= Tax Rate= 30.00%), are known, then it's (A= After Tax Income planned) is:

$$A = [1-t][\$-\$v-\$'f[1+s]-\$i\}$$
$$= [1-t]\{\$[1-v-i]-\$'f[1+s]\}$$
$$= [1-0.3000]\{51,294[1-0.1900-0.0100] - 48,851*0.5800][1+0.0500]\}$$
$$= 7,899 \text{ millions USD}$$

Law-1226:
If (A= After Tax Income= 7,899 millions USD), (f= F/S=Fixed Portion= 58.00%), (v= V/S= Variable Portion= 19.00%), ($'= Sales of Past Year= 48,851 millions USD), (s= [S/S']-1= Sales Growth= 5.00%), (i= I/S= Interest Portion= 1.00%), and (t= T/B= Tax Rate= 30.00%), are known, then it's ($= Sales or Revenues planned) is:

$$\$ = \{A/[1-t]+\$'f[1+s]\}/[1-v-i]$$
$$= \{7.899/[1-0.3000] + 48,851*0.5800][1+0.0500]\}/[1-0.1900-0.0100]$$
$$= 51,294 \text{ millions USD}$$

Law-1227:

If (A= After Tax Income= 7,899 millions USD), (f= F/S=Fixed Portion= 58.00%), (S= Sales or Revenues= 51,294 millions USD), (S'= Sales of Past Year= 48,851 millions USD), (s= [S/S']-1= Sales Growth= 5.00%), (i= I/S= Interest Portion= 1.00%), and (t= T/B= Tax Rate= 30.00%), are known, then it's (v= Variable Portion planned) is:

v= 1-i-{$S'f$[1+s]+A/[1-t]}/S
$\quad$ = 1-0.0100-{48,851*0.5800][1+0.0500]
$\quad\quad$ +7.899/[1-0.3000]}/51,294
$\quad$ = 19.00%

Law-1228:

If (A= After Tax Income= 7,899 millions USD), (f= F/S=Fixed Portion= 58.00%), (S= Sales or Revenues= 51,294 millions USD), (S'= Sales of Past Year= 48,851 millions USD), (s= [S/S']-1= Sales Growth= 5.00%), (v= V/S= Variable Portion= 19.00%), and (t= T/B= Tax Rate= 30.00%), are known, then it's (i= Interest Portion planned) is:

i= 1-v-{$S'f$[1+s]+A/[1-t]}/S
$\quad$ = 1-0.1900-{48,851*0.5800][1+0.0500]
$\quad\quad$ +{7.899/[1-0.3000]}/51,294
$\quad$ = 1.00%

Law-1229:

If (**A**= After Tax Income= 7,899 millions USD), (**f**= F/S=Fixed Portion= 58.00%), (**$**= Sales or Revenues= 51,294 millions USD), (**i**= I/S= Interest Portion= 1.00%), (**s**= [S/S']-1= Sales Growth= 5.00%), (**v**= V/S= Variable Portion= 19.00%), and (**t**= T/B= Tax Rate= 30.00%), are known, then it's (**$'**= Sales Past) must be:

$'= ${[1-**v**-**i**]-**A**/[1-**t**]}/{**f**[1+**s**]}
    = 51,294[1-0.1900-0.01]-7.899
      /[1-0.3000]}
      /{0.5800[1+0.0500]}
    = 48,851 millions USD

Law-1230:

If (**A**= After Tax Income= 7,899 millions USD), (**$**= Sales or Revenues= 51,294 millions USD), (**i**= I/S= Interest Portion= 1.00%), (**s**= [S/S']-1= Sales Growth= 5.00%), (**$'**= Sales of Past Year= 48,851 millions USD), (**v**= V/S= Variable Portion= 19.00%), and (**t**= T/B= Tax Rate= 30.00%), are known, then it's (**f**= Fixed Portion planned) is:

f= {${[1-**v**-**i**]-**A**/[1-**t**]}/{$'[1+**s**]}
    = 51,294[1-0.1900-0.01]-7.899
      /[1-0.3000]}
      /{48,851[1+0.0500]}
    = 58.00%

Law-1231:

If (**A**= After Tax Income= 7,899 millions USD), (**$**= Sales or Revenues= 51,294 millions USD), (**i**= I/S= Interest Portion= 1.00%), (**f**= F/S=Fixed Portion= 58.00%), (**$'**= Sales of Past Year= 48,851 millions USD), (**v**= V/S= Variable Portion= 19.00%), and (**t**= T/B= Tax Rate= 30.00%), are known, then it's (**s**= Sales Growth planned) is:

$$s= \{S[1-v-i]-A/[1-t]\}/[S'f]-1$$
$$= 51,294[1-0.1900-0.0100]-7.899$$
$$/[1-0.3000]\}/[48,851*0.5800]-1$$
$$= 5.00\%$$

Law-1232:

If (**A**= After Tax Income= 7,899 millions USD), (**$**= Sales or Revenues= 51,294 millions USD), (**i**= I/S= Interest Portion= 1.00%), (**f**= F/S=Fixed Portion= 58.00%), (**$'**= Sales of Past Year= 48,851 millions USD), (**v**= V/S= Variable Portion= 19.00%), and (**s**= [S/S']-1= Sales Growth= 5.00%), are known, then it's (**t**= Tax Rate planned) is:

$$t= 1-A/\{S-Sv-S'f[1+s]-Si\}$$
$$= 1-A/\{S[1-v-i]-S'f[1+s]\}$$
$$= 1-7,899/\{51,294[1-0.1900-0.0100]$$
$$-48,851*0.5800[1+0.0500]\}$$
$$= 30.00\%$$

Law-1233:

If (**t**= T/B= Tax Rate= 30.00%), (**$**= Sales or Revenues= 51,294 millions USD), (**i**= I/S= Interest Portion= 1.00%), (**f**= F/S=Fixed Portion= 58.00%), (**$'**= Sales of Past Year= 48,851 millions USD), (**v**= V/S= Variable Portion= 19.00%), and (**s**= [S/S']-1= Sales Growth= 5.00%), are known, then it's (**A**= After Tax Income planned) is:

$$A = [1-t]\{\$ - \$v - \$'f[1+s] - \$'i[1+s]\}$$
$$= [1-t]\{\$[1-v] - \$'[1+s][f+i]\}$$
$$= [1-0.3000]\{51,294[1-0.1900] - 48,851$$
$$[1+0.0500][0.5800+0.0100]\}$$
$$= 7,899 \text{ millions USD}$$

Law-1234:

If (**t**= T/B= Tax Rate= 30.00%), (**A**= After Tax Income= 7,899 millions USD), (**i**= I/S= Interest Portion= 1.00%), (**f**= F/S=Fixed Portion= 58.00%), (**$'**= Sales of Past Year= 48,851 millions USD), (**v**= V/S= Variable Portion= 19.00%), and (**s**= [S/S']-1= Sales Growth= 5.00%), are known, then it's (**$**= Sales or Revenues planned) is:

$$\$ = \{\$'[1+s][f+i] + A/[1-t]\}/[1-v]$$
$$= \{48,851[1+0.0500][0.5800+0.0100]$$
$$+ 7,899/[1-0.3000]\}/[1-0.1900]$$
$$= 51,294 \text{ millions USD}$$

Law-1235:

If (t= T/B= Tax Rate= 30.00%), (A= After Tax Income= 7,899 millions USD), (i= I/S= Interest Portion= 1.00%), (f= F/S=Fixed Portion= 58.00%), ($\$'$= Sales of Past Year= 48,851 millions USD), ($\$$= Sales or Revenues= 51,294 millions USD), and (s= [S/S']-1= Sales Growth= 5.00%), are known, then it's (v= Variable Portion planned) is:

$$v = 1-\{\$'[1+s][f+i]+A/[1-t]\}/\$$$
$$= 1-48,851[1+0.0500][0.5800+0.0100]$$
$$+7,899/[1-0.3000]\}/51,294$$
$$= 19.00\%$$

Law-1236:

If (t= T/B= Tax Rate= 30.00%), (A= After Tax Income= 7,899 millions USD), (i= I/S= Interest Portion= 1.00%), (f= F/S=Fixed Portion= 58.00%), (v= V/S= Variable Portion= 19.00%), ($\$$= Sales or Revenues= 51,294 millions USD), and (s= [S/S']-1= Sales Growth= 5.00%), are known, then it's ($\$'$= Sales Past) must be:

$$\$' = \{\$[1-v]-A/[1-t]\}/\{[1+s][f+i]\}$$
$$= \{51,294[1-0.1900]-7,899/[1-0.3000]\}$$
$$/\{[1+0.0500][0.5800+0.0100]\}$$
$$= 48,851 \text{ millions USD}$$

Law-1237:

If (**t**= T/B= Tax Rate= 30.00%), (**A**= After Tax Income= 7,899 millions USD), (**i**= I/S= Interest Portion= 1.00%), (**f**= F/S=Fixed Portion= 58.00%), (**v**= V/S= Variable Portion= 19.00%), (**S**= Sales or Revenues= 51,294 millions USD), and (**S'**= Sales of Past Year= 48,851 millions USD), are known, then it's (**s**=Sales Growth planned) is:

$$s = \{S[1-v]-A/[1-t]\}/\{S'[f+i]\}-1$$
$$= \{51,294[1-0.1900]-7,899/[1-0.3000]\}$$
$$/\{48,851[0.5800+0.0100]\}-1$$
$$= 5.00\%$$

Law-1238:

If (**t**= T/B= Tax Rate= 30.00%), (**A**= After Tax Income= 7,899 millions USD), (**i**= I/S= Interest Portion= 1.00%), (**s**= [S/S']-1= Sales Growth= 5.00%), (**v**= V/S= Variable Portion= 19.00%), (**S**= Sales or Revenues= 51,294 millions USD), and (**S'**= Sales of Past Year= 48,851 millions USD), are known, then it's (**f**= Fixed Portion planned) is:

$$f = \{S[1-v]-A/[1-t]\}/\{S'[1+s]\}-i$$
$$= \{51,294[1-0.1900]-7,899/[1-0.3000]\}$$
$$/\{48,851[1+0.0500]\}-0.0100$$
$$= 58.00\%$$

Law-1239:
If (**t**= T/B= Tax Rate= 30.00%), (**A**= After Tax Income= 7,899 millions USD), (**f**= F/S=Fixed Portion= 58.00%), (**s**= [S/S']-1= Sales Growth= 5.00%), (**v**= V/S= Variable Portion= 19.00%), (**$**= Sales or Revenues= 51,294 millions USD), and (**$'**= Sales of Past Year= 48,851 millions USD), are known, then it's (**i**= Interest Portion planned) is:

$$i = \{\$[1-v]-A/[1-t]\}/\{\$'[1+s]\}-f$$
$$= \{51,294[1-0.1900]-7,899/[1-0.3000]\}$$
$$/\{48,851[1+0.0500]\}-0.5800$$
$$= 1.00\%$$

Law-1240:
If (**i**= I/S= Interest Portion= 1.00%), (**A**= After Tax Income= 7,899 millions USD), (**f**= F/S=Fixed Portion= 58.00%), (**s**= [S/S']-1= Sales Growth= 5.00%), (**v**= V/S= Variable Portion= 19.00%), (**$**= Sales or Revenues= 51,294 millions USD), and (**$'**= Sales of Past Year= 48,851 millions USD), are known, then it's (**t**= Tax Rate planned) is:

$$t = 1-A/\{\$-\$v-\$'f[1+s]-\$'i[1+s]\}$$
$$= 1-A/\{\$[1-v]-\$'[1+s][f+i]\}$$
$$= 1-7,899/\{\{51,294[1-0.1900]-48,851$$
$$[1+0.0500][0.5800+0.0100]\}$$
$$= 30.00\%$$

Law-1241:

If (t= T/B= Tax Rate= 30.00%), (F= Fixed Cost= 29,740 millions USD), (I= Interest Expense= 513 millions USD), (s= [S/S']-1= Sales Growth= 5.00%), (v= V/S= Variable Portion= 19.00%), (S= Sales or Revenues= 51,294 millions USD), and (S'= Sales of Past Year= 48,851 millions USD), are known, then it's (A= After Tax Income planned) is:

$A = [1-t]\{S-F-I-S'v[1+s]\}$
$= [1-0.3000]\{51,294-29,750-513$
$-48,851*0.1900[1+0.0500]\}$
$= 7,899$ millions USD

Law-1242:

If (t= T/B= Tax Rate= 30.00%), (F= Fixed Cost= 29,740 millions USD), (I= Interest Expense= 513 millions USD), (s= [S/S']-1= Sales Growth= 5.00%), (v= V/S= Variable Portion= 19.00%), (A= After Tax Income= 7,899 millions USD), and (S'= Sales of Past Year= 48,851 millions USD), are known, then it's (S= Sales or Revenues planned) is:

$S = S'v[1+s]+F+I+A/[1-t]\}$
$= 48,851*0.1900[1+0.0500] +29,750$
$+513+7,899/[1-0.3000]\}$
$= 51,294$ millions USD

Law-1243:
If (**t**= T/B= Tax Rate= 30.00%), (**F**= Fixed Cost= 29,740 millions USD), (**I**= Interest Expense= 513 millions USD), (**s**= [S/S']-1= Sales Growth= 5.00%), (**v**= V/S= Variable Portion= 19.00%), (**A**= After Tax Income= 7,899 millions USD), and (**S**= Sales or Revenues= 51,294 millions USD), are known, then it's (**S'**= Sales Past) must be:

$$S' = \{S\text{-}F\text{-}I\text{-}A/[1\text{-}t]\}/\{v[1+s]\}$$
$$= 51{,}294\text{-}29{,}750\text{-}513\text{-}7{,}899/[1\text{-}0.3000]\}$$
$$/\{0.1900[1+0.0500]\}$$
$$= 48{,}851 \text{ millions USD}$$

Law-1244:
If (**t**= T/B= Tax Rate= 30.00%), (**F**= Fixed Cost= 29,740 millions USD), (**I**= Interest Expense= 513 millions USD), (**s**= [S/S']-1= Sales Growth= 5.00%), (**S'**= Sales of Past Year= 48,851 millions USD), (**A**= After Tax Income=7,899 millions USD), and (**S**= Sales or Revenues= 51,294 millions USD), are known, then it's (**v**= Variable Portion planned) is:

$$v = \{S\text{-}F\text{-}I\text{-}A/[1\text{-}t]\}/\{S'[1+s]\}$$
$$= \{51{,}294\text{-}29{,}750\text{-}513\text{-}7{,}899/[1\text{-}0.3000]\}$$
$$/\{48{,}851[1+0.0500]\}$$
$$= 19.00\%$$

Law-1245:

If (**t**= T/B= Tax Rate= 30.00%), (**F**= Fixed Cost= 29,740 millions USD), (**I**= Interest Expense= 513 millions USD), (**v**= V/S= Variable Portion= 19.00%), (**S'**= Sales of Past Year= 48,851 millions USD), (**A**= After Tax Income=7,899 millions USD), and (**S**= Sales or Revenues= 51,294 millions USD), are known, then it's (**s**= Sales Growth planned) is:

$$s = \{S-F-I-A/[1-t]\}/\{S'v\} - 1$$
$$= \{51,294-29,750-513-7,899/[1-0.3000]\}/[48,851*0.1900]-1$$
$$= 5.00\%$$

Law-1246:

If (**t**= T/B= Tax Rate= 30.00%), (**s**= [S/S']-1= Sales Growth= 5.00%), (**I**= Interest Expense= 513 millions USD), (**v**= V/S= Variable Portion= 19.00%), (**S'**= Sales of Past Year= 48,851 millions USD), (**A**= After Tax Income=7,899 millions USD), and (**S**= Sales or Revenues= 51,294 millions USD), are known, then it's (**F**= Fixed Cost planned) is:

$$F = \{S-I-S'v[1+s]-A/[1-t]\}$$
$$= \{51,294-513-48,851*0.1900][1+0.0500]-7,899/[1-0.3000]\}$$
$$= 29,750 \text{ millions USD}$$

Law-1247:

If (t= T/B= Tax Rate= 30.00%), (s= [S/S']-1= Sales Growth= 5.00%), (F= Fixed Cost= 29,740 millions USD), (v= V/S= Variable Portion= 19.00%), (S'= Sales of Past Year= 48,851 millions USD), (A= After Tax Income=7,899 millions USD), and (S= Sales or Revenues= 51,294 millions USD), are known, then it's (I= Interest Portion planned) is:

I = {S-F-$S'v$[1+s]-A/[1-t]}
 ={51,294-29,750-48,851*0.1900]
  [1+0.0500] -7,899/[1-0.3000]}
 = 513 millions USD

Law-1248:

If (I= Interest Expense= 513 millions USD), (s= [S/S']-1= Sales Growth= 5.00%), (F= Fixed Cost= 29,740 millions USD), (v= V/S= Variable Portion= 19.00%), (S'= Sales of Past Year= 48,851 millions USD), (A= After Tax Income=7,899 millions USD), and (S= Sales or Revenues= 51,294 millions USD), are known, then it's (t=Tax Rate planned) is:

t= 1-A/{S -$S'v$[1+s]-F-I}
 =1-7,899/{51,294 -48,851*0.1900]
  [1+0.0500] -29,750-513}
 = 30.00%

Law-1249:
If (i= I/S= Interest Portion= 1.00%), (s= [S/S']-1= Sales Growth= 5.00%), (F= Fixed Cost= 29,740 millions USD), (v= V/S= Variable Portion= 19.00%), (S'= Sales of Past Year= 48,851 millions USD), (t= T/B= Tax Rate= 30.00%), and (S= Sales or Revenues= 51,294 millions USD), are known, then it's (A= After Tax Income planned) is:

A=[1-t]{S -S'v[1+s]-F-Si}
=[1-0.3000]{51,294 -48,851*0.1900]
[1+0.0500] -29,750
-51,294*0.0100}
= 7,899 millions USD

Law-1250:
If (i= I/S= Interest Portion= 1.00%), (s= [S/S']-1= Sales Growth= 5.00%), (F= Fixed Cost= 29,740 millions USD), (v= V/S= Variable Portion= 19.00%), (S'= Sales of Past Year= 48,851 millions USD), (t= T/B= Tax Rate= 30.00%), and (A= After Tax Income= 7,899 millions USD), are known, then it's (S= Sales or Revenues planned) is:

S={F+$S'v$[1+s]+A/[1-t]}/[1-i]
= {29,750+48,851*0.1900][1+0.0500]
+7,899/[1-0.3000]}/[1-0.0100]
= 51,294 millions USD

Law-1251:

If ($= Sales or Revenues= 51,294 millions USD), ($= [S/S']-1= Sales Growth= 5.00%), (F= Fixed Cost= 29,740 millions USD), (v= V/S= Variable Portion= 19.00%), ($'= Sales of Past Year= 48,851 millions USD), (t= T/B= Tax Rate= 30.00%), and (A= After Tax Income= 7,899 millions USD), are known, then it's (i= Interest Portion planned) is:

$$i = 1-\{F+\$'v[1+s]+A/[1-t]\}/\$$$
$$= 1-\{29,750+48,851*0.1900][1+0.0500]$$
$$+7,899/[1-0.3000]\}/51,294$$
$$= \underline{1.00\%}$$

Law-1252:

If ($= Sales or Revenues= 51,294 millions USD), ($= [S/S']-1= Sales Growth= 5.00%), (F= Fixed Cost= 29,740 millions USD), (v= V/S= Variable Portion= 19.00%), (i= I/S= Interest Portion= 1.00%), (t= T/B= Tax Rate= 30.00%), and (A= After Tax Income= 7,899 millions USD), are known, then it's ($'= Sales Past) must be:

$$\$' = \{\$[1-i]-F-A/[1-t]\}/\{v[1+s]\}$$
$$= \{51,294[1-0.0100]- 29,750-7,899$$
$$/[1-0.3000]\}$$
$$/\{0.1900[1+0.0500]\}$$
$$= \underline{48,851} \text{ millions USD}$$

Law-1253:

If ($\$$= Sales or Revenues= 51,294 millions USD), ($\pmb{s}$= [S/S']-1= Sales Growth= 5.00%), ($\pmb{F}$= Fixed Cost= 29,740 millions USD), ($\$'$= Sales of Past Year= 48,851 millions USD), ($\pmb{i}$= I/S= Interest Portion= 1.00%), ($\pmb{t}$= T/B= Tax Rate= 30.00%), and ($\pmb{A}$= After Tax Income= 7,899 millions USD), are known, then it's ($\pmb{v}$= Variable Portion planned) is:

$\pmb{v}$= {$\$$[1-$\pmb{i}$]-$\pmb{F}$-$\pmb{A}$/[1-$\pmb{t}$]}/{$\$'$[1+$\pmb{s}$]}
 = {51,294[1-0.0100]- 29,750-7,899
           /[1-0.3000]}
           /{48,851[1+0.0500]}
 = 19.00%

Law-1254:

If ($\$$= Sales or Revenues= 51,294 millions USD), ($\pmb{v}$= V/S= Variable Portion= 19.00%), ($\pmb{F}$= Fixed Cost= 29,740 millions USD), ($\$'$= Sales of Past Year= 48,851 millions USD), ($\pmb{i}$= I/S= Interest Portion= 1.00%), ($\pmb{t}$= T/B= Tax Rate= 30.00%), and ($\pmb{A}$= After Tax Income= 7,899 millions USD), are known, then it's ($\pmb{s}$= Sales Growth planned) is:

$\pmb{s}$= {$\$$[1-$\pmb{i}$]-$\pmb{F}$-$\pmb{A}$/[1-$\pmb{t}$]}/[$\$'\pmb{v}$]-1
 = {51,294[1-0.0100]- 29,750-7,899
           /[1-0.3000]}/[48,851*0.1900]-1
 = 5.00%

Law-1255:

If ($=Sales or Revenues= 51,294 millions USD), (υ= V/S= Variable Portion= 19.00%), (s= [S/S']-1= Sales Growth= 5.00%), ($'= Sales of Past Year= 48,851 millions USD), (i= I/S= Interest Portion= 1.00%), (t= T/B= Tax Rate= 30.00%), and (A= After Tax Income= 7,899 millions USD), are known, then it's (F= Fixed Cost planned) is:

F = $[1-i]-$'υ[1+s]-A/[1-t]
= 51,294[1-0.0100]-48,851*0.1900]
    [1+0.0500]-7,899/[1-0.3000]
= 29,750 millions USD

Law-1256:

If ($= Sales or Revenues= 51,294 millions USD), (υ= V/S= Variable Portion= 19.00%), (s= [S/S']-1= Sales Growth= 5.00%), ($'= Sales of Past Year= 48,851 millions USD), (i= I/S= Interest Portion= 1.00%), (F= Fixed Cost= 29,740 millions USD), and (A= After Tax Income= 7,899 millions USD), are known, then it's (t= Tax Rate planned) is:

t = 1-A/{$-$'υ[1+s]-F-$i}
= 1-A/{$[1-i]-$'υ[1+s]-F}
= 1-7,899/{ 51,294[1-0.0100]-48,851
        *0.1900][1+0.0500] -29,750}
= 30.00%

Law-1257:

If ($= Sales or Revenues= 51,294 millions USD), (v= V/S= Variable Portion= 19.00%), (s= [S/S']-1= Sales Growth= 5.00%), ($'= Sales of Past Year= 48,851 millions USD), (i= I/S= Interest Portion= 1.00%), (F= Fixed Cost= 29,740 millions USD), and (t= T/B= Tax Rate= 30.00%), are known, then it's (A= After Tax Income planned) is:

$$A = [1-t]\{\$-\$'v[1+s]-F-\$'i[1+s]\}$$
$$= [1-t]\{\$-F-\$'[1+s][v+i]\}$$
$$= [1-0.3000]\{51,294-29,750-48,851$$
$$[1+0.0500][0.1900+0.0100]\}$$
$$= 7,899 \text{ millions USD}$$

Law-1258:

If (A= After Tax Income= 7,899 millions USD), (v= V/S= Variable Portion= 19.00%), (s= [S/S']-1= Sales Growth= 5.00%), ($'= Sales of Past Year= 48,851 millions USD), (i= I/S= Interest Portion= 1.00%), (F= Fixed Cost= 29,740 millions USD), and (t= T/B= Tax Rate= 30.00%), are known, then it's ($= Sales or Revenues planned) is:

$$\$ = F+\$'[v+i][1+s]+A/[1-t]$$
$$= 29,750+48,851[0.1900+0.0100][1+.0500]$$
$$+7,899/[1-0.3000]$$
$$= 51,294 \text{ millions USD}$$

Law-1260:
If (**A**= After Tax Income= 7,899 millions USD), (**$'**= Sales of Past Year= 48,851 millions USD), (**s**= [S/S']-1= Sales Growth= 5.00%), (**$**= Sales or Revenues= 51,294 millions USD), (**i**= I/S= Interest Portion= 1.00%), (**F**= Fixed Cost= 29,740 millions USD), and (**t**= T/B= Tax Rate= 30.00%), are known, then it's (**v**= Variable Portion planned) is:

$$v = \{\$-F-A/[1-t]\}/\{\$'[1+s]\}-i$$
$$= \{51,294-29,750-7,899/[1-0.3000]\}/48,851[1+0.0500]\}-0.0100$$
$$= 19.00\%$$

Law-1261:
If (**A**= After Tax Income= 7,899 millions USD), (**$'**= Sales of Past Year= 48,851 millions USD), (**s**= [S/S']-1= Sales Growth= 5.00%), (**$**= Sales or Revenues= 51,294 millions USD), (**v**= V/S= Variable Portion= 19.00%), (**F**= Fixed Cost= 29,740 millions USD), and (**t**= T/B= Tax Rate= 30.00%), are known, then it's (**i**= Interest Portion planned) is:

$$i = \{\$-F-A/[1-t]\}/\{\$'[1+s]\}-v$$
$$= \{51,294-29,750-7,899/[1-0.3000]\}/48,851[1+0.0500]\}-0.1900$$
$$= 1.00\%$$

Law-1262:

If (**A**= After Tax Income= 7,899 millions USD), (**S'**= Sales of Past Year= 48,851 millions USD), (**i**= I/S= Interest Portion= 1.00%), (**S**= Sales or Revenues= 51,294 millions USD), (**v**= V/S= Variable Portion= 19.00%), (**F**= Fixed Cost= 29,740 millions USD), and (**t**= T/B= Tax Rate= 30.00%), are known, then it's (**s**= Sales Growth planned) is:

$$s = \{S-F-A/[1-t]\}/\{S'[v+i]\} - 1$$
$$= \{51,294 - 29,750 - 7,899/[1-0.3000]\}/48,851[0.1900+0.0100]\} - 1$$
$$= 5.00\%$$

Law-1263:

If (**A**= After Tax Income= 7,899 millions USD), (**S'**= Sales of Past Year= 48,851 millions USD), (**i**= I/S= Interest Portion= 1.00%), (**S**= Sales or Revenues= 51,294 millions USD), (**v**= V/S= Variable Portion= 19.00%), (**s**= [S/S']-1= Sales Growth= 5.00%), and (**t**= T/B= Tax Rate= 30.00%), are known, then it's (**F**= Fixed Cost planned) is:

$$F = \{S - S'[v+i][1+s] - A/[1-t]\}$$
$$= \{51294 - 48,851[0.1900+0.0100][1+0.0500] - 7,899/[1-0.3000]\}$$
$$= 29,750 \text{ millions USD}$$

Law-1264:
If (**A**= After Tax Income= 7,899 millions USD), (**$'**= Sales of Past Year= 48,851 millions USD), (**i**= I/S= Interest Portion= 1.00%), (**$**= Sales or Revenues= 51,294 millions USD), (**v**= V/S= Variable Portion= 19.00%), (**s**= [S/S']-1= Sales Growth= 5.00%), and (**F**= Fixed Cost= 29,740 millions USD), are known, then it's (**F**= Tax Rate planned) is:

$$t = 1-A\{\$-\$'v[1+s]-F-\$'i[1+s]\}$$
$$= 1-A/\{\$-\$'[v+i][1+s]-F\}$$
$$= 1-7,899\{51,294-48,851[0.1900$$
$$+0.0100][1+0.0500] -29,750\}$$
$$= 30.00\%$$

Law-1265:
If (**$'**= Sales of Past Year= 48,851 millions USD), (**I**= Interest Expense= 513 millions USD), (**t**= T/B= Tax Rate= 30.00%), (**$**= Sales or Revenues= 51,294 millions USD), (**v**= V/S= Variable Portion= 19.00%), (**s**= [S/S']-1= Sales Growth= 5.00%), and (**f**= F/S= Fixed Portion= 58.00%), are known, then it's (**A**= After Tax Income planned) is:

$$A = [1-t]\{\$-\$'v[1+s]-\$f-I\}$$
$$= [1-t]\{\$[1-f]-\$'v[1+s]-I\}$$
$$= [1-0.3000]\{51294[1-0.5800]-48,851$$
$$*0.1900[1+0.0500] -513\}$$
$$= 7,899 \text{ millions USD}$$

Law-1266:
If ($'= Sales of Past Year= 48,851 millions USD), (I= Interest Expense= 513 millions USD), (t= T/B= Tax Rate= 30.00%), (A= After Tax Income= 7,899 millions USD), (v= V/S= Variable Portion= 19.00%), (s= [S/S']-1= Sales Growth= 5.00%), and (f= F/S= Fixed Portion= 58.00%), are known, then it's ($= Sales or Revenues planned) is:

$$\$ = \{I+\$'v[1+s]+A/1-t\}/[1-f]$$
$$= \{513+48,851*0.1900[1+0.0500]$$
$$+7,899/[1-0.3000]\}/[1-0.5800]$$
$$= \underline{51,294} \text{ millions USD}$$

Law-1267:
If ($'= Sales of Past Year= 48,851 millions USD), (I= Interest Expense= 513 millions USD), (t= T/B= Tax Rate= 30.00%), (A= After Tax Income= 7,899 millions USD), (v= V/S= Variable Portion= 19.00%), (s= [S/S']-1= Sales Growth= 5.00%), and ($= Sales or Revenues= 51,294 millions USD), are known, then it's (f= Fixed Portion planned) is:

$$f = 1-\{I+\$'v[1+s]+A/1-t\}/\$$$
$$= 1-\{513+48,851*0.1900[1+0.0500]$$
$$+7,899/[1-0.3000]\}/51,294$$
$$= \underline{58.00\%}$$

Law-1268:
If (**f**= F/S=Fixed Portion= 58.00%), (**I**= Interest Expense= 513 millions USD), (**t**= T/B= Tax Rate= 30.00%), (**A**= After Tax Income= 7,899 millions USD), (**v**= V/S= Variable Portion= 19.00%), (**s**= [S/S']-1= Sales Growth= 5.00%), and (**$**= Sales or Revenues= 51,294 millions USD), are known, then it's (**$'**= Sales Past) must be:

$$\$' = \$[1-f]-\{I-A/[1-t]\}/\{v[1+s]\}$$
$$= 51,294[1-0.5800]-\{513-7,899$$
$$/[1-0.3000]\}$$
$$/\{0.1900[1+0.0500]\}$$
$$= 48,851 \text{ millions USD}$$

Law-1269:
If (**f**= F/S=Fixed Portion= 58.00%), (**I**= Interest Expense= 513 millions USD), (**t**= T/B= Tax Rate= 30.00%), (**A**= After Tax Income= 7,899 millions USD), (**$'**= Sales of Past Year= 48,851 millions USD), (**s**= [S/S']-1= Sales Growth= 5.00%), and (**$**= Sales or Revenues= 51,294 millions USD), are known, then it's (**v**= Variable Portion planned) is:

$$v = \$[1-f]-\{I-A/[1-t]\}/\{\$'[1+s]\}$$
$$= 51,294[1-0.5800]-\{513-7,899$$
$$/[1-0.3000]\}$$
$$/\{48,851[1+0.0500]\}$$
$$= 19.00\%$$

Law-1270:

If (f= F/S=Fixed Portion= 58.00%), (I= Interest Expense= 513 millions USD), (t= T/B= Tax Rate= 30.00%), (A= After Tax Income= 7,899 millions USD), (S'= Sales of Past Year= 48,851 millions USD), (v= V/S= Variable Portion= 19.00%), and (S= Sales or Revenues= 51,294 millions USD), are known, then it's (s= Sales Growth planned) is:

$s = \{S[1-f]-I-A/[1-t]\}/[S'v]-1$
$= \{51,294[1-0.5800]-513-7,899$
$\qquad /[1-0.3000]\}/[48,851*0.1900]-1$
$= 5.00\%$

Law-1271:

If (f= F/S=Fixed Portion= 58.00%), (s= [S/S']-1= Sales Growth= 5.00%), (t= T/B= Tax Rate= 30.00%), (A= After Tax Income= 7,899 millions USD), (S'= Sales of Past Year= 48,851 millions USD), (v= V/S= Variable Portion= 19.00%), and (S= Sales or Revenues= 51,294 millions USD), are known, then it's (I= Interest Expense planned) is:

$I = \{S[1-f]-S'v[1+s]-A/[1-t]\}$
$= \{51,294[1-0.5800]- 48,851*0.1900$
$\qquad [1+0.0500]-7,899/[1-0.3000]\}$
$= 513$ millions USD

Finance Construction-2, *Tim Asikin, Steve Asikin, Indra Senihardja*

Law-1272:
If (f= F/S=Fixed Portion= 58.00%), (s= [S/S']-1= Sales Growth= 5.00%), (I= Interest Expense= 513 millions USD), (A= After Tax Income= 7,899 millions USD), (S'= Sales of Past Year= 48,851 millions USD), (v= V/S= Variable Portion= 19.00%), and (S= Sales or Revenues= 51,294 millions USD), are known, then it's (t= Tax Rate planned) is:

$$t = 1-A/\{S-S'v[1+s]-Sf-I\}$$
$$= 1-A/\{S[1-f]-S'v[1+s]-I\}$$
$$= 1-7,899/\{51,294[1-0.5800]- 48,851$$
$$*0.1900[1+0.0500]-513\}$$
$$= 30.00\%$$

Law-1273:
If (f= F/S=Fixed Portion= 58.00%), (s= [S/S']-1= Sales Growth= 5.00%), (i= I/S= Interest Portion= 1.00%), (t= T/B= Tax Rate= 30.00%), (S'= Sales of Past Year= 48,851 millions USD), (v= V/S= Variable Portion= 19.00%), and (S= Sales or Revenues= 51,294 millions USD), are known, then it's (A= After Tax Income planned) is:

$$A = [1-t]\{S-S'v[1+s]-Sf-Si\}$$
$$= [1-t]\{S[1-f-i]-S'v[1+s]\}$$
$$= [1-0.3000]\{51,294[1-0.5800-0.0100]$$
$$- 48,851*0.1900[1+0.0500]\}$$
$$= 7,899 \text{ millions USD}$$

Law-1274:

If (f= F/S=Fixed Portion= 58.00%), (s= [S/S']-1= Sales Growth= 5.00%), (i= I/S= Interest Portion= 1.00%), (t= T/B= Tax Rate= 30.00%), (S'= Sales of Past Year= 48,851 millions USD), (v= V/S= Variable Portion= 19.00%), and (A= After Tax Income= 7,899 millions USD), are known, then it's (S= Sale or Revenues planned) is:

S = {A/[1-t]+$S'v$[1+s]}/[1-f-i]
  = {7,899/[1-0.3000] + 48,851*0.1900
      [1+0.0500]/[1-0.5800-0.0100]
  = 51,294 millions USD

Law-1275:

If (S= Sales or Revenues= 51,294 millions USD), (s= [S/S']-1= Sales Growth= 5.00%), (i= I/S= Interest Portion= 1.00%), (t= T/B= Tax Rate= 30.00%), (S'= Sales of Past Year= 48,851 millions USD), (v= V/S= Variable Portion= 19.00%), and (A= After Tax Income= 7,899 millions USD), are known, then it's (f= Fixed Portion planned) is:

f= 1-i-{A/[1-t]+$S'v$[1+s]}/S
  = 1-0.0100-{7,899/[1-0.3000]+ 48,851
      *0.1900[1+0.0500]/51,294
  = 58.00%

Law-1276:

If ($= Sales or Revenues= 51,294 millions USD), ($= [S/S']-1= Sales Growth= 5.00%), (f= F/S=Fixed Portion= 58.00%), (t= T/B= Tax Rate= 30.00%), ($'= Sales of Past Year= 48,851 millions USD), (v= V/S= Variable Portion= 19.00%), and (A= After Tax Income= 7,899 millions USD), are known, then it's (i= Interest Portion planned) is:

$$i = 1-f-\{A/[1-t]+\$'v[1+s]\}/\$$$
$$= 1-0.5800-\{7,899/[1-0.3000]+48,851$$
$$*0.1900[1+0.0500]\}/51,294$$
$$= 1.00\%$$

Law-1277:

If ($= Sales or Revenues= 51,294 millions USD), ($= [S/S']-1= Sales Growth= 5.00%), (f= F/S=Fixed Portion= 58.00%), (t= T/B= Tax Rate= 30.00%), (i= I/S= Interest Portion= 1.00%), (v= V/S= Variable Portion= 19.00%), and (A= After Tax Income= 7,899 millions USD), are known, then it's ($'= Sales Past) must be:

$$\$' = \{\$[1-f-i]-A/[1-t]\}/\{v[1+s]\}$$
$$= \{51,294[1-0.5800-0.0100]- 7,899$$
$$/[1-0.3000]\}$$
$$/\{0.1900[1+0.0500]\}$$
$$= 48,851 \text{ millions USD}$$

Law-1278:

If ($= Sales or Revenues= 51,294 millions USD), (s= [S/S']-1= Sales Growth= 5.00%), (f= F/S=Fixed Portion= 58.00%), (t= T/B= Tax Rate= 30.00%), (i= I/S= Interest Portion= 1.00%), ($'= Sales of Past Year= 48,851 millions USD), and (A= After Tax Income= 7,899 millions USD), are known, then it's (v= Variable Portion planned) is:

$$v = \{\$[1-f-i]-A/[1-t]\}/\{\$'[1+s]\}$$
$$= \{51,294[1-0.5800-0.0100]-7,899$$
$$/[1-0.3000]\}$$
$$/\{48,851[1+0.0500]\}$$
$$= \underline{19.00\%}$$

Law-1279:

If ($= Sales or Revenues= 51,294 millions USD), (v= V/S= Variable Portion= 19.00%), (f= F/S=Fixed Portion= 58.00%), (t= T/B= Tax Rate= 30.00%), (i= I/S= Interest Portion= 1.00%), ($'= Sales of Past Year= 48,851 millions USD), and (A= After Tax Income= 7,899 millions USD), are known, then it's (s= Sales Growth planned) is:

$$s = \{\$[1-f-i]-A/[1-t]\}/[\$'v]-1$$
$$= \{51,294[1-0.5800-0.0100]-7,899$$
$$/[1-0.3000]\}/[48,851*0.1900]-1$$
$$= \underline{5.00\%}$$

Law-1280:
If ($\$$= Sales or Revenues= 51,294 millions USD), (v= V/S= Variable Portion= 19.00%), (f= F/S=Fixed Portion= 58.00%), (s= [S/S']-1= Sales Growth= 5.00%), (i= I/S= Interest Portion= 1.00%), ($\$'$= Sales of Past Year= 48,851 millions USD), and (A= After Tax Income= 7,899 millions USD), are known, then it's (t= Tax Rate planned) is:

$t = 1-A/\{\$-\$'v[1+s]-\$f-\$i\}$
$= 1-A/\{\$[1-f-i]-\$'v[1+s]\}$
$= 1-7,899/\{51,294[1-0.5800-0.0100]$
$\qquad -48,851*0.1900[1+0.0500]\}$
$= 30.00\%$

Law-1281:
If ($\$$= Sales or Revenues= 51,294 millions USD), (v= V/S= Variable Portion= 19.00%), (f= F/S=Fixed Portion= 58.00%), (s= [S/S']-1= Sales Growth= 5.00%), (i= I/S= Interest Portion= 1.00%), ($\$'$= Sales of Past Year= 48,851 millions USD), and (t= T/B= Tax Rate= 30.00%), are known, then it's (A= After Tax Income planned) is:

$A = [1-t]\{\$-\$'v[1+s]-\$f-\$'i[1+s]\}$
$= [1-t]\{\$[1-f]-\$'[1+s][v+i]\}$
$= [1-0.3000]\{51,294[1-0.5800]-48,851$
$\qquad [1+0.0500][0.1900+0.0100]\}$
$= 7,899$ millions USD

Law-1282:

If (A= After Tax Income= 7,899 millions USD), (v= V/S= Variable Portion= 19.00%), (f= F/S=Fixed Portion= 58.00%), (s= [S/S']-1= Sales Growth= 5.00%), (i= I/S= Interest Portion= 1.00%), ($\$'$= Sales of Past Year= 48,851 millions USD), and (t= T/B= Tax Rate= 30.00%), are known, then it's ($\$$= Sales or Revenues planned) is:

$\$ = \{\$'[1+s][v+i]+A/[1-t]\}/[1-f]$
$= \{48,851[1+0.0500][0.1900+0.0100]$
$+7,899/[1-0.3000]/[1-0.5800]$
= <u>51,294</u> millions USD

Law-1283:

If (A= After Tax Income= 7,899 millions USD), (v= V/S= Variable Portion= 19.00%), ($\$$= Sales or Revenues= 51,294 millions USD), (s= [S/S']-1= Sales Growth= 5.00%), (i= I/S= Interest Portion= 1.00%), ($\$'$= Sales of Past Year= 48,851 millions USD), and (t= T/B= Tax Rate= 30.00%), are known, then it's (f= Fixed Cost planned) is:

$f = 1-\{\$'[1+s][v+i]+A/1-t]\}/\$$
$= \{48,851[1+0.0500][0.1900+0.0100]$
$+7,899/[1-0.3000]/51,294$
= <u>58.00</u>%

Law-1284:
If (A= After Tax Income= 7,899 millions USD), (v= V/S= Variable Portion= 19.00%), (S= Sales or Revenues= 51,294 millions USD), (s= [S/S']-1= Sales Growth= 5.00%), (i= I/S= Interest Portion= 1.00%), (f= F/S=Fixed Portion= 58.00%), and (t= T/B= Tax Rate= 30.00%), are known, then it's (S'=Sales Past) must be:

$S' = \{S[1-f] - A/[1-t]\} / \{[1+s][v+i]\}$
$= 51,294[1-0.5800] - 7,899/[1-0.3000]\}$
$/\{[1+0.0500][0.1900+0.0100]\}$
$= \underline{48,851}$ millions USD

Law-1285:
If (A= After Tax Income= 7,899 millions USD), (v= V/S= Variable Portion= 19.00%), (S'= Sales of Past Year= 48,851 millions USD), (S= Sales or Revenues= 51,294 millions USD), (i= I/S= Interest Portion= 1.00%), (f= F/S=Fixed Portion= 58.00%), and (t= T/B= Tax Rate= 30.00%), are known, then it's (s= Sales Growth planned) is:

$s = \{S[1-f] - A/[1-t]\} / \{S'[v+i]\} - 1$
$= 51,294[1-0.5800] - 7,899/[1-0.3000]\}$
$/\{48,851[0.1900+0.0100]\} - 1$
$= \underline{5.00}\%$

Law-1286:

If (A= After Tax Income= 7,899 millions USD), (s= [S/S']-1= Sales Growth= 5.00%), (S'= Sales of Past Year= 48,851 millions USD), (S= Sales or Revenues= 51,294 millions USD), (i= I/S= Interest Portion= 1.00%), (f= F/S=Fixed Portion= 58.00%), and (t= T/B= Tax Rate= 30.00%), are known, then it's (v=Variable Portion planned) is:

$v = \{S[1-f] - A/[1-t]\} / \{S'[1+s]\} - i$
$= \{51,294[1-0.5800] - 7,899/[1-0.3000]\}$
$/\{48,851[1+0.0500]\} - 1$
$= \underline{19.00\%}$

Law-1287:

If (A= After Tax Income= 7,899 millions USD), (s= [S/S']-1= Sales Growth= 5.00%), (S'= Sales of Past Year= 48,851 millions USD), (S= Sales or Revenues= 51,294 millions USD), (v= V/S= Variable Portion= 19.00%), (f= F/S=Fixed Portion= 58.00%), and (t= T/B= Tax Rate= 30.00%), are known, then it's (i= Interest Portion planned) is:

$i = \{S[1-f] - A/[1-t]\} / \{S'[1+s]\} - v$
$= \{51,294[1-0.5800] - 7,899/[1-0.3000]\}$
$/\{48,851[1+0.0500]\} - 0.1900$
$= \underline{1.00\%}$

Law-1288:

If (**A**= After Tax Income= 7,899 millions USD), (**s**= [S/S']-1= Sales Growth= 5.00%), (**$'**= Sales of Past Year= 48,851 millions USD), (**$**= Sales or Revenues= 51,294 millions USD), (**v**= V/S= Variable Portion= 19.00%), (**f**= F/S=Fixed Portion= 58.00%), and (**i**= I/S= Interest Portion= 1.00%), are known, then it's (**t**= Tax Rate planned) is:

$$\begin{aligned}\mathbf{t} &= 1\text{-}\mathbf{A}/\{\mathbf{\$}\text{-}\mathbf{\$'v}[1+\mathbf{s}]\text{-}\mathbf{\$f}\text{-}\mathbf{\$'i}\{1+\mathbf{s}]\}\\ &= 1\text{-}\mathbf{A}/\{\mathbf{\$}[1\text{-}\mathbf{f}]\text{-}\mathbf{\$'}[1+\mathbf{s}][\mathbf{v}+\mathbf{i}]\}\\ &= 1\text{-}7{,}899\{51{,}294[1\text{-}0.5800]\text{ -}48{,}851\\ &\qquad[1+0.0500][0.1900+0.0100]\}\\ &= 30.00\%\end{aligned}$$

Law-1289:

If (**t**= T/B= Tax Rate= 30.00%), (**s**= [S/S']-1= Sales Growth= 5.00%), (**$'**= Sales of Past Year= 48,851 millions USD), (**$**= Sales or Revenues= 51,294 millions USD), (**v**= V/S= Variable Portion= 19.00%), (**f**= F/S=Fixed Portion= 58.00%), and (**i**= I/S= Interest Portion= 1.00%), are known, then it's (**A**= After Tax Income planned) is:

$$\begin{aligned}\mathbf{A} &= [1\text{-}\mathbf{t}]/\{\mathbf{\$}\text{-}\mathbf{\$'v}[1+\mathbf{s}]\text{-}\mathbf{\$'f}[1+\mathbf{s}]\text{-}\mathbf{I}\}\\ &= [1\text{-}\mathbf{t}]\{\mathbf{\$}\text{-}\mathbf{I}\text{-}\mathbf{\$'}[1+\mathbf{s}][\mathbf{v}+\mathbf{f}]\}\\ &= [1\text{-}0.3000]\{51{,}294\text{-}513\text{ -}48{,}851\\ &\qquad[1+0.0500][0.1900+0.5800]\}\\ &= 7{,}899 \text{ millions USD}\end{aligned}$$

Law-1290:

If (t= T/B= Tax Rate= 30.00%), (s= [S/S']-1= Sales Growth= 5.00%), (S'= Sales of Past Year= 48,851 millions USD), (A= After Tax Income= 7,899 millions USD), (v= V/S= Variable Portion= 19.00%), (f= F/S=Fixed Portion= 58.00%), and (i= I/S= Interest Portion= 1.00%), are known, then it's (S= Sales or Revenues planned) is:

$$S= I+S'[1+s][v+f]+A/[1-t]$$
$$= 513 +48,851[1+0.0500][0.1900 +0.5800] +7,899/[1-0.3000]$$
$$=\underline{51,294} \text{ millions USD}$$

Law-1291:

If (t= T/B= Tax Rate= 30.00%), (s= [S/S']-1= Sales Growth= 5.00%), (S= Sales or Revenues= 51,294 millions USD), (A= After Tax Income= 7,899 millions USD), (v= V/S= Variable Portion= 19.00%), (f= F/S=Fixed Portion= 58.00%), and (i= I/S= Interest Portion= 1.00%), are known, then it's (S'= Sales Past) must be:

$$S'= \{S-I-A/[1-t]\}/\{[1+s][v+f]\}$$
$$= \{51,294-513- 7,899/ [1-0.3000]\} /\{[1+0.0500][0.1900+0.5800]\}$$
$$=\underline{48,851} \text{ millions USD}$$

Law-1292:
If (t= T/B= Tax Rate= 30.00%), (S'= Sales of Past Year= 48,851 millions USD), (S= Sales or Revenues= 51,294 millions USD), (A= After Tax Income= 7,899 millions USD), (v= V/S= Variable Portion= 19.00%), (f= F/S=Fixed Portion= 58.00%), and (i= I/S= Interest Portion= 1.00%), are known, then it's (s= Sales Growth planned) is:

$$s = \{S-I-A/[1-t]\}/\{S'[v+f]\}-1$$
$$= \{51,294-513-7,899/[1-0.3000]\}/48,851\{[0.1900+0.5800]\}-1$$
$$= 5.00\%$$

Law-1293:
If (t= T/B= Tax Rate= 30.00%), (S'= Sales of Past Year= 48,851 millions USD), (S= Sales or Revenues= 51,294 millions USD), (A= After Tax Income= 7,899 millions USD), (s= [S/S']-1= Sales Growth= 5.00%), (f= F/S=Fixed Portion= 58.00%), and (i= I/S= Interest Portion= 1.00%), are known, then it's (v= Variable Portion planned) is:

$$v = \{S-I-A/[1-t]\}/\{S'[1+s]\}-f$$
$$= \{51,294-513-7,899/[1-0.3000]\}/48,851\{[1+0.0500]\}-0.5800$$
$$= 19.00\%$$

Law-1294:

If (**t**= T/B= Tax Rate= 30.00%), (**$'**= Sales of Past Year= 48,851 millions USD), (**$**= Sales or Revenues= 51,294 millions USD), (**A**= After Tax Income= 7,899 millions USD), (**v**= V/S= Variable Portion= 19.00%), (**s**= [S/S']-1= Sales Growth= 5.00%), and (**i**= I/S= Interest Portion= 1.00%), are known, then it's (**f**= Fixed Portion planned) is:

$$f = \{\$-I-A/[1-t]\}/\{\$'[1+s]\}-v$$
$$= \{51{,}294-513-7{,}899/[1-0.3000]\}$$
$$/48{,}851\{[1+0.0500]\}-0.1900$$
$$= \underline{58.00\%}$$

Law-1295:

If (**t**= T/B= Tax Rate= 30.00%), (**$'**= Sales of Past Year= 48,851 millions USD), (**$**= Sales or Revenues= 51,294 millions USD), (**A**= After Tax Income= 7,899 millions USD), (**v**= V/S= Variable Portion= 19.00%), (**s**= [S/S']-1= Sales Growth= 5.00%), and (**f**= F/S= Fixed Portion= 58.00%), are known, then it's (**I**= Interest Expense planned) is:

$$I = \{\$-\$'[1+s][v+f]-A/[1-t]\}$$
$$= \{51{,}294-48{,}851[1+0.0500][0.1900$$
$$+0.5800]-7{,}899/[1-0.3000]\}$$
$$= \underline{513} \text{ millions USD}$$

Law-1296:
If (**I**= Interest Expense= 513 millions USD), (**S'**= Sales of Past Year= 48,851 millions USD), (**S**= Sales or Revenues= 51,294 millions USD), (**A**= After Tax Income= 7,899 millions USD), (**v**= V/S= Variable Portion= 19.00%), (**s**= [S/S']-1= Sales Growth= 5.00%), and (**f**= F/S= Fixed Portion= 58.00%), are known, then it's (**t**= Tax Rate planned) is:
$$t= 1-\mathbf{A}/\{\mathbf{S}-\mathbf{S'v}[1+\mathbf{s}]-\mathbf{S'f}[1+\mathbf{s}]-\mathbf{I}\}$$
$$= 1-\mathbf{A}\{\mathbf{S}-\mathbf{S'}[1+\mathbf{s}][\mathbf{v}+\mathbf{f}]-\mathbf{I}\}$$
$$= 1-7,899/\{51,294-48,851[1+0.0500]$$
$$[0.1900+0.5800]-513\}$$
$$=\underline{30.00}\%$$

Law-1297:
If (**i**= I/S= Interest Portion= 1.00%), (**S'**= Sales of Past Year= 48,851 millions USD), (**S**= Sales or Revenues= 51,294 millions USD), (**t**= T/B= Tax Rate= 30.00%), (**v**= V/S= Variable Portion= 19.00%), (**s**= [S/S']-1= Sales Growth= 5.00%), and (**f**= F/S=Fixed Portion= 58.00%), are known, then it's (**A**= After Tax Income planned) is:
$$\mathbf{A}= [1-\mathbf{t}]\{\mathbf{S}-\mathbf{S'v}[1+\mathbf{s}]-\mathbf{S'f}[1+\mathbf{s}]-\mathbf{Si}\}$$
$$= 1-\mathbf{A}\{\mathbf{S}[1-\mathbf{i}]-\mathbf{S'}[1+\mathbf{s}][\mathbf{v}+\mathbf{f}]\}$$
$$= [1-0.3000]\{51,294[1-0.0100]-48,851$$
$$[1+0.0500][0.1900+0.5800]\}$$
$$=\underline{7,899} \text{ millions USD}$$

Law-1298:

If (i= I/S= Interest Portion= 1.00%), ($\$'$= Sales of Past Year= 48,851 millions USD), (A= After Tax Income= 7,899 millions USD), (t= T/B= Tax Rate= 30.00%), (v= V/S= Variable Portion= 19.00%), (s= [S/S']-1= Sales Growth= 5.00%), and (f= F/S= Fixed Portion= 58.00%), are known, then it's ($\$$= Sales or Revenues planned) is:

$$\$ = \{\$'[1+s][v+f]+A/[1-t]\}/[1-i]$$
$$= \{48,851[1+0.0500][0.1900+0.5800]$$
$$+7,899/[1-0.3000]\}/[1-0.0100]$$
$$= \underline{51,294} \text{ millions USD}$$

Law-1299:

If ($\$$= Sales or Revenues= 51,294 millions USD), ($\$'$= Sales of Past Year= 48,851 millions USD), (A= After Tax Income= 7,899 millions USD), (t= T/B= Tax Rate= 30.00%), (v= V/S= Variable Portion= 19.00%), (s= [S/S']-1= Sales Growth= 5.00%), and (f= F/S= Fixed Portion= 58.00%), are known, then it's (i= Interest Portion planned) is:

$$i = 1-\{\$'[1+s][v+f]+A/[1-t]\}/\$$$
$$= 1-\{48,851[1+0.0500][0.1900+0.5800]$$
$$+7,899/[1-0.3000]\}/51,294$$
$$= \underline{1.00}\%$$

Law-1300:

If ($= Sales or Revenues= 51,294 millions USD), (i= I/S= Interest Portion= 1.00%), (A= After Tax Income= 7,899 millions USD), (t= T/B= Tax Rate= 30.00%), (v= V/S= Variable Portion= 19.00%), (s= [S/S']-1= Sales Growth= 5.00%), and (f= F/S=Fixed Portion= 58.00%), are known, then it's ($'= Sales Past) must be:

$$\$' = \{\$[1-i]-A/[1-t]\}/\{[1+s][v+f]\}$$
$$= \{51,294[1-0.0100]-7,899/[1-0.3000]\}$$
$$/\{[1+0.0500][0.1900+0.5800]\}$$
$$= 48,851 \text{ millions USD}$$

Law-1301:

If ($= Sales or Revenues= 51,294 millions USD), (i= I/S= Interest Portion= 1.00%), (A= After Tax Income= 7,899 millions USD), (t= T/B= Tax Rate= 30.00%), (v= V/S= Variable Portion= 19.00%), ($'= Sales of Past Year= 48,851 millions USD), and (f= F/S=Fixed Portion= 58.00%), are known, then it's (s= Sales Growth planned) is: (v= V/S= Variable Portion= 19.00%),

$$s = \{\$[1-i]-A/[1-t]\}/\{\$'[v+f]\}-1$$
$$= \{51,294[1-0.0100]-7,899/[1-0.3000]\}$$
$$/\{48,851 [0.1900+0.5800]\}-1$$
$$= 5.00\%$$

Law-1302:

If ($ = Sales or Revenues = 51,294 millions USD), (i = I/S = Interest Portion = 1.00%), (A = After Tax Income = 7,899 millions USD), (t = T/B = Tax Rate = 30.00%), (s = [S/S']-1 = Sales Growth = 5.00%), ($' = Sales of Past Year = 48,851 millions USD), and (f = F/S = Fixed Portion = 58.00%), are known, then it's (v = Variable Portion planned) is:

$$v = \{\$[1-i] - A/[1-t]\} / \{\$'[1+s]\} - f$$
$$= \{51,294[1-0.0100] - 7,899/[1-0.3000]\} / \{48,851[1+0.0500]\} - 0.5800$$
$$= \underline{19.00\%}$$

Law-1303:

If ($ = Sales or Revenues = 51,294 millions USD), (i = I/S = Interest Portion = 1.00%), (A = After Tax Income = 7,899 millions USD), (t = T/B = Tax Rate = 30.00%), (s = [S/S']-1 = Sales Growth = 5.00%), ($' = Sales of Past Year = 48,851 millions USD), and (v = V/S = Variable Portion = 19.00%), are known, then it's (f = Fixed Portion planned) is:

$$f = \{\$[1-i] - A/[1-t]\} / \{\$'[1+s]\} - v$$
$$= \{51,294[1-0.0100] - 7,899/[1-0.3000]\} / \{48,851[1+0.0500]\} - 0.1900$$
$$= \underline{58.00\%}$$

Law-1304:

If ($= Sales or Revenues= 51,294 millions USD), (i= I/S= Interest Portion= 1.00%), (A= After Tax Income= 7,899 millions USD), (f= F/S=Fixed Portion= 58.00%), (s= [S/S']-1= Sales Growth= 5.00%), ($'= Sales of Past Year= 48,851 millions USD), and (v= V/S= Variable Portion= 19.00%), are known, then it's (t= Tax Rate planned) is:

$t = 1-A/\{\$-\$'v[1+s]-\$'f\{1+s\}-\$i\}$
$= 1-A/\{\$[1-i]-\$'[1+s][v+f]\}$
$= 1-7,899/\{51,294[1-0.0100]- 48,851$
$[1+0.0500][0.1900+0.5800]\}$
$= \underline{30.00\%}$

Law-1305:

If ($= Sales or Revenues= 51,294 millions USD), (i= I/S= Interest Portion= 1.00%), (t= T/B= Tax Rate= 30.00%), (f= F/S=Fixed Portion= 58.00%), (s= [S/S']-1= Sales Growth= 5.00%), ($'= Sales of Past Year= 48,851 millions USD), and (v= V/S= Variable Portion= 19.00%), are known, then it's (A= After Tax Income planned) is:

$A = [1-t]\{\$-\$'v[1+s]-\$'f\{1+s\}-\$'i[1+s]\}$
$= [1-t]\{\$-\$'[1+s][v+f+i]\}$
$= [1-0.3000]\{51,294- 48,851[1+0.0500]$
$[0.1900+0.5800+0.0100]\}$
$= \underline{7,899}$ millions USD

Law-1306:

If (**A**= After Tax Income= 7,899 millions USD), (**i**= I/S= Interest Portion= 1.00%), (**t**= T/B= Tax Rate= 30.00%), (**f**= F/S=Fixed Portion= 58.00%), (**s**= [S/S']-1= Sales Growth= 5.00%), (**S'**= Sales of Past Year= 48,851 millions USD), and (**v**= V/S= Variable Portion= 19.00%), are known, then it's (**S**= Sales or Revenues planned) is:

$$S = S'[1+s][v+f+i] + A/[1-t]$$
$$= 48,851 [1+0.0500][0.1900+0.5800 +0.0100]\} + 7,899/[1-0.3000]$$
$$= \underline{51,294} \text{ millions USD}$$

Law-1307:

If (**A**= After Tax Income= 7,899 millions USD), (**i**= I/S= Interest Portion= 1.00%), (**t**= T/B= Tax Rate= 30.00%), (**f**= F/S=Fixed Portion= 58.00%), (**s**= [S/S']-1= Sales Growth= 5.00%), (**S**= Sales or Revenues= 51,294 millions USD), and (**v**= V/S= Variable Portion= 19.00%), are known, then it's (**S'**= Sales Past) must be:

$$S' = \{S-A/[1-t]\}/\{[1+s][v+f+i]\}$$
$$= \{51,294-7,899/[1-0.3000]\} /\{[1+0.0500][0.1900+0.5800 +0.0100]\}$$
$$= \underline{48,851} \text{ millions USD}$$

Law-1308:

If (**A**= After Tax Income= 7,899 millions USD), (**i**= I/S= Interest Portion= 1.00%), (**t**= T/B= Tax Rate= 30.00%), (**f**= F/S=Fixed Portion= 58.00%), (**$'**= Sales of Past Year= 48,851 millions USD), (**$**= Sales or Revenues= 51,294 millions USD), and (**v**= V/S= Variable Portion= 19.00%), are known, then it's (**s**= Sales Growth planned) is:

$s = \{\$-A/[1-t]\}/\{\$'[v+f+i]\} - 1$
$= \{51,294-7,899/[1-0.3000]\}$
$\quad /\{48,851[0.1900+0.5800$
$\quad +0.0100]\} - 1$
$= 5.00\%$

Law-1309:

If (**A**= After Tax Income= 7,899 millions USD), (**i**= I/S= Interest Portion= 1.00%), (**t**= T/B= Tax Rate= 30.00%), (**f**= F/S=Fixed Portion= 58.00%), (**$'**= Sales of Past Year= 48,851 millions USD), (**$**= Sales or Revenues= 51,294 millions USD), and (**s**= [S/S']-1= Sales Growth= 5.00%), are known, then it's (**v**= Variable Portion planned) is:

$v = 1-f-i-\{\$-A/[1-t]\}/\{\$'[1+s]\}$
$= 1-0.5800-0.0100-\{51,294-7,899$
$\quad /[1-0.3000]\}$
$\quad /\{48,851[1+0.0500]\}$
$= 19.00\%$

Law-1310 :

If (A= After Tax Income= 7,899 millions USD), (i= I/S= Interest Portion= 1.00%), (t= T/B= Tax Rate= 30.00%), (v= V/S= Variable Portion= 19.00%), (S'= Sales of Past Year= 48,851 millions USD), (S= Sales or Revenues= 51,294 millions USD), and (s= [S/S']-1= Sales Growth= 5.00%), are known, then it's (f= Fixed Portion planned) is:

$f = \{S-A/[1-t]\}/\{S'[1+s]\}-[v+i]$
  = {51,294-7,899/ [1-0.3000] }
    /{48,851[1+0.0500]}
    -[0.1900+0.0100]
  = 58.00%

Law-1311:

If (A= After Tax Income= 7,899 millions USD), (t= T/B= Tax Rate= 30.00%), (v= V/S= Variable Portion= 19.00%), (f= F/S=Fixed Portion= 58.00%), (S'= Sales of Past Year= 48,851 millions USD), (S= Sales or Revenues= 51,294 millions USD), and (s= [S/S']-1= Sales Growth= 5.00%), are known, then it's (i= Interest Portion planned) is:

$i = \{S-A/[1-t]\}/\{S'[1+s]\}-[v+f]$
  = {51,294-7,899/ [1-0.3000]}
    /{48,851[1+0.0500]}
    -[0.1900+0.5800]
  = 1.00%

Law-1312:
If (A= After Tax Income= 7,899 millions USD), (i= I/S= Interest Portion= 1.00%), (v= V/S= Variable Portion= 19.00%), (f= F/S=Fixed Portion= 58.00%), (S'= Sales of Past Year= 48,851 millions USD), (S= Sales or Revenues= 51,294 millions USD), and (s= [S/S']-1= Sales Growth= 5.00%), are known, then it's (t= Tax Rate planned) is:

$$t = 1-A\{S-S'v[1+s]-S'f[1+s]-S'i[1+s]\}$$
$$= 1-A/\{S-S'[1+s][1-v-f-i]\}$$
$$= 1-7,899/\{51,294- 48,851[1+0.0500] [1-0.1900-0.5800-0.0100]\}$$
$$= 30.00\%$$

Law-1313:
If (t= T/B= Tax Rate= 30.00%), (V= Variable Cost= 9,746 millions USD), (I= Interest Expense= 513 millions USD), (F= Fixed Cost= 29,740 millions USD), (S'= Sales of Past Year= 48,851 millions USD), and (s= [S/S']-1= Sales Growth= 5.00%), are known, then it's (A=After Tax Income planned) is:

$$A = [1-t]\{S'[1+s]-V-F-I\}$$
$$= [1-0.3000]\{48,851[1+0.0500]-9,746 -29,750-513\}$$
$$= 7,899 \text{ millions USD}$$

Law-1314:

If (**t**= T/B= Tax Rate= 30.00%), (**V**= Variable Cost= 9,746 millions USD), (**I**= Interest Expense= 513 millions USD), (**F**= Fixed Cost= 29,740 millions USD),(**A**=After Tax Income= 7,899 millions USD), and (**s**= [S/S']-1= Sales Growth= 5.00%), are known, then it's (**S'**= Sales Past) must be:

$$S' = \{V+F+I+A/[1-t]\}/[1+s]$$
$$= \{9,746+29,750+513+7,899/[1-0.3000]\}/[1+0.0500]$$
$$= \underline{48,851} \text{ millions USD}$$

Law-1315:

If (**t**= T/B= Tax Rate= 30.00%), (**V**= Variable Cost= 9,746 millions USD), (**I**= Interest Expense= 513 millions USD), (**F**= Fixed Cost= 29,740 millions USD), (**A**=After Tax Income= 7,899 millions USD), and (**S'**= Sales of Past Year= 48,851 millions USD), are known, then it's (**s**= Sales Growth planned) is:

$$s = \{V+F+I+A/[1-t]\}/S' - 1$$
$$= \{9,746+29,750+513+7,899/[1-0.3000]\}/48,851 - 1$$
$$= \underline{5.00\%}$$

Law-1316:

If (**t**= T/B= Tax Rate= 30.00%), (**s**= [S/S']-1= Sales Growth= 5.00%), (**I**= Interest Expense= 513 millions USD), (**F**= Fixed Cost= 29,740 millions USD), (**A**= After Tax Income= 7,899 millions USD), and (**$'**= Sales of Past Year= 48,851 millions USD), are known, then it's (**V**= Variable Cost planned) is:

V= **$'**[1+**s**]-**F**-**I**-**A**/[1-**t**]
= 48,851[1+0.0500] -29,750-513
-7,899/[1-0.3000]
=9,746 millions USD

Law-1317:

If (**t**= T/B= Tax Rate= 30.00%), (**s**= [S/S']-1= Sales Growth= 5.00%), (**I**= Interest Expense= 513 millions USD), (**V**= Variable Cost= 9,746 millions USD), (**A**= After Tax Income= 7,899 millions USD), and (**$'**= Sales of Past Year= 48,851 millions USD), are known, then it's (**F**= Fixed Cost planned) is:

F= **$'**[1+**s**]-**V**-**A**/[1-**t**]-**I**
= 48,851[1+0.0500] -9,746-7,899
/[1-0.3000] -513
=29,750 millions USD

Law-1318:
If (t= T/B= Tax Rate= 30.00%), (s= [S/S']-1= Sales Growth= 5.00%), (F= Fixed Cost= 29,740 millions USD), (V= Variable Cost= 9,746 millions USD), (A= After Tax Income= 7,899 millions USD), and (S'= Sales of Past Year= 48,851 millions USD), are known, then it's (I= Interest Expense planned) is:

$I = S'[1+s] - V - F - A/[1-t]$
= 48,851[1+0.0500] -9,746-29,750
  -7,899/[1-0.3000]
= 513 millions USD

Law-1319:
If (I= Interest Expense= 513 millions USD), (s= [S/S']-1= Sales Growth= 5.00%), (F= Fixed Cost= 29,740 millions USD), (V= Variable Cost= 9,746 millions USD), (A= After Tax Income= 7,899 millions USD), and (S'= Sales of Past Year= 48,851 millions USD), are known, then it's (t= Tax Rate planned) is:

$t = 1 - A/\{S'[1+s] - V - F - I\}$
= 1-7,899/{48,851[1+0.0500]-9,746
  -29,750- 513}
= 30.00%

Law-1320:

If (**i**= I/S= Interest Portion= 1.00%), (**s**= [S/S']-1= Sales Growth= 5.00%), (**F**= Fixed Cost= 29,740 millions USD), (**V**= Variable Cost= 9,746 millions USD), (**t**= T/B= Tax Rate= 30.00%), and (**S'**= Sales of Past Year= 48,851 millions USD), are known, then it's (**A**= After Tax Income planned) is:

$$A = [1-t]\{S'[1+s] - V - F - Si\}$$
$$= [1-0.3000]\{48,851[1+0.0500] - 9,746 - 29,750 - 51,294 \cdot 0.0100\}$$
$$= 7,899 \text{ millions USD}$$

Law-1321:

If (**i**= I/S= Interest Portion= 1.00%), (**s**= [S/S']-1= Sales Growth= 5.00%), (**F**= Fixed Cost= 29,740 millions USD), (**V**= Variable Cost= 9,746 millions USD), (**t**= T/B= Tax Rate= 30.00%), and (**A**= After Tax Income= 7,899 millions USD), are known, then it's (**S'**= Sales Past) must be:

$$S' = \{V + F + Si + A/[1-t]\}/[1+s]$$
$$= \{9,746 + 29,750 + 51,294 \cdot 0.0100 + 7,899/[1-0.3000]\}/[1+0.0500]$$
$$= 48,851 \text{ millions USD}$$

Law-1322:
   If (i= I/S= Interest Portion= 1.00%), (S'= Sales of Past Year= 48,851 millions USD), (F= Fixed Cost= 29,740 millions USD), (V= Variable Cost= 9,746 millions USD), (t= T/B= Tax Rate= 30.00%), and (A= After Tax Income= 7,899 millions USD), are known, then it's (s= Sales Growth planned) is:

$$s = \{V+F+Si+A/[1-t]\}/S'-1$$
$$= \{9,746+29,750+51,294*0.0100 +7,899/[1-0.3000]\}/48,851-1$$
$$= 5.00\%$$

Law-1323:
   If (i= I/S= Interest Portion= 1.00%), (S'= Sales of Past Year= 48,851 millions USD), (F= Fixed Cost= 29,740 millions USD), (s= [S/S']-1= Sales Growth= 5.00%), (t= T/B= Tax Rate= 30.00%), and (A= After Tax Income= 7,899 millions USD), are known, then it's (V= Variable Cost planned) is:

$$V = F-Si-S'[1+s]-A/[1-t]$$
$$= 29,750-51,294*0.0100-48,851 [1+0.0500]-7,899/[1-0.3000]$$
$$= 9,746 \text{ millions USD}$$

Law-1324 :
  If (i= I/S= Interest Portion= 1.00%), ($\$'$= Sales of Past Year= 48,851 millions USD), (V= Variable Cost= 9,746 millions USD), (s= [S/S']-1= Sales Growth= 5.00%), (t= T/B= Tax Rate= 30.00%), and (A= After Tax Income= 7,899 millions USD), are known, then it's (F= Fixed Cost planned) is:
  $F = \$'[1+s] - V - \$i - A/[1-t]$
  $= 48,851[1+0.0500] - 9,746 - 51,294$
      $*0.0100 - 7,899/[1-0.3000]$
  $= 29,750$ millions USD

Law-1325:
  If (i= I/S= Interest Portion= 1.00%), ($\$'$= Sales of Past Year= 48,851 millions USD), (V= Variable Cost= 9,746 millions USD), (s= [S/S']-1= Sales Growth= 5.00%), (t= T/B= Tax Rate= 30.00%), (F= Fixed Cost= 29,740 millions USD), and (A= After Tax Income= 7,899 millions USD), are known, then it's ($\$$= Sales or Revenues planned) is:
  $\$ = \{\$'[1+s] - V - F - A/[1-t]\}/i$
  $= \{48,851[1+0.0500] - 9,746 - 29,750$
      $-7,899/[1-0.3000]\}/0.0100$
  $= 51,294$ millions USD

Law-1326:

If ($ = Sales or Revenues = 51,294 millions USD), ($' = Sales of Past Year = 48,851 millions USD), (V = Variable Cost = 9,746 millions USD), (s = [S/S']-1 = Sales Growth = 5.00%), (t = T/B = Tax Rate = 30.00%), (F = Fixed Cost = 29,740 millions USD), and (A = After Tax Income = 7,899 millions USD), are known, then it's (i = Interest Portion planned) is:

$$i = \{\$'[1+s]-V-F-A/[1-t]\}/\$$$
$$= \{48,851[1+0.0500] -9,746-29,750$$
$$-7,899/[1-0.3000]\}/51,294$$
$$= \underline{1.00\%}$$

Law-1327:

If ($' = Sales of Past Year = 48,851 millions USD), (V = Variable Cost = 9,746 millions USD), (s = [S/S']-1 = Sales Growth = 5.00%), (i = I/S = Interest Portion = 1.00%), (F = Fixed Cost = 29,740 millions USD), and (A = After Tax Income = 7,899 millions USD), are known, then it's (t = Tax Rate planned) is:

$$t = 1-A/\{\$'[1+s]-V-F-\$i\}$$
$$= 1-7,899/\{48,851[1+0.0500] -9,746$$
$$-29,750-7,899/[1-0.3000]$$
$$-51,294*0.0100\}$$
$$= \underline{30.00\%}$$

Law-1328:
If ($= Sales of Past Year= 48,851 millions USD), (V= Variable Cost= 9,746 millions USD), (s= [S/S']-1= Sales Growth= 5.00%), (i= I/S= Interest Portion= 1.00%), (F= Fixed Cost= 29,740 millions USD), and (t= T/B= Tax Rate= 30.00%), are known, then it's (A= After Tax Income planned) is:
A= [1-t]{$'[1+s]-V-F-$'i[1+s]}
  = [1-t]{$'[1+s][1-i]-V-F}
  = [1-0.3000]{48,851[1+0.0500]
       [1-0.0100] -9,746-29,750}
  = 7,899 millions USD

Law-1329:
If (A= After Tax Income= 7,899 millions USD), (V= Variable Cost= 9,746 millions USD), (s= [S/S']-1= Sales Growth= 5.00%), (i= I/S= Interest Portion= 1.00%), (F= Fixed Cost= 29,740 millions USD), and (t= T/B= Tax Rate= 30.00%), are known, then it's ($'= Sales Past) must be:
$'= {V+F+A/[1-t]}/{[1+s][1-i]}
  ={9,746+29,750+7,899/ [1-0.3000]}
       /{[1+0.0500][1-0.0100]}
  = 48,851 millions USD

Law-1330:

If (A= After Tax Income= 7,899 millions USD), (V= Variable Cost= 9,746 millions USD), (S'= Sales of Past Year= 48,851 millions USD), (i= I/S= Interest Portion= 1.00%), (F= Fixed Cost= 29,740 millions USD), and (t= T/B= Tax Rate= 30.00%), are known, then it's (s= Sales Growth planned) is:

$s = \{V+F+A/[1-t]\}/\{S'[1-i]\}-1$
$= \{9,746+29,750+7,899/[1-0.3000]\}$
$/\{48,851[1-0.0100]\}-1$
$= 5.00\%$

Law-1331:

If (A= After Tax Income= 7,899 millions USD), (V= Variable Cost= 9,746 millions USD), (S'= Sales of Past Year= 48,851 millions USD), (s= [S/S']-1= Sales Growth= 5.00%), (F= Fixed Cost= 29,740 millions USD), and (t= T/B= Tax Rate= 30.00%), are known, then it's (i= Internet Portion planned) is:

$i = 1-\{V+F+A/[1-t]\}/\{S'[1+s]\}$
$= 1-\{9,746+29,750+7,899/[1-0.3000]\}$
$/\{48,851[1+0.0500]\}-1$
$= 1.00\%$

<u>Law-1332</u>:

If (**A**= After Tax Income= <u>7,899</u> millions USD), (**i**= I/S= Interest Portion= <u>1.00</u>%), (**$'**= Sales of Past Year= <u>48,851</u> millions USD), (**s**= [S/S']-1= Sales Growth= <u>5.00</u>%), (**F**= Fixed Cost= <u>29,740</u> millions USD), and (**t**= T/B= Tax Rate= <u>30.00</u>%), are known, then it's (**V**= Variable Cost planned) is:

$V = \$'[1+s]\}[1-i]-F-A/[1-t]$
$= 48,851[1+0.0500][1-0.0100]- 29,750$
$-7,899/[1-0.3000]\}$
$= \underline{9,746}$ millions USD

<u>Law-1333</u>:

If (**A**= After Tax Income= <u>7,899</u> millions USD), (**i**= I/S= Interest Portion= <u>1.00</u>%), (**$'**= Sales of Past Year= <u>48,851</u> millions USD), (**s**= [S/S']-1= Sales Growth= <u>5.00</u>%), (**V**= Variable Cost= <u>9,746</u> millions USD), and (**t**= T/B= Tax Rate= <u>30.00</u>%), are known, then it's (**F**= Fixed Cost planned) is:

$F = \$'[1+s]\{[1-i]-V-A/[1-t]$
$= 48,851[1+0.0500][1-0.0100]-9,746$
$-7,899/[1-0.3000]\}$
$= \underline{29,750}$ millions USD

Law-1334:
If (A= After Tax Income= 7,899 millions USD), (i= I/S= Interest Portion= 1.00%), (S'= Sales of Past Year= 48,851 millions USD), (s= [S/S']-1= Sales Growth= 5.00%), (s= [S/S']-1= Sales Growth= 5.00%), (V= Variable Cost= 9,746 millions USD), and (F= Fixed Cost= 29,740 millions USD), are known, then it's (t= Tax Rate planned) is:

t= 1-A/{S'[1+s]-V-F-$S'i$[1+s]}
 = 1-A/{S'[1+s][1-i]-V-F}
 = 1-7.899/{48,851[1+0.0500][1-0.0100] -9,746-29,750}
 = 30.00%

Law-1335:
If (t= T/B= Tax Rate= 30.00%), (I= Interest Expense= 513 millions USD), (S'= Sales of Past Year= 48,851 millions USD), (s= [S/S']-1= Sales Growth= 5.00%), (S= Sales or Revenues= 51,294 millions USD), (V= Variable Cost= 9,746 millions USD), and (f= F/S=Fixed Portion= 58.00%), are known, then it's (A= After Tax Income planned) is:

A= [1-t]{S'[1+s]-V-Sf-I}
 = [1-0.3000]{ 48,851[1+0.0500] -9,746 -51,294*0.5800-513}
 =7,899 millions USD

Law-1336:
If (**t**= T/B= Tax Rate= 30.00%), (**I**= Interest Expense= 513 millions USD), (**A**= After Tax Income= 7,899 millions USD), (**s**= [S/S']-1= Sales Growth= 5.00%), (**$**= Sales or Revenues= 51,294 millions USD), (**V**= Variable Cost= 9,746 millions USD), and (**f**= F/S=Fixed Portion= 58.00%), are known, then it's (**$'**= Sales Past) must be:

$$\$' = \{V+\$f+I+A/[1-t]\}/[1+s]$$
$$= \{9,746+51,294*0.5800+513+7,899\allowbreak/[1-0.3000]\}/[1+0.0500]$$
$$= 48,851 \text{ millions USD}$$

Law-1337:
If (**t**= T/B= Tax Rate= 30.00%), (**I**= Interest Expense= 513 millions USD), (**$'**= Sales of Past Year= 48,851 millions USD), (**A**= After Tax Income= 7,899 millions USD), (**$**= Sales or Revenues= 51,294 millions USD), (**V**= Variable Cost= 9,746 millions USD), and (**f**= F/S=Fixed Portion= 58.00%), are known, then it's (**s**= Sales Growth planned) is:

$$s = \{V+\$f+I+A/[1-t]\}/\$'-1$$
$$= \{9,746+51,294*0.5800+513\}+7,899\allowbreak/[1-0.3000]\}/48,851 -1$$
$$= 5.00\%$$

Law-1338:

If (t= T/B= Tax Rate= 30.00%), (I= Interest Expense= 513 millions USD), (S'= Sales of Past Year= 48,851 millions USD), (A= After Tax Income= 7,899 millions USD), (S= Sales or Revenues= 51,294 millions USD), (s= [S/S']-1= Sales Growth= 5.00%), and (f= F/S=Fixed Portion= 58.00%), are known, then it's (V= Variable Cost planned) is:

$V = S'[1+s]-Sf-I-A/[1-t]$
$= 48,851[1+0.0500]- 51,294*0.5800$
$-513\}-7,899/[1-0.3000]$
$= 9,746$ millions USD

Law-1339:

If (t= T/B= Tax Rate= 30.00%), (I= Interest Expense= 513 millions USD), (S'= Sales of Past Year= 48,851 millions USD), (A= After Tax Income= 7,899 millions USD), (V= Variable Cost= 9,746 millions USD), (s= [S/S']-1= Sales Growth= 5.00%), and (f= F/S=Fixed Portion= 58.00%), are known, then it's (S= Sales or Revenues planned) is:

$S = \{S'[1+s]-V-A/[1-t]-I\}/f$
$= \{48,851[1+0.0500]- 9,746-7,899$
$/[1-0.3000]-513\}/0.5800$
$= 51,294$ millions USD

Law-1340:

If (t= T/B= Tax Rate= 30.00%), (I= Interest Expense= 513 millions USD), (S'= Sales of Past Year= 48,851 millions USD), (A= After Tax Income= 7,899 millions USD), (V= Variable Cost= 9,746 millions USD), (s= [S/S']-1= Sales Growth= 5.00%), and (S= Sales or Revenues= 51,294 millions USD), are known, then it's (f= Fixed Portion planned) is:

$f = \{S'[1+s]-V-I-A/[1-t]\}/S$
$= \{48,851[1+0.0500]- 9,746-513-7,899 /[1-0.3000]\}/51,294$
$= \underline{58.00\%}$

Law-1341:

If (t= T/B= Tax Rate= 30.00%), (f= F/S=Fixed Portion= 58.00%), (S'= Sales of Past Year= 48,851 millions USD), (A= After Tax Income= 7,899 millions USD), (V= Variable Cost= 9,746 millions USD), (s= [S/S']-1= Sales Growth= 5.00%), and (S= Sales or Revenues= 51,294 millions USD), are known, then it's (I= Interest Expense planned) is:

$I = S'[1+s]-V-Sf-A/[1-t]$
$= \{48,851[1+0.0500]- 9,746-51,294 *0.5800-7,899/ [1-0.3000]$
$= \underline{513}$ millions USD

Law-1342:

If (**I**= Interest Expense= 513 millions USD), (**f**= F/S=Fixed Portion= 58.00%), (**S'**= Sales of Past Year= 48,851 millions USD), (**A**= After Tax Income= 7,899 millions USD), (**V**= Variable Cost= 9,746 millions USD), (**s**= [S/S']-1= Sales Growth= 5.00%), and (**S**= Sales or Revenues= 51,294 millions USD), are known, then it's (**t**= Tax Rate planned) is:

$$\mathbf{t} = 1-\mathbf{A}/\{\mathbf{S'}[1+\mathbf{s}]-\mathbf{V}-\mathbf{Sf}-\mathbf{I}\}$$
$$= 1-7,899/\{48,851[1+0.0500]- 9,746 -51,294*0.5800-513\}$$
$$= 30.00\%$$

Law-1343:

If (**i**= I/S= Interest Portion= 1.00%), (**f**= F/S=Fixed Portion= 58.00%), (**S'**= Sales of Past Year= 48,851 millions USD), (**t**= T/B= Tax Rate= 30.00%), (**V**= Variable Cost= 9,746 millions USD), (**s**= [S/S']-1= Sales Growth= 5.00%), and (**S**= Sales or Revenues= 51,294 millions USD), are known, then it's (**A**= After Tax Income planned) is:

$$\mathbf{A} = [1-\mathbf{t}]\{\mathbf{S'}[1+\mathbf{s}]-\mathbf{V}-\mathbf{Sf}-\mathbf{Si}\}$$
$$= [1-\mathbf{t}]\{\mathbf{S'}[1+\mathbf{s}]-\mathbf{V}-\mathbf{S}[\mathbf{f}+\mathbf{i}]\}$$
$$= [1-0.3000]\{48,851[1+0.0500]-9,746 -51,294[0.5800+0.0100]\}$$
$$= 7,899 \text{ millions USD}$$

Law-1344:

If (**i**= I/S= Interest Portion= 1.00%), (**f**= F/S=Fixed Portion= 58.00%), (**A**= After Tax Income= 7,899 millions USD), (**t**= T/B= Tax Rate= 30.00%), (**V**= Variable Cost= 9,746 millions USD), (**s**= [S/S']-1= Sales Growth= 5.00%), and (**S**= Sales or Revenues= 51,294 millions USD), are known, then it's (**S'**= Sales Past) must be:

$$S' = \{V + S[f+i] + A/[1-t]\}/[1+s]$$
$$= \{9,746 + 51,294[0.5800 + 0.0100] + 7,899 /[1-0.3000]\}/[1+0.0500]$$
$$= 48,851 \text{ millions USD}$$

Law-1345:

If (**i**= I/S= Interest Portion= 1.00%), (**f**= F/S=Fixed Portion= 58.00%), (**A**= After Tax Income= 7,899 millions USD), (**t**= T/B= Tax Rate= 30.00%), (**V**= Variable Cost= 9,746 millions USD), (**S'**= Sales of Past Year= 48,851 millions USD), and (**S**= Sales or Revenues= 51,294 millions USD), are known, then it's (**s**= Sales Growth planned) is:

$$s = \{V + S[f+i] + A/[1-t]\}/S' - 1$$
$$= \{9,746 + 51,294[0.5800 + 0.0100] + 7,899 /[1-0.3000]\}/48,851 - 1$$
$$= 5.00\%$$

Law-1346:
If (i= I/S= Interest Portion= 1.00%), (f= F/S=Fixed Portion= 58.00%), (A= After Tax Income= 7,899 millions USD), (t= T/B= Tax Rate= 30.00%), (s= [S/S']-1= Sales Growth= 5.00%), ($\$'$= Sales of Past Year= 48,851 millions USD), and ($\$$= Sales or Revenues= 51,294 millions USD), are known, then it's (V= Variable Cost planned) is:
$V = \{\$'[1+s] - \$[f+i] - A/[1-t]\}$
$= \{48,851[1+0.0500] - 51,294[0.5800 + 0.0100] - 7,899/[1-0.3000]\}$
$= 9,746$ millions USD

Law-1347:
If (i= I/S= Interest Portion= 1.00%), (f= F/S=Fixed Portion= 58.00%), (A= After Tax Income= 7,899 millions USD), (t= T/B= Tax Rate= 30.00%), (s= [S/S']-1= Sales Growth= 5.00%), ($\$'$= Sales of Past Year= 48,851 millions USD), and (V= Variable Cost= 9,746 millions USD), are known, then it's ($\$$= Sales or Revenues planned) is:
$\$ = \{\$'[1+s] - V - A/[1-t]\}/[f+i]$
$= \{48,851[1+0.0500] - 9,746 - 7,899 /[1-0.3000]\}/[0.5800+0.0100]$
$= 51,294$ millions USD

Law-1348:
If (i= I/S= Interest Portion= 1.00%), (S= Sales or Revenues= 51,294 millions USD), (A= After Tax Income= 7,899 millions USD), (t= T/B= Tax Rate= 30.00%), (s= [S/S']-1= Sales Growth= 5.00%), (S'= Sales of Past Year= 48,851 millions USD), and (V= Variable Cost= 9,746 millions USD), are known, then it's (f= Fixed Portion planned) is:

f = {S'[1+s]-V-A/[1-t]}/S-i
= {48,851[1+0.0500]-9,746 -7,899
/[1-0.3000]}/51,294- 0.0100
= 58.00%

Law-1349:
If (f= F/S=Fixed Portion= 58.00%), (S= Sales or Revenues= 51,294 millions USD), (A= After Tax Income= 7,899 millions USD), (t= T/B= Tax Rate= 30.00%), (s= [S/S']-1= Sales Growth= 5.00%), (S'= Sales of Past Year= 48,851 millions USD), and (V= Variable Cost= 9,746 millions USD), are known, then it's (i= Interest Portion planned) is:

i = {S'[1+s]-V-A/[1-t]}/S-f
= {48,851[1+0.0500]-9,746 -7,899
/[1-0.3000]}/51,294- 0.5800
= 1.00%

Law-1350:
If (**f**= F/S=Fixed Portion= 58.00%), (**S**= Sales or Revenues= 51,294 millions USD), (**A**= After Tax Income= 7,899 millions USD), (**i**= I/S= Interest Portion= 1.00%), (**s**= [S/S']-1= Sales Growth= 5.00%), (**S'**= Sales of Past Year= 48,851 millions USD), and (**V**= Variable Cost= 9,746 millions USD), are known, then it's (**t**= Tax Rate planned) is:

$$\begin{aligned}
\mathbf{t} &= 1-\mathbf{A}/\{\mathbf{S'}[1+\mathbf{s}]-\mathbf{V}-\mathbf{Sf}-\mathbf{Si}\} \\
&= 1-\mathbf{A}/\{\mathbf{S'}[1+\mathbf{s}]-\mathbf{V}-\mathbf{S}[\mathbf{f}+\mathbf{i}]\} \\
&= 1-7,899/\{48,851[1+0.0500]-9,746 \\
&\qquad -51,294[\,0.5800+0.0100]\} \\
&= 30.00\%
\end{aligned}$$

Law-1351:
If (**f**= F/S=Fixed Portion= 58.00%), (**S**= Sales or Revenues= 51,294 millions USD), (**t**= T/B= Tax Rate= 30.00%), (**i**= I/S= Interest Portion= 1.00%), (**s**= [S/S']-1= Sales Growth= 5.00%), (**S'**= Sales of Past Year= 48,851 millions USD), and (**V**= Variable Cost= 9,746 millions USD), are known, then it's (**A**= After Tax Income planned) is:

$$\begin{aligned}
\mathbf{A} &= [1-\mathbf{t}]\{\mathbf{S'}[1+\mathbf{s}]-\mathbf{V}-\mathbf{Sf}-\mathbf{S'i}[1+\mathbf{s}]\} \\
&= [1-\mathbf{t}]\{\mathbf{S'}[1+\mathbf{s}][1-\mathbf{i}]-\mathbf{V}-\mathbf{Sf}\} \\
&= [1-0.3000]\{48,851[1+0.0500] \\
&\qquad [1-0.0100]-9,746 \\
&\qquad -51,294*0.5800\} \\
&= 7,899 \text{ millions USD}
\end{aligned}$$

Law-1352:

If (**f**= F/S=Fixed Portion= 58.00%), (**S**= Sales or Revenues= 51,294 millions USD), (**t**= T/B= Tax Rate= 30.00%), (**i**= I/S= Interest Portion= 1.00%), (**s**= [S/S']-1= Sales Growth= 5.00%), (**A**= After Tax Income= 7,899 millions USD), and (**V**= Variable Cost= 9,746 millions USD), are known, then it's (**S'**= Sales Past) must be:

$S' = \{V+Sf+A/1-t]\}/\{[1+s][1-i]\}$
$= \{9,746+51,294* 0.5800+7,899$
$/[1-0.3000]\}/\{[1+0.0500]$
$[1-0.0100]\}$
$= 48,851$ millions USD

Law-1353:

If (**f**= F/S=Fixed Portion= 58.00%), (**S**= Sales or Revenues= 51,294 millions USD), (**t**= T/B= Tax Rate= 30.00%), (**i**= I/S= Interest Portion= 1.00%), (**S'**= Sales of Past Year= 48,851 millions USD), (**A**= After Tax Income= 7,899 millions USD), and (**V**= Variable Cost= 9,746 millions USD), are known, then it's (**s**= Sales Growth planned) is:

$s = \{V+Sf+A/1-t]\}/\{S'[1-i]\}-1$
$= \{9,746+51,294* 0.5800+7,899$
$/[1-0.3000]\}$
$/\{48,851[1-0.0100]\}-1$
$= 5.00\%$

Law-1354:
   If (**f**= F/S=Fixed Portion= 58.00%), (**$**= Sales or Revenues= 51,294 millions USD), (**t**= T/B= Tax Rate= 30.00%), (**s**= [S/S']-1= Sales Growth= 5.00%), (**$'**= Sales of Past Year= 48,851 millions USD), (**A**= AfterTax Income= 7,899 millions USD), and (**V**= Variable Cost= 9,746 millions USD), are known, then it's (**i**= Interest Portion planned) is:

   i = 1-{**V+$f+A**/1-**t**]}/{**$'**[1+**s**]}
   = 1- {9,746+51,294* 0.5800+7,899
       /[1-0.3000]}
       /{48,851[1+0.0500]}
   = 1.00%

Law-1355:
   If (**f**= F/S=Fixed Portion= 58.00%), (**$**= Sales or Revenues= 51,294 millions USD), (**t**= T/B= Tax Rate= 30.00%), (**s**= [S/S']-1= Sales Growth= 5.00%), (**$'**= Sales of Past Year= 48,851 millions USD), (**A**= AfterTax Income= 7,899 millions USD), and (**i**= I/S= Interest Portion= 1.00%), are known, then it's (**V**= Variable Cost planned) is:

   V= **$'**[1+**s**][1-**i**]-**$f-A**/1-**t**]
   = 48,851[1+0.0500][1-0.0100]-51,294
       *0.5800-7,899/[1-0.3000]}
   = 9,746 millions USD

Law-1356:
If (**f**= F/S=Fixed Portion= 58.00%), (**V**= Variable Cost= 9,746 millions USD), (**t**= T/B= Tax Rate= 30.00%), (**s**= [S/S']-1= Sales Growth= 5.00%), (**$'**= Sales of Past Year= 48,851 millions USD), (**A**= AfterTax Income= 7,899 millions USD), and (**i**= I/S= Interest Portion= 1.00%), are known, then it's (**$**= Sales or Revenues planned) is:

$= {$'[1+s][1-i]-**V**-**A**/1-**t**]}/**f**
  = {48,851[1+0.0500][1-0.0100]-9,746
      -7,899/[1-0.3000]}/0.5800
  = 51,294 millions USD

Law-1357:
If (**$**= Sales or Revenues= 51,294 millions USD), (**V**= Variable Cost= 9,746 millions USD), (**t**= T/B= Tax Rate= 30.00%), (**s**= [S/S']-1= Sales Growth= 5.00%), (**$'**= Sales of Past Year= 48,851 millions USD), (**A**= AfterTax Income= 7,899 millions USD), and (**i**= I/S= Interest Portion= 1.00%), are known, then it's (**f**= Fixed Portion planned) is:

f= {$'[1+s][1-i]-**V**-**A**/1-**t**]}/$
  = {48,851[1+0.0500][1-0.0100]-9,746
      -7,899/[1-0.3000]}/51,294
  = 58.00%

Law-1358:

If ($= Sales or Revenues= 51,294 millions USD), (V= Variable Cost= 9,746 millions USD), (f= F/S=Fixed Portion= 58.00%), (s= [S/S']-1= Sales Growth= 5.00%), ($'= Sales of Past Year= 48,851 millions USD), (A= AfterTax Income= 7,899 millions USD), and (i= I/S= Interest Portion= 1.00%), are known, then it's (t= Tax Rate planned) is:

$t = 1 - A/\{\$'[1+s] - V - \$f - \$'i[1+s]\}$
$= 1 - A/\{\$'[1+s][1-i] - V - \$f\}$
$= 1 - 7,899/\{48,851[1+0.0500][1-0.0100] - 9,746 - 51,294*0.5800\}$
$= 30.00\%$

Law-1359:

If ($= Sales or Revenues= 51,294 millions USD), (V= Variable Cost= 9,746 millions USD), (f= F/S=Fixed Portion= 58.00%), (s= [S/S']-1= Sales Growth= 5.00%), ($'= Sales of Past Year= 48,851 millions USD), (t= T/B= Tax Rate= 30.00%), and (i= I/S= Interest Portion= 1.00%), are known, then it's (A= After Tax Income planned) is:

$A = [1-t]\{\$'[1+s] - V - \$'f[1+s] - \$'i[1+s]\}$
$= [1-t]\{\$'[1+s][1-f-i] - V\}$
$= [1-0.3000]\{48,851[1+0.0500][1-0.5800-0.0100] - 9,746\}$
$= 7,899$ millions USD

Law-1360:
If ($= Sales or Revenues= 51,294 millions USD), (V= Variable Cost= 9,746 millions USD), (f= F/S=Fixed Portion= 58.00%), (s= [S/S']-1= Sales Growth= 5.00%), (A= After Tax Income= 7,899 millions USD), (t= T/B= Tax Rate= 30.00%), and (i= I/S= Interest Portion= 1.00%), are known, then it's ($'= Sales Past) must be:

$$\$' = \{V+A/[1-t]\}/\{[1+s][1-f-i]\}$$
$$= \{9,746 +7,899/[1-0.3000]\}/\{[1+0.0500][1-0.5800-0.0100]\}$$
$$= 48,851 \text{ millions USD}$$

Law-1361:
If ($'= Sales of Past Year= 48,851 millions USD), (V= Variable Cost= 9,746 millions USD), (f= F/S=Fixed Portion= 58.00%), ($= Sales or Revenues= 51,294 millions USD), (A= After Tax Income= 7,899 millions USD), (t= T/B= Tax Rate= 30.00%), and (i= I/S= Interest Portion= 1.00%), are known, then it's (s= Sales Growth Planned) is:

$$s = \{V+A/[1-t]\}/\{\$'[1-f-i]\}-1$$
$$= \{9,746 +7,899/[1-0.3000]\}/\{48,851[1-0.5800-0.0100]\}-1$$
$$= 5.00\%$$

Law-1362:
If ($'= Sales of Past Year= 48,851 millions USD), (V= Variable Cost= 9,746 millions USD), ($= Sales or Revenues= 51,294 millions USD), (s= [S/S']-1= Sales Growth= 5.00%), (A= After Tax Income= 7,899 millions USD), (t= T/B= Tax Rate= 30.00%), and (i= I/S= Interest Portion= 1.00%), are known, then it's (f= Fixed Portion Planned) is:

$$f = 1-i-\{V+A/[1-t]\}/\{\$'[1+s]\}$$
$$= 1-0.0100-\{9,746 +7,899/[1-0.3000]\}/\{48,851[1 +0.0500]\}$$
$$= 58.00\%$$

Law-1363:
If ($'= Sales of Past Year= 48,851 millions USD), (V= Variable Cost= 9,746 millions USD), ($= Sales or Revenues= 51,294 millions USD), (s= [S/S']-1= Sales Growth= 5.00%), (A= After Tax Income= 7,899 millions USD), (t= T/B= Tax Rate= 30.00%), and (f= F/S=Fixed Portion= 58.00%), are known, then it's (i= Interest Portion Planned) is:

$$i = 1-f-\{V+A/[1-t]\}/\{\$'[1+s]\}$$
$$= 1-0.5800-\{9,746 +7,899/[1-0.3000\}/\{48,851[1 +0.0500]\}$$
$$= 1.00\%$$

Law-1364:
If ($'= Sales of Past Year= 48,851 millions USD), (i= I/S= Interest Portion= 1.00%), ($= Sales or Revenues= 51,294 millions USD), (s= [S/S']-1= Sales Growth= 5.00%), (A= After Tax Income= 7,899 millions USD), (t= T/B= Tax Rate= 30.00%), and (f= F/S=Fixed Portion= 58.00%), are known, then it's (V= Variable Cost Planned) is:

$$V = \$'[1+s][1-f-i] - A/[1-t]$$
$$= 48,851[1+0.0500][1-0.5800-0.0100]$$
$$-7,899/[1-0.3000]$$
$$= \underline{9,746} \text{ millions USD}$$

Law-1365:
If ($'= Sales of Past Year= 48,851 millions USD), (i= I/S= Interest Portion= 1.00%), ($= Sales or Revenues= 51,294 millions USD), (s= [S/S']-1= Sales Growth= 5.00%), (A= After Tax Income= 7,899 millions USD), (V= Variable Cost= 9,746 millions USD), and (f= F/S=Fixed Portion= 58.00%), are known, then it's (t= Tax Rate Planned) is:

$$t = 1 - A/\{\$'[1+s] - V - \$'f[1+s] - \$'i[1+s]\}$$
$$= 1 - A/\{\$'[1+s][1-f-i] - V\}$$
$$= 1 - 7,899/\{48,851[1+0.0500][1-0.5800$$
$$-0.0100] - 9,746\}$$
$$= \underline{30.00\%}$$

Law-1366:
If ($$= Sales of Past Year= 48,851 millions USD), (I= Interest Expense= 513 millions USD), ($= Sales or Revenues= 51,294 millions USD), (s= [S/S']-1= Sales Growth= 5.00%), (t= T/B= Tax Rate= 30.00%), (v= V/S= Variable Portion= 19.00%), and (F= Fixed Cost= 29,740 millions USD), are known, then it's (A= After Tax Income Planned) is:

A= [1-t]{$'[1+s]-$v-F-I}
  = [1-0.3000]{48,851[1 +0.0500]
      -51,294*0.1900-29,750-513}
  = 7,899 millions USD

Law-1367:
If (A= After Tax Income= 7,899 millions USD), (I= Interest Expense= 513 millions USD), ($= Sales or Revenues= 51,294 millions USD), (s= [S/S']-1= Sales Growth= 5.00%), (t= T/B= Tax Rate= 30.00%), (v= V/S= Variable Portion= 19.00%), and (F= Fixed Cost= 29,740 millions USD), are known, then it's ($'= Sales Past) is:

$'= {$v+F+I+A/[1-t]}/[1+s]
  = {51,294*0.1900+29,750+513+7,899
      /[1-0.3000]}/[1 +0.0500]
  = 48,851 millions USD

Law-1368:

If (A= After Tax Income= 7,899 millions USD), (I= Interest Expense= 513 millions USD), (S= Sales or Revenues= 51,294 millions USD), (S'= Sales of Past Year= 48,851 millions USD), (t= T/B= Tax Rate= 30.00%), (v= V/S= Variable Portion= 19.00%), and (F= Fixed Cost= 29,740 millions USD), are known, then it's (s= Sales Growth planned) is:

$s = \{Sv+F+I+A/[1-t]\}/S'-1$
$= \{51,294*0.1900+29,750+513+7,899 /[1-0.3000]\}/48,851-1$
$= 5.00\%$

Law-1369:

If (A= After Tax Income= 7,899 millions USD), (I= Interest Expense= 513 millions USD), (s= [S/S']-1= Sales Growth= 5.00%), (S'= Sales of Past Year= 48,851 millions USD), (t= T/B= Tax Rate= 30.00%), (v= V/S= Variable Portion= 19.00%), and (F= Fixed Cost= 29,740 millions USD), are known, then it's (S= Sales or Revenues planned) is:

$S = S'[1+s]-F-I-A/[1-t]\}/v$
$= 48,851[1+0.0500]-29,750-513-7,899 /[1-0.3000]\}/0.1900$
$= 51,294$ millions USD

Law-1370:

If (A= After Tax Income= 7,899 millions USD), (I= Interest Expense= 513 millions USD), (s= [S/S']-1= Sales Growth= 5.00%), (S'= Sales of Past Year= 48,851 millions USD), (t= T/B= Tax Rate= 30.00%), (S= Sales or Revenues= 51,294 millions USD), and (F= Fixed Cost= 29,740 millions USD), are known, then it's (v= Variable Portion planned) is:

v = {S'[1+s]-F-I-A/[1-t]}/S
= 48,851[1+0.0500]-29,750-513-7,899
/[1-0.3000]}/51,294
= 19.00%

Law-1371:

If (v= V/S= Variable Portion= 19.00%), (I= Interest Expense= 513 millions USD), (A= After Tax Income= 7,899 millions USD), (s= [S/S']-1= Sales Growth= 5.00%), (S'= Sales of Past Year= 48,851 millions USD), (t= T/B= Tax Rate= 30.00%), (S= Sales or Revenues= 51,294 millions USD), and are known, then it's (F= Fixed Cost planned) is:

F = S'[1+s]-Sv-I-A/[1-t]
= 48,851[1+0.0500]-51,294*0.1900
-513-7,899/[1-0.3000]
= 29,750 millions USD

Law-1372:
If (v= V/S= Variable Portion= 19.00%), (A= After Tax Income= 7,899 millions USD), (s= [S/S']-1= Sales Growth= 5.00%), (S'= Sales of Past Year= 48,851 millions USD), (t= T/B= Tax Rate= 30.00%), (S= Sales or Revenues= 51,294 millions USD), and (F= Fixed Cost= 29,740 millions USD), are known, then it's (I= Interest Expense planned) is:
I= S'[1+s]-Sv-F-A/[1-t]
   = 48,851[1+0.0500]-51,294* 0.1900
     -29,750-7,899/[1-0.3000]
   = 513 millions USD

Law-1373:
If (v= V/S= Variable Portion= 19.00%), (A= After Tax Income= 7,899 millions USD), (s= [S/S']-1= Sales Growth= 5.00%), (S'= Sales of Past Year= 48,851 millions USD), (I= Interest Expense= 513 millions USD), (S= Sales or Revenues= 51,294 millions USD), and (F= Fixed Cost= 29,740 millions USD), are known, then it's (t= Tax Rate planned) is:
t= 1-A/{S'[1+s]-Sv-F-I}
   = 1-7,899/{48,851[1+0.0500]-51,294
     *0.1900-29,750-513}
   = 30.00%

Law-1374:

If (v= V/S= Variable Portion= 19.00%), (i= I/S= Interest Portion= 1.00%), (s= [S/S']-1= Sales Growth= 5.00%), (S'= Sales of Past Year= 48,851 millions USD), (t= T/B= Tax Rate= 30.00%), (S= Sales or Revenues= 51,294 millions USD), and (F= Fixed Cost= 29,740 millions USD), are known, then it's (A= After Tax Income planned) is:

$$A = [1-t]\{S'[1+s]-Sv-F-Si\}$$
$$= [1-t]\{S'[1+s]-S[v+i]-F\}$$
$$= [1-0.3000]\{48,851[1+0.0500]-51,294$$
$$*[0.1900+0.0100]-29,750\}$$
$$= 7,899 \text{ millions USD}$$

Law-1375:

If (v= V/S= Variable Portion= 19.00%), (i= I/S= Interest Portion= 1.00%), (s= [S/S']-1= Sales Growth= 5.00%), (A= After Tax Income= 7,899 millions USD), (t= T/B= Tax Rate= 30.00%), (S= Sales or Revenues= 51,294 millions USD), and (F= Fixed Cost= 29,740 millions USD), are known, then it's (S'= Sales Past) must be:

$$S' = \{F+S[v+i]+A/[1-t]\}/[1+s]$$
$$= \{29,750+51,294[0.1900+0.0100]$$
$$+7,899/[1-0.3000]\}/[1+0.0500]$$
$$= 48,851 \text{ millions USD}$$

Law-1376:
If (v= V/S= Variable Portion= 19.00%), (i= I/S= Interest Portion= 1.00%), (S'= Sales of Past Year= 48,851 millions USD), (A= After Tax Income= 7,899 millions USD), (t= T/B= Tax Rate= 30.00%), (S= Sales or Revenues= 51,294 millions USD), and (F= Fixed Cost= 29,740 millions USD), are known, then it's (s= Sales Growth planned) is:

$s = \{F+S[v+i]+A/[1-t]\}/S'-1$
$= \{29,750+51,294[0.1900+0.0100]$
$\quad +7,899/[1-0.3000]\}/48,851-1$
$= \underline{5.00\%}$

Law-1377:
If (v= V/S= Variable Portion= 19.00%), (i= I/S= Interest Portion= 1.00%), (S'= Sales of Past Year= 48,851 millions USD), (A= After Tax Income= 7,899 millions USD), (t= T/B= Tax Rate= 30.00%), (s= [S/S']-1= Sales Growth= 5.00%), and (F= Fixed Cost= 29,740 millions USD), are known, then it's (S= Sales Growth planned) is:

$S = \{S'[1+s]-F-A/[1-t]\}/[v+i]$
$= \{48,851[1+0.0500]-29,750-7,899$
$\quad /[1-0.3000]\}/[0.1900+0.0100]$
$= \underline{51,294}$ millions USD

Law-1378:

If ($= Sales or Revenues= 51,294 millions USD), (i= I/S= Interest Portion= 1.00%), ($'= Sales of Past Year= 48,851 millions USD), (A= After Tax Income= 7,899 millions USD), (t= T/B= Tax Rate= 30.00%), (s= [S/S']-1= Sales Growth= 5.00%), and (F= Fixed Cost= 29,740 millions USD), are known, then it's (v= Variable Portion planned) is:

$$v = \{\$'[1+s]-F-A/[1-t]\}/\$-i$$
$$= \{48,851[1+0.0500]-29,750-7,899 /[1-0.3000]\}/51,294-0.0100$$
$$= 19.00\%$$

Law-1379:

If ($= Sales or Revenues= 51,294 millions USD), (v= V/S= Variable Portion= 19.00%), ($'= Sales of Past Year= 48,851 millions USD), (A= After Tax Income= 7,899 millions USD), (t= T/B= Tax Rate= 30.00%), (s= [S/S']-1= Sales Growth= 5.00%), and (F= Fixed Cost= 29,740 millions USD), are known, then it's (i= Interest Portion planned) is:

$$i = \{\$'[1+s]-F-A/[1-t]\}/\$-v$$
$$= \{48,851[1+0.0500]-29,750-7,899 /[1-0.3000]\}/51,294-0.1900$$
$$= 1.00\%$$

Law-1380:

If ($= Sales or Revenues= 51,294 millions USD), (v= V/S= Variable Portion= 19.00%), ($'= Sales of Past Year= 48,851 millions USD), (A= After Tax Income= 7,899 millions USD), (t= T/B= Tax Rate= 30.00%), (s= [S/S']-1= Sales Growth= 5.00%), and (i= I/S= Interest Portion= 1.00%), are known, then it's (F= Fixed Cost planned) is:

$$F = \$'[1+s] - \$[v+i] - A/[1-t]$$
$$= 48,851[1+0.0500] - 51,294[0.1900 + 0.0100] - 7,899/[1-0.3000]$$
$$= \underline{29,750} \text{ millions USD}$$

Law-1381:

If ($= Sales or Revenues= 51,294 millions USD), (v= V/S= Variable Portion= 19.00%), ($'= Sales of Past Year= 48,851 millions USD), (A= After Tax Income= 7,899 millions USD), (F= Fixed Cost= 29,740 millions USD), (s= [S/S']-1= Sales Growth= 5.00%), and (i= I/S= Interest Portion= 1.00%), are known, then it's (t= Tax Rate planned) is:

$$t = 1 - A/\{\$'[1+s] - \$v - F - \$i\}$$
$$= 1 - A/\{\$'[1+s] - \$[v+i] - F\}$$
$$= 1 - 7,899/\{48,851[1+0.0500] - 51,294[0.1900 + 0.0100] - 29,750\}$$
$$= \underline{30.00\%}$$

Law-1382:

If ($= Sales or Revenues= 51,294 millions USD), (v= V/S= Variable Portion= 19.00%), ($'= Sales of Past Year= 48,851 millions USD), (t= T/B= Tax Rate= 30.00%), (F= Fixed Cost= 29,740 millions USD), (s= [S/S']-1= Sales Growth= 5.00%), and (i= I/S= Interest Portion= 1.00%), are known, then it's (A= Tax Rate planned) is:

$$\begin{aligned}
A &= [1\text{-}t]\{\$'[1+s]\text{-}\$v\text{-}F\text{-}\$'i[1+s]\} \\
&= [1\text{-}t]\{\$'[1+s][1\text{-}i]\text{-}\$v\text{-}F\} \\
&= [1\text{-}0.3000]\{48,851[1+0.0500] \\
&\quad [1\text{-}0.0100]\text{-}51,294 \\
&\quad *0.1900\text{-}29,750\} \\
&= 7,899 \text{ millions USD}
\end{aligned}$$

Law-1383:

If ($= Sales or Revenues= 51,294 millions USD), (v= V/S= Variable Portion= 19.00%), (A= After Tax Income= 7,899 millions USD), (t= T/B= Tax Rate= 30.00%), (F= Fixed Cost= 29,740 millions USD), (s= [S/S']-1= Sales Growth= 5.00%), and (i= I/S= Interest Portion= 1.00%), are known, then it's ($'= Sales Past) must be:

$$\begin{aligned}
\$' &= \{\$v+F+A/[1\text{-}t]\}/\{[1+s][1\text{-}i]\} \\
&= \{51,294*0.1900+29,750+7,899 \\
&\quad /[1\text{-}0.3000]\} \\
&\quad /\{[1+0.0500][1\text{-}0.0100]\} \\
&= 48,851 \text{ millions USD}
\end{aligned}$$

Law-1384:
If ($= Sales or Revenues= 51,294 millions USD), (v= V/S= Variable Portion= 19.00%), (A= After Tax Income= 7,899 millions USD), (t= T/B= Tax Rate= 30.00%), (F= Fixed Cost= 29,740 millions USD), ($'= Sales of Past Year= 48,851 millions USD), and (i= I/S= Interest Portion= 1.00%), are known, then it's (s= Sales Growth planned) is:

$s = \{\$v+F+A/[1-t]\}/\{\$'[1-i]\} - 1$
$ = \{51,294*0.1900+29,750+7,899$
$\phantom{s = \{} /[1-0.3000]\}$
$\phantom{s = \{} /\{48,851[1-0.0100]\} - 1$
$ = 5.00\%$

Law-1385:
If ($= Sales or Revenues= 51,294 millions USD), (v= V/S= Variable Portion= 19.00%), (A= After Tax Income= 7,899 millions USD), (t= T/B= Tax Rate= 30.00%), (F= Fixed Cost= 29,740 millions USD), ($'= Sales of Past Year= 48,851 millions USD), and (s= [S/S']-1= Sales Growth= 5.00%), are known, then it's (i= Interest Portion planned) is:

$i = 1 - \{\$v+F+A/[1-t]\}/\{\$'[1+s]\}$
$ = 1 - \{51,294*0.1900+29,750+7,899$
$\phantom{i = 1 - \{} /[1-0.3000]\}$
$\phantom{i = 1 - \{} /\{48,851[1+0.0500]\}$
$ = 1.00\%$

Law-1387:

If (i= I/S= Interest Portion= $\underline{1.00}$%), ($\$$= Sales or Revenues= $\underline{51,294}$ millions USD), (A= After Tax Income= $\underline{7,899}$ millions USD), (t= T/B= Tax Rate= $\underline{30.00}$%), (F= Fixed Cost= $\underline{29,740}$ millions USD), ($\$'$= Sales of Past Year= $\underline{48,851}$ millions USD), and (s= [S/S']-1= Sales Growth= $\underline{5.00}$%), are known, then it's (v= Variable Portion planned) is:

v= {$\$'$[1+$s$][1-$i$]-$F$-$A$/[1-$t$]}/$\$$
  = {48,851[1+0.0500][1-0.0100]-29,750
     -7,899/[1-0.3000]}/51,294
  = $\underline{19.00}$%

Law-1388:

If (i= I/S= Interest Portion= $\underline{1.00}$%), ($\$$= Sales or Revenues= $\underline{51,294}$ millions USD), (A= After Tax Income= $\underline{7,899}$ millions USD), (t= T/B= Tax Rate= $\underline{30.00}$%), (v= V/S= Variable Portion= $\underline{19.00}$%), ($\$'$= Sales of Past Year= $\underline{48,851}$ millions USD), and (s= [S/S']-1= Sales Growth= $\underline{5.00}$%), are known, then it's (F= Fixed Cost planned) is:

F= $\$'$[1+$s$][1-$i$]-$\$v$-$A$/[1-$t$]}
  = 48,851[1+0.0500][1-0.0100]-51,294
    *0.1900 -7,899/[1-0.3000]}
  = $\underline{29,750}$ millions USD

Law-1389:

If (**i**= I/S= Interest Portion= 1.00%), (**$**= Sales or Revenues= 51,294 millions USD), (**A**= After Tax Income= 7,899 millions USD), (**F**= Fixed Cost= 29,740 millions USD), (**v**= V/S= Variable Portion= 19.00%), (**$'**= Sales of Past Year= 48,851 millions USD), and (**s**= [S/S']-1= Sales Growth= 5.00%), are known, then it's (**t**= Tax Rate planned) is:

$$\begin{aligned}
\mathbf{t} &= 1\text{-}\mathbf{A}/\{\mathbf{\$'}[1\text{+}\mathbf{s}]\text{-}\mathbf{\$v}\text{-}\mathbf{F}\text{-}\mathbf{\$'i}[1\text{+}\mathbf{s}]\} \\
&= 1\text{-}\mathbf{A}/\{\mathbf{\$'}[1\text{+}\mathbf{s}][1\text{-}\mathbf{i}]\text{-}\mathbf{\$v}\text{-}\mathbf{F}\} \\
&= 1\text{-}7{,}899/\{\ 48{,}851[1\text{+}0.0500][1\text{-}0.0100] \\
&\qquad \text{-}51{,}294*0.1900 \text{ -}29{,}750\} \\
&= 30.00\%
\end{aligned}$$

Law-1390:

If (**I**= Interest Expense= 513 millions USD), (**$**= Sales or Revenues= 51,294 millions USD), (**t**= T/B= Tax Rate= 30.00%), (**f**= F/S=Fixed Portion= 58.00%), (**v**= V/S= Variable Portion= 19.00%), (**$'**= Sales of Past Year= 48,851 millions USD), and (**s**= [S/S']-1= Sales Growth= 5.00%), are known, then it's (**A**= After Tax Income planned) is:

$$\begin{aligned}
\mathbf{A} &= [1\text{-}\mathbf{t}]\{\mathbf{\$'}[1\text{+}\mathbf{s}]\text{-}\mathbf{\$v}\text{-}\mathbf{\$f}\text{-}\mathbf{I}\} \\
&= [1\text{-}0.3000]\{\ 48{,}851[1\text{+}0.0500]\text{-}51{,}294 \\
&\qquad *0.1900\ \text{-}51{,}294*0.5800\text{- }513\} \\
&= 7{,}899 \text{ millions USD}
\end{aligned}$$

Law-1391:
　If (I= Interest Expense= 513 millions USD), (S= Sales or Revenues= 51,294 millions USD), (t= T/B= Tax Rate= 30.00%), (f= F/S=Fixed Portion= 58.00%), (v= V/S= Variable Portion= 19.00%), (A= After Tax Income= 7,899 millions USD), and (s= [S/S']-1= Sales Growth= 5.00%), are known, then it's (S'= Sales Past) must be:

S' = {S[v+f]+I+A/[1-t]}/[1+s]
　　= {51,294[0.1900+0.5800]+513+7,899/[1-0.3000]}/[1+0.0500]
　　= 48,851 millions USD

Law-1392:
　If (I= Interest Expense= 513 millions USD), (S= Sales or Revenues= 51,294 millions USD), (t= T/B= Tax Rate= 30.00%), (f= F/S=Fixed Portion= 58.00%), (v= V/S= Variable Portion= 19.00%), (A= After Tax Income= 7,899 millions USD), and (S'= Sales of Past Year= 48,851 millions USD), are known, then it's (s= Sales Growth planned) is:

s = {A/[1-t]+S[v+f]+I}/S'-1
　　= {7,899/[1-0.3000]+{51,294[0.1900+0.5800]+513}/48,851-1
　　= 5.00%

Law-1393:
If (**I**= Interest Expense= 513 millions USD), (**s**= [S/S']-1= Sales Growth= 5.00%), (**t**= T/B= Tax Rate= 30.00%), (**f**= F/S=Fixed Portion= 58.00%), (**v**= V/S= Variable Portion= 19.00%), (**A**= After Tax Income= 7,899 millions USD), and (**S'**= Sales of Past Year= 48,851 millions USD), are known, then it's (**S**= Sales or Revenues planned) is:

$$S = \{S'[1+s]-I-A/[1-t]\}/[v+f]$$
$$= \{48,851[1+0.0500]-513-7,899 /[1-0.3000]\}/[0.1900+0.5800]$$
$$= 51,294 \text{ millions USD}$$

Law-1394:
If (**I**= Interest Expense= 513 millions USD), (**s**= [S/S']-1= Sales Growth= 5.00%), (**t**= T/B= Tax Rate= 30.00%), (**f**= F/S=Fixed Portion= 58.00%), (**S**= Sales or Revenues= 51,294 millions USD), (**A**= After Tax Income= 7,899 millions USD), and (**S'**= Sales of Past Year= 48,851 millions USD), are known, then it's (**v**=Variable Portion planned) is:

$$v = \{S'[1+s]-I-A/[1-t]\}/S-f$$
$$= \{48,851[1+0.0500]-513-7,899 /[1-0.3000]\}/51,294-0.5800$$
$$= 19.00\%$$

Law-1395:

If (**I**= Interest Expense= 513 millions USD), (**s**= [S/S']-1= Sales Growth= 5.00%), (**t**= T/B= Tax Rate= 30.00%), (**v**= V/S= Variable Portion= 19.00%), (**$**= Sales or Revenues= 51,294 millions USD), (**A**= After Tax Income= 7,899 millions USD), and (**$'**= Sales of Past Year= 48,851 millions USD), are known, then it's (**f**= Fixed Portion planned) is:

$$f = \{\$'[1+s]-I-A/[1-t]\}/\$-v$$
$$= \{48,851[1+0.0500]-513-7,899 \\ /[1-0.3000]\}/51,294-0.1900$$
$$= 58.00\%$$

Law-1396:

If (**f**= F/S=Fixed Portion= 58.00%), (**s**= [S/S']-1= Sales Growth= 5.00%), (**t**= T/B= Tax Rate= 30.00%), (**v**= V/S= Variable Portion= 19.00%), (**$**= Sales or Revenues= 51,294 millions USD), (**A**= After Tax Income= 7,899 millions USD), and (**$'**= Sales of Past Year= 48,851 millions USD), are known, then it's (**I**= Interest Expense planned) is:

$$I = \$'[1+s]-\$[v+f]-A/[1-t]$$
$$= 48,851[1+0.0500]-51,294[0.1900 \\ +0.5800]-7,899/[1-0.3000]$$
$$= 513 \text{ millions USD}$$

Law-1397:
 If (f= F/S=Fixed Portion= 58.00%), (s= [S/S']-1= Sales Growth= 5.00%), (I= Interest Expense= 513 millions USD), (v= V/S= Variable Portion= 19.00%), ($\$$= Sales or Revenues= 51,294 millions USD), (A= After Tax Income= 7,899 millions USD), and ($\$'$= Sales of Past Year= 48,851 millions USD), are known, then it's (t= Tax Rate planned) is:
 t= 1-A/{$\$'$[1+$s$]-$\$v$-$\$f$-$I$}
   = 1-A/{$\$'$[1+$s$]-$\$$[v+f]-I}
   = 1-7,899/{48,851[1+0.0500]-51,294
         [0.1900+0.5800]-513
   = 30.00%

Law-1398:
 If (f= F/S=Fixed Portion= 58.00%), (s= [S/S']-1= Sales Growth= 5.00%), (i= I/S= Interest Portion= 1.00%), (v= V/S= Variable Portion= 19.00%), ($\$$= Sales or Revenues= 51,294 millions USD), (t= T/B= Tax Rate= 30.00%), and ($\$'$= Sales of Past Year= 48,851 millions USD), are known, then it's (A= After Tax Income planned) is:
 A= [1-t]{$\$'$[1+$s$]-$\$v$-$\$f$-$\$i$}
   = 1-A/{$\$'$[1+$s$]-$\$$[v+f+i]}
   = [1-0.3000]{48,851[1+0.0500]-51,294
         [0.1900+0.5800+0.0100]}
   = 7,899 millions USD

Law-1399:

If (**f**= F/S=Fixed Portion= 58.00%), (**s**= [S/S']-1= Sales Growth= 5.00%), (**i**= I/S= Interest Portion= 1.00%), (**v**= V/S= Variable Portion= 19.00%), (**$**= Sales or Revenues= 51,294 millions USD), (**t**= T/B= Tax Rate= 30.00%), and (**A**= After Tax Income= 7,899 millions USD), are known, then it's (**$'** =Sales Past) must be:

$'= {**$**[**v**+**f**+**i**]+**A**/[1-**t**]}/[1+**s**]
  = {51,294[0.1900+0.5800+0.0100]
       +7,899/[1-0.3000]}/[1+0.0500]
  = 48,851 millions USD

Law-1400:

If (**f**= F/S=Fixed Portion= 58.00%), (**$'**= Sales of Past Year= 48,851 millions USD), (**i**= I/S= Interest Portion= 1.00%), (**v**= V/S= Variable Portion= 19.00%), (**$**= Sales or Revenues= 51,294 millions USD), (**t**= T/B= Tax Rate= 30.00%), and (**A**= After Tax Income= 7,899 millions USD), are known, then it's (**s**= Sales Growth planned) is:

s= {**$**[**v**+**f**+**i**]+**A**/[1-**t**]}/**$'**-1
  = {51,294[0.1900+0.5800+0.0100]
       +7,899/[1-0.3000]}/48,851-1
  = 5.00%

Law-1401:
If (**f**= F/S=Fixed Portion= 58.00%), (**$'**= Sales of Past Year= 48,851 millions USD), (**i**= I/S= Interest Portion= 1.00%), (**v**= V/S= Variable Portion= 19.00%), (**s**= [S/S']-1= Sales Growth= 5.00%), (**t**= T/B= Tax Rate= 30.00%), and (**A**= After Tax Income= 7,899 millions USD), are known, then it's (**$**= Sales or Revenues planned) is:
$\$ = \{\$'[1+s]-A/[1-t]\}/[v+f+i]$
    = 48,851[1+0.0500]-7,899/[1-0.3000]}
       /[0.1900+0.5800+0.0100]
    = 51,294 millions USD

Law-1402:
If (**f**= F/S=Fixed Portion= 58.00%), (**$'**= Sales of Past Year= 48,851 millions USD), (**i**= I/S= Interest Portion= 1.00%), (**$**= Sales or Revenues= 51,294 millions USD), (**s**= [S/S']-1= Sales Growth= 5.00%), (**t**= T/B= Tax Rate= 30.00%), and (**A**= After Tax Income= 7,899 millions USD), are known, then it's (**v**= Variable Portion planned) is:
$v = \{\$'[1+s]-A/[1-t]\}/\$-f-i$
    = 48,851[1+0.0500]-7,899/[1-0.3000]}
       /51,294 -0.5800-0.0100
    = 19.00%

Law-1403:

If (v= V/S= Variable Portion= 19.00%), (S'= Sales of Past Year= 48,851 millions USD), (i= I/S= Interest Portion= 1.00%), (S= Sales or Revenues= 51,294 millions USD), (s= [S/S']-1= Sales Growth= 5.00%), (t= T/B= Tax Rate= 30.00%), and (A= After Tax Income= 7,899 millions USD), are known, then it's (f= Fixed Portion planned) is:

f = {S'[1+s]-A/[1-t]}/S-v-i
    = {48,851[1+0.0500]-7,899/[1-0.3000]}
       /51,294-0.1900-0.0100
    = 58.00%

Law-1404:

If (v= V/S= Variable Portion= 19.00%), (S'= Sales of Past Year= 48,851 millions USD), (f= F/S=Fixed Portion= 58.00%), (S= Sales or Revenues= 51,294 millions USD), (s= [S/S']-1= Sales Growth= 5.00%), (t= T/B= Tax Rate= 30.00%), and (A= After Tax Income= 7,899 millions USD), are known, then it's (i= Interest Portion planned) is:

i = {S'[1+s]-A/[1-t]}/S-f-v
    = {48,851[1+0.0500]-7,899/[1-0.3000]}
       /51,294 -0.5800-0.1900
    = 1.00%

Law-1405:
If (v= V/S= Variable Portion= 19.00%), ($\$'$= Sales of Past Year= 48,851 millions USD), (f= F/S=Fixed Portion= 58.00%), ($\$$= Sales or Revenues= 51,294 millions USD), (s= [S/S']-1= Sales Growth= 5.00%), (i= I/S= Interest Portion= 1.00%), and (A= After Tax Income= 7,899 millions USD), are known, then it's (t= Tax Rate planned) is:

t= 1-A/{$\$'$[1+$s$]-$\$v$-$\$f$-$\$i$}
 = 1-A/{$\$'$[1+$s$]-$\$$[v+f+i]}
 = 1-7,899/{48,851[1+0.0500]-51,294
   [0.1900+0.5800+0.0100]}
 = 30.00%

Law-1406:
If (v= V/S= Variable Portion= 19.00%), ($\$'$= Sales of Past Year= 48,851 millions USD), (f= F/S=Fixed Portion= 58.00%), ($\$$= Sales or Revenues= 51,294 millions USD), (s= [S/S']-1= Sales Growth= 5.00%), (i= I/S= Interest Portion= 1.00%), and (t= T/B= Tax Rate= 30.00%), are known, then it's (A= After Tax Income planned) is:

A= [1-t]{$\$'$[1+$s$]-$\$v$-$\$f$-$\$'i$[1+$s$]}
 = [1-t]{$\$'$[1+$s$][1-$i$]-$\$$[v+f]}
 = [1-0.3000]{48,851[1+0.0500]
   [1-0.0100]-51,294
   [0.1900+0.5800]}
 = 7,899 millions USD

Law-1407:
If (v= V/S= Variable Portion= 19.00%), (A= After Tax Income= 7,899 millions USD), (f= F/S=Fixed Portion= 58.00%), (S= Sales or Revenues= 51,294 millions USD), (s= [S/S']-1= Sales Growth= 5.00%), (i= I/S= Interest Portion= 1.00%), and (t= T/B= Tax Rate= 30.00%), are known, then it's (S'= Sales Past) must be:

$$S' = \{S[v+f]+A/[1-t]\}/[1+s][1-i]$$
$$= \{51,294\ [0.1900+0.5800]+7,899 /[1-0.3000]\}/\{[1+0.0500][1-0.0100]\}$$
$$= 48,851 \text{ millions USD}$$

Law-1408:
If (v= V/S= Variable Portion= 19.00%), (A= After Tax Income= 7,899 millions USD), (f= F/S=Fixed Portion= 58.00%), (S= Sales or Revenues= 51,294 millions USD), (S'= Sales of Past Year= 48,851 millions USD), (i= I/S= Interest Portion= 1.00%), and (t= T/B= Tax Rate= 30.00%), are known, then it's (s= Sales Growth planned) is:

$$s = \{S[v+f]+A/[1-t]\}/\{S'[1-i]\}-1$$
$$= \{51,294\ [0.1900+0.5800]+7,899 /[1-0.3000]\} /\{48,851[1-0.0100]\}-1$$
$$= 5.00\%$$

Law-1409:
If (v= V/S= Variable Portion= 19.00%), (A= After Tax Income= 7,899 millions USD), (f= F/S=Fixed Portion= 58.00%), (S= Sales or Revenues= 51,294 millions USD), (S'= Sales of Past Year= 48,851 millions USD), (s= [S/S']-1= Sales Growth= 5.00%), and (t= T/B= Tax Rate= 30.00%), are known, then it's (i= Interest Portion planned) is:

$i = 1 - \{S[v+f] + A/[1-t]\} / \{S'[1+s]\}$
$= 1 - \{51,294 [0.1900+0.5800] + 7,899 /[1-0.3000]\}$
$/\{48,851[1+0.0500]\}$
$= 1.00\%$

Law-1410:
If (v= V/S= Variable Portion= 19.00%), (A= After Tax Income= 7,899 millions USD), (f= F/S=Fixed Portion= 58.00%), (i= I/S= Interest Portion= 1.00%), (S'= Sales of Past Year= 48,851 millions USD), (s= [S/S']-1= Sales Growth= 5.00%), and (t= T/B= Tax Rate= 30.00%), are known, then it's (S= Sales or Revenues planned) is:

$S = \{S'[1+s][1-i] - A/[1-t]\} / [v+f]$
$= \{48,851[1+0.0500][1-0.0100] - 7,899 /[1-0.3000]\}/[0.1900+0.5800]$
$= 51,294$ millions USD

Law-1411:
If ($= Sales or Revenues= 51,294 millions USD), (A= After Tax Income= 7,899 millions USD), (f= F/S=Fixed Portion= 58.00%), (i= I/S= Interest Portion= 1.00%), ($'= Sales of Past Year= 48,851 millions USD), (s= [S/S']-1= Sales Growth= 5.00%), and (t= T/B= Tax Rate= 30.00%), are known, then it's (v= Variable Portion planned) is:

$$v = \{\$'[1+s][1-i] - A/[1-t]\}/\$ - f$$
$$= \{48,851[1+0.0500][1-0.0100] - 7,899 /[1-0.3000]\}/51,294 - 0.5800$$
$$= 19.00\%$$

Law-1412:
If ($= Sales or Revenues= 51,294 millions USD), (A= After Tax Income= 7,899 millions USD), (v= V/S= Variable Portion= 19.00%), (i= I/S= Interest Portion= 1.00%), ($'= Sales of Past Year= 48,851 millions USD), (s= [S/S']-1= Sales Growth= 5.00%), and (t= T/B= Tax Rate= 30.00%), are known, then it's (f= Fixed Portion planned) is:

$$f = \{\$'[1+s][1-i] - A/[1-t]\}/\$ - v$$
$$= \{48,851[1+0.0500][1-0.0100] - 7,899 /[1-0.3000]\}/51,294 - 0.1900$$
$$= 58.00\%$$

Law-1413:
 If ($= Sales or Revenues= 51,294 millions USD), (A= After Tax Income= 7,899 millions USD), (v= V/S= Variable Portion= 19.00%), (i= I/S= Interest Portion= 1.00%), ($'= Sales of Past Year= 48,851 millions USD), (s= [S/S']-1= Sales Growth= 5.00%), and (f= F/S=Fixed Portion= 58.00%), are known, then it's (t= Tax Rate planned) is:

$$t = 1-A/\{\$'[1+s]-\$v-\$f-\$'i[1+s]\}$$
$$= 1-A/\{\$'[1+s][1-i]-\$[v+f]\}$$
$$= 1-7,899/\{48,851[1+0.0500][1-0.0100]$$
$$\qquad -51,294\,[0.1900+0.5800]\}$$
$$= \underline{30.00}\%$$

Law-1414:
 If ($= Sales or Revenues= 51,294 millions USD), (t= T/B= Tax Rate= 30.00%), (v= V/S= Variable Portion= 19.00%), (I= Interest Expense= 513 millions USD), ($'= Sales of Past Year= 48,851 millions USD), (s= [S/S']-1= Sales Growth= 5.00%), and (f= F/S=Fixed Portion= 58.00%), are known, then it's (A= After Tax Income planned) is:

$$A = [1-t]\{\$'[1+s]-\$v-\$'f[1+s]-I\}$$
$$= [1-t]\{\$'[1+s][1-f]-\$v-I\}$$
$$= [1-0.3000]\{\,48,851[1+0.0500]$$
$$\qquad [1-0.5800]-51,294*0.1900-513\}$$
$$= \underline{7,899}\text{ millions USD}$$

Law-1415:
If ($ = Sales or Revenues = 51,294 millions USD), (t = T/B = Tax Rate = 30.00%), (v = V/S = Variable Portion = 19.00%), (I = Interest Expense = 513 millions USD), (A = After Tax Income = 7,899 millions USD), (s = [S/S']-1 = Sales Growth = 5.00%), and (f = F/S = Fixed Portion = 58.00%), are known, then it's ($' = Sales Past) must be:

$$\$' = \{\$v + I + A/[1-t]\}\{[1+s][1-f]\}$$
$$= 51{,}294 * 0.1900 + 513 + 7{,}899$$
$$/[1-0.3000]\}$$
$$/[1+0.0500][1-0.5800]$$
$$= 48{,}851 \text{ millions USD}$$

Law-1416:
If ($ = Sales or Revenues = 51,294 millions USD), (t = T/B = Tax Rate = 30.00%), (v = V/S = Variable Portion = 19.00%), (I = Interest Expense = 513 millions USD), (A = After Tax Income = 7,899 millions USD), ($' = Sales of Past Year = 48,851 millions USD), and (f = F/S = Fixed Portion = 58.00%), are known, then it's (s = Sales Growth planned) is:

$$s = \{A/[1-t] + \$v + I\}/\{\$'[1-f]\} - 1$$
$$= \{7{,}899/[1-0.3000] + 51{,}294*0.1900$$
$$+ 513\}/\{48{,}851[1-0.5800]\} - 1$$
$$= 5.00\%$$

Law-1417:

If ($= Sales or Revenues= 51,294 millions USD), (t= T/B= Tax Rate= 30.00%), (v= V/S= Variable Portion= 19.00%), (I= Interest Expense= 513 millions USD), (A= After Tax Income= 7,899 millions USD), ($'= Sales of Past Year= 48,851 millions USD), and (s= [S/S']-1= Sales Growth= 5.00%), are known, then it's (f= Fixed Portion planned) is:

$$f = 1-\{A/[1-t]+\$v+I\}/\{\$'[1+s]\}$$
$$= 1-\{7,899/[1-0.3000]+51,294*0.1900+513\}/\{48,851[1+0.0500]\}$$
$$= \underline{58.00}\%$$

Law-1418:

If (f= F/S=Fixed Portion= 58.00%), (t= T/B= Tax Rate= 30.00%), (v= V/S= Variable Portion= 19.00%), (I= Interest Expense= 513 millions USD), (A= After Tax Income= 7,899 millions USD), ($'= Sales of Past Year= 48,851 millions USD), and (s= [S/S']-1= Sales Growth= 5.00%), are known, then it's ($= Sales or Revenues planned) is:

$$\$ = \{\$'[1+s][1-f]-I-A/[1-t]\}/v$$
$$= \{48,851[1+0.0500][1-0.5800]-513-\{7,899/[1-0.3000]\}/0.1900$$
$$= \underline{51,294} \text{ millions USD}$$

Law-1419:

If (f= F/S=Fixed Portion= 58.00%), (t= T/B= Tax Rate= 30.00%), (S= Sales or Revenues= 51,294 millions USD), (I= Interest Expense= 513 millions USD), (A= After Tax Income= 7,899 millions USD), (S'= Sales of Past Year= 48,851 millions USD), and (s= [S/S']-1= Sales Growth= 5.00%), are known, then it's (v= Variable Portion planned) is:

$v = \{S'[1+s][1-f]-Si-A/[1-t]\}/S$
$= \{48,851[1+0.0500][1-0.5800]-51,294$
$\quad *0.0100-\{7,899/[1-0.3000]\}$
$\quad /51,294$
$= \underline{19.00\%}$

Law-1420:

If (f= F/S=Fixed Portion= 58.00%), (t= T/B= Tax Rate= 30.00%), (S= Sales or Revenues= 51,294 millions USD), (v= V/S= Variable Portion= 19.00%), (A= After Tax Income= 7,899 millions USD), (S'= Sales of Past Year= 48,851 millions USD), and (s= [S/S']-1= Sales Growth= 5.00%), are known, then it's (I= Interest Expense planned) is:

$I = \{S'[1+s][1-f]-Sv-A/[1-t]\}$
$= \{48,851[1+0.0500][1-0.5800]- 51,294$
$\quad *0.1900-\{7,899/[1-0.3000]\}$
$= \underline{513}$ millions USD

Law-1421:

If (**f**= F/S=Fixed Portion= 58.00%), (**I**= Interest Expense= 513 millions USD), (**$**= Sales or Revenues= 51,294 millions USD), (**v**= V/S= Variable Portion= 19.00%), (**A**= After Tax Income= 7,899 millions USD), (**$'**= Sales of Past Year= 48,851 millions USD), and (**s**= [S/S']-1= Sales Growth= 5.00%), are known, then it's (**t**= Tax Rate planned) is:

$$t = 1 - A/\{\$'[1+s] - \$v - \$'f[1+s] - I\}$$
$$= 1 - A/\{\$'[1+s][1-f] - \$v - I\}$$
$$= 1 - 7{,}899/\{48{,}851[1+0.0500][1-0.5800] - 51{,}294*0.1900 - 513\}$$
$$= \underline{30.00\%}$$

Law-1422:

If (**f**= F/S=Fixed Portion= 58.00%), (**i**= I/S= Interest Portion= 1.00%), (**$**= Sales or Revenues= 51,294 millions USD), (**v**= V/S= Variable Portion= 19.00%), (**t**= T/B= Tax Rate= 30.00%), (**$'**= Sales of Past Year= 48,851 millions USD), and (**s**= [S/S']-1= Sales Growth= 5.00%), are known, then it's (**A**= After Tax Income planned) is:

$$A = [1-t]\{\$'[1+s] - \$v - \$'f[1+s] - \$i\}$$
$$= [1-t]\{\$'[1+s][1-f] - \$[v+i]\}$$
$$= [1-0.3000]\{48{,}851[1+0.0500][1-0.5800] - 51{,}294[0.1900+0.0100]\}$$
$$= \underline{7{,}899} \text{ millions USD}$$

Law-1423:
If (f= F/S=Fixed Portion= 58.00%), (i= I/S= Interest Portion= 1.00%), (S= Sales or Revenues= 51,294 millions USD), (v= V/S= Variable Portion= 19.00%), (t= T/B= Tax Rate= 30.00%), (A= After Tax Income= 7,899 millions USD), and (s= [S/S']-1= Sales Growth= 5.00%), are known, then it's (S'= Sales Past) is:

S'= {S[v+i]+A/[1-t]}/{[1+s][1-f]}
 = 51,294[0.1900+0.0100]+7,899
 /[1-0.3000]/{ [1+0.0500]
 [1-0.5800]}
 = 48,851 millions USD

Law-1424:
If (f= F/S=Fixed Portion= 58.00%), (i= I/S= Interest Portion= 1.00%), (S= Sales or Revenues= 51,294 millions USD), (v= V/S= Variable Portion= 19.00%), (t= T/B= Tax Rate= 30.00%), (A= After Tax Income= 7,899 millions USD), and (S'= Sales of Past Year= 48,851 millions USD), are known, then it's (s= Sales Growth planned) is:

s= {S[v+i]+A/[1-t]}/{S'[1-f]}-1
 = {51,294[0.1900+0.0100]}+7,899
 /[1-0.3000]}/{48,851
 [1-0.5800]}-1
 = 5.00%

Law-1425:

If (s= [S/S']-1= Sales Growth= 5.00%), (i= I/S= Interest Portion= 1.00%), (S= Sales or Revenues= 51,294 millions USD), (v= V/S= Variable Portion= 19.00%), (t= T/B= Tax Rate= 30.00%), (A= After Tax Income= 7,899 millions USD), and (S'= Sales of Past Year= 48,851 millions USD), are known, then it's (f= Fixed Portion planned) is:

$$f = 1 - \{S[v+i] + A/[1-t]\} / \{S'[1+s]\}$$
$$= 1 - \{51,294[0.1900+0.0100]\} + 7,899$$
$$/[1-0.3000]\}$$
$$/\{48,851[1+0.0500]\}$$
$$= \underline{58.00\%}$$

Law-1426:

If (s= [S/S']-1= Sales Growth= 5.00%), (i= I/S= Interest Portion= 1.00%), (f= F/S=Fixed Portion= 58.00%), (v= V/S= Variable Portion= 19.00%), (t= T/B= Tax Rate= 30.00%), (A= After Tax Income= 7,899 millions USD), and (S'= Sales of Past Year= 48,851 millions USD), are known, then it's (S= Sales or Revenues planned) is:

$$S = \{S'[1+s][1-f] - A/[1-t]\} / [v+i]\}$$
$$= \{48,851[1+0.0500][1-0.5800] - 7,899$$
$$/[1-0.3000]\} / [0.1900+0.0100]\}$$
$$= \underline{51,294} \text{ millions USD}$$

Law-1427:

If (s= [S/S']-1= Sales Growth= 5.00%), (i= I/S= Interest Portion= 1.00%), (f= F/S=Fixed Portion= 58.00%), ($\$$= Sales or Revenues= 51,294 millions USD), (t= T/B= Tax Rate= 30.00%), (A= After Tax Income= 7,899 millions USD), and ($\$'$= Sales of Past Year= 48,851 millions USD), are known, then it's (v= Variable Portion planned) is:

$$v = \{\$'[1+s][1-f]-A/[1-t]\}/\$ - i$$
$$= \{48,851[1+0.0500][1-0.5800]-7,899/[1-0.3000]\}/51,294 - 0.0100$$
$$= \underline{19.00\%}$$

Law-1428:

If (s= [S/S']-1= Sales Growth= 5.00%), (v= V/S= Variable Portion= 19.00%), (f= F/S=Fixed Portion= 58.00%), ($\$$= Sales or Revenues= 51,294 millions USD), (t= T/B= Tax Rate= 30.00%), (A= After Tax Income= 7,899 millions USD), and ($\$'$= Sales of Past Year= 48,851 millions USD), are known, then it's (i= Interest Portion planned) is:

$$i = \{\$'[1+s][1-f]-A/[1-t]\}/\$ - v$$
$$= \{48,851[1+0.0500][1-0.5800]-7,899/[1-0.3000]\}/51,294 - 0.1900$$
$$= \underline{1.00\%}$$

Law-1429:

If (s= [S/S']-1= Sales Growth= 5.00%), (v= V/S= Variable Portion= 19.00%), (f= F/S=Fixed Portion= 58.00%), ($\$$= Sales or Revenues= 51,294 millions USD), (i= I/S= Interest Portion= 1.00%), (**A**= After Tax Income= 7,899 millions USD), and ($\$'$= Sales of Past Year= 48,851 millions USD), are known, then it's (**t**= Tax Rate planned) is:

$$t = 1-\mathbf{A}/\{\$'[1+s]-\$v-\$'f[1+s]-\$i\}$$
$$= 1-\mathbf{A}/\{\$'[1+s][1-f]-\$[v+i]\}$$
$$= 1-7,899/\{48,851[1+0.0500][1-0.5800]$$
$$- 51,294[0.1900 +0.0100]\}$$
$$= \underline{30.00\%}$$

Law-1430:

If (s= [S/S']-1= Sales Growth= 5.00%), (v= V/S= Variable Portion= 19.00%), (f= F/S=Fixed Portion= 58.00%), ($\$$= Sales or Revenues= 51,294 millions USD), (i= I/S= Interest Portion= 1.00%), (**t**= T/B= Tax Rate= 30.00%), and ($\$'$= Sales of Past Year= 48,851 millions USD), are known, then it's (**A**= After Tax Income planned) is:

$$\mathbf{A} = [1-t]\{\$'[1+s]-\$v-\$'f[1+s]-\$i[1+s]\}$$
$$= 1-\mathbf{A}/\{\$'[1+s][1-f-i]-\$v]\}$$
$$= [1-0.3000]\{48,851[1+0.0500]$$
$$[1-0.5800-0.0100]$$
$$-51,294*0.1900\}$$
$$= \underline{7,899} \text{ milions USD}$$

Law-1431:
If (s= [S/S']-1= Sales Growth= 5.00%), (v= V/S= Variable Portion= 19.00%), (f= F/S=Fixed Portion= 58.00%), ($\$$= Sales or Revenues= 51,294 millions USD), (i= I/S= Interest Portion= 1.00%), (t= T/B= Tax Rate= 30.00%), and (A= After Tax Income= 7,899 millions USD), are known, then it's ($\$'$= Sales Past) must be:

$$\$' = \{\$v + A/[1-t]\}/\{[1+s][1-f-i]\}$$
$$= 51,294*0.1900 + 7,899 [1-0.3000]] /\{[1+0.0500][1-0.5800-0.0100]\}$$
$$= \underline{48,851} \text{ milions USD}$$

Law-1432:
If ($\$'$= Sales of Past Year= 48,851 millions USD), (v= V/S= Variable Portion= 19.00%), (f= F/S=Fixed Portion= 58.00%), ($\$$= Sales or Revenues= 51,294 millions USD), (i= I/S= Interest Portion= 1.00%), (t= T/B= Tax Rate= 30.00%), and (A= After Tax Income= 7,899 millions USD), are known, then it's (s= Sales Growth planned) is:

$$s = \{\$v + A/[1-t]\}/\{\$'[1-f-i]\} - 1$$
$$= \{51,294*0.1900 + 7,899 [1-0.3000] /48,851[1-0.5800-0.0100]\} - 1$$
$$= \underline{5.00\%}$$

Law-1433:
If ($'= Sales of Past Year= 48,851 millions USD), (v= V/S= Variable Portion= 19.00%), (s= [S/S']-1= Sales Growth= 5.00%), ($= Sales or Revenues= 51,294 millions USD), (i= I/S= Interest Portion= 1.00%), (t= T/B= Tax Rate= 30.00%), and (A= After Tax Income= 7,899 millions USD), are known, then it's (f= Fixed Portion planned) is:

f= 1-{$v+A/[1-t]}/{$'[1+s]}-i
= 1-{51,294*0.1900+7,899[1-0.3000]}
/{48,851[1+0.0500]}-0.0100}
= 58.00%

Law-1434:
If ($'= Sales of Past Year= 48,851 millions USD), (v= V/S= Variable Portion= 19.00%), (s= [S/S']-1= Sales Growth= 5.00%), ($= Sales or Revenues= 51,294 millions USD), (f= F/S=Fixed Portion= 58.00%), (t= T/B= Tax Rate= 30.00%), and (A= After Tax Income= 7,899 millions USD), are known, then it's (i= Interest Portion planned) is:

i= 1-{$v+A/[1-t]}/{$'[1+s]}-f
=1-{51,294*0.1900+7,899[1-0.3000]}
/{48,851[1+0.0500]}-0.5800
= 1.00%

Law-1435:

If ($\$'$= Sales of Past Year= 48,851 millions USD), (v= V/S= Variable Portion= 19.00%), (s= [S/S']-1= Sales Growth= 5.00%), (i= I/S= Interest Portion= 1.00%), (f= F/S=Fixed Portion= 58.00%), (t= T/B= Tax Rate= 30.00%), and (A= After Tax Income= 7,899 millions USD), are known, then it's ($\$$= Sales or Revenues planned) is:

$\$ = \{\$'[1+s][1-f-i] - A/[1-t]\}/v$
$ = \{48,851[1+0.0500][1-0.5800-0.0100]$
$\phantom{\$ = \{} -7,899[1-0.3000]\}/0.1900$
$ = \underline{51,294}$ millions USD

Law-1436:

If ($\$'$= Sales of Past Year= 48,851 millions USD), ($\$$= Sales or Revenues= 51,294 millions USD), (s= [S/S']-1= Sales Growth= 5.00%), (i= I/S= Interest Portion= 1.00%), (f= F/S=Fixed Portion= 58.00%), (t= T/B= Tax Rate= 30.00%), and (A= After Tax Income= 7,899 millions USD), are known, then it's (v= Variable Portion planned) is:

$v = \{\$'[1+s][1-f-i] - A/[1-t]\}/\$$
$ = \{48,851[1+0.0500] - [1-0.5800-0.0100]$
$\phantom{v = \{} -7,899[1-0.3000]\}/51,294$
$ = \underline{19.00}\%$

Law-1437:
If ($'$= Sales of Past Year= 48,851 millions USD), ($= Sales or Revenues= 51,294 millions USD), ($= [S/S']-1= Sales Growth= 5.00%), (i= I/S= Interest Portion= 1.00%), (f= F/S=Fixed Portion= 58.00%), (v= V/S= Variable Portion= 19.00%), and (A= After Tax Income= 7,899 millions USD), are known, then it's (t= Tax Rate planned) is:

t = 1-A/{$'$[1+s]-$ v$-$'$$f$[1+$s$]-$'$$i$[1+$s$]}
  = 1-A/{$'$[1+s][1-f-i]-$ v$}
  = 1-7,899 {48,851[1+0.0500][1-0.5800
      -0.0100]-51,294*0.1900}
  = 30.00%

Law-1438:
If ($'$= Sales of Past Year= 48,851 millions USD), ($= Sales or Revenues= 51,294 millions USD), ($= [S/S']-1= Sales Growth= 5.00%), (I= Interest Expense= 513 millions USD), (F= Fixed Cost= 29,750 millions USD), (v= V/S= Variable Portion= 19.00%), and (t= T/B= Tax Rate= 30.00%), are known, then it's (A= After Tax Income planned) is:

A= [1-t]{$'$[1+s]-$'$$v$[1+$s$]-$F$-$I$}
  = [1-t]{$'$[1+s][1-v]- F-I}
  = [1-0.3000]{48,851[1+0.0500]
      [1-0.1900]-29,750-513}
  = 7,899 millions USD

Law-1439:

If (A= After Tax Income= 7,899 millions USD), (S= Sales or Revenues= 51,294 millions USD), (s= [S/S']-1= Sales Growth= 5.00%), (I= Interest Expense= 513 millions USD), (F= Fixed Cost= 29,750 millions USD), (v= V/S= Variable Portion= 19.00%), and (t= T/B= Tax Rate= 30.00%), are known, then it's (S'= Sales Past) must be:

$$S' = \{F+I+A/[1-t]\}/\{[1+s][1-v]\}$$
$$= \{29,750+513+7,899\,[1-0.3000]\}$$
$$/\{[1+0.0500]\,[1-0.1900\,]\}$$
$$= \underline{48,851} \text{ millions USD}$$

Law-1440:

If (A= After Tax Income= 7,899 millions USD), (S= Sales or Revenues= 51,294 millions USD), (S'= Sales of Past Year= 48,851 millions USD), (I= Interest Expense= 513 millions USD), (F= Fixed Cost= 29,750 millions USD), (v= V/S= Variable Portion= 19.00%), and (t= T/B= Tax Rate= 30.00%), are known, then it's (s= Sales or Revenues planned) is:

$$s = \{F+I+A/[1-t]\}/\{S'[1-v]\}-1$$
$$= \{29,750+513+7,899\,[1-0.3000]\}$$
$$/\{[1-0.1900\,]\}-1$$
$$= \underline{51,294} \text{ millions USD}$$

Law-1441:
If (A= After Tax Income= 7,899 millions USD), (S= Sales or Revenues= 51,294 millions USD), (S'= Sales of Past Year= 48,851 millions USD), (I= Interest Expense= 513 millions USD), (s= [S/S']-1= Sales Growth= 5.00%), (F= Fixed Cost= 29,750 millions USD), and (t= T/B= Tax Rate= 30.00%), are known, then it's (v= Variable Portion planned) is:

$v = 1 - \{F+I+A/[1-t]\}/\{S'[1+s]\}$
$= 1 - \{29,750+513+7,899\,[1-0.3000]\}$
$\quad /\{48,851[1+0.0500\,]\}$
$= \underline{19.00\%}$

Law-1442:
If (A= After Tax Income= 7,899 millions USD), (S= Sales or Revenues= 51,294 millions USD), (S'= Sales of Past Year= 48,851 millions USD), (I= Interest Expense= 513 millions USD), (v= V/S= Variable Portion=19.00%), (s= [S/S']-1= Sales Growth= 5.00%), and (t= T/B= Tax Rate= 30.00%), are known, then it's (F= Fixed Cost planned) is:

$F = \{S'[1+s][1-v]-I-A\}/[1-t]\}$
$= \{48,851\,[1+0.0500[1-0.1900]-513$
$\quad -7,899/\,[1-0.3000]\}$
$= \underline{29,750}$ millions USD

Law-1443:
If (**A**= After Tax Income= 7,899 millions USD), (**S**= Sales or Revenues= 51,294 millions USD), (**S'**= Sales of Past Year= 48,851 millions USD), (**v**= V/S= Variable Portion=19.00%), (**F**= Fixed Cost= 29,750 millions USD), (**s**= [S/S']-1= Sales Growth= 5.00%), and (**t**= T/B= Tax Rate= 30.00%), are known, then it's (**I**= Interest Expense planned) is:
I= {**S'**[1+**s**][1-**v**]-**F**-**A**}/[1-**t**]}
 = {48,851[1+0.0500[1-0.1900]-29,750
    -7,899/ [1-0.3000]}
 = 513 millions USD

Law-1444:
If (**A**= After Tax Income= 7,899 millions USD), (**S**= Sales or Revenues= 51,294 millions USD), (**S'**= Sales of Past Year= 48,851 millions USD), (**v**= V/S= Variable Portion=19.00%), (**F**= Fixed Cost= 29,750 millions USD), (**s**= [S/S']-1= Sales Growth= 5.00%), and (**I**= Interest Expense= 513 millions USD), are known, then it's (**t**= Tax Rate planned) is:
t= 1-**A**/{**S'**[1+**s**]-**S'v**[1+**s**]-**F**-**I**}
 = 1-**A**/{**S'**[1+**s**][1-**v**]-**F**-**I**}
 = 1-7,899/{48,851[1+0.0500[1-0.1900]
    -29,750-513}
 = 30.00%

Law-1445:

If (t= T/B= Tax Rate= 30.00%), (S= Sales or Revenues= 51,294 millions USD), (S'= Sales of Past Year= 48,851 millions USD), (v= V/S= Variable Portion=19.00%), (F= Fixed Cost= 29,750 millions USD), (s= [S/S']-1= Sales Growth= 5.00%), and (i= I/S= Interest Portion= 1.00%), are known, then it's (A= After Tax Income planned) is:

A= [1-t]{S'[1+s]-$S'v$[1+s]-F-Si}
 = [1-t]{S'[1+s][1-v]-F-Si}
 = 1-7,899/{48,851[1+0.0500[1-0.1900]
   -29,750-51,294*0.0100}
 = 7,899 millions USD

Law-1446:

If (t= T/B= Tax Rate= 30.00%), (S= Sales or Revenues= 51,294 millions USD), (A= After Tax Income= 7,899 millions USD), (v= V/S= Variable Portion=19.00%), (F= Fixed Cost= 29,750 millions USD), (s= [S/S']-1= Sales Growth= 5.00%), and (i= I/S= Interest Portion= 1.00%), are known, then it's (S'= Sales Past) must be:

S'= {F+Si+A/[1-t]}/{[1+s][1-v]}
 = {29,750+51,294*0.0100+ 7,899
   /[1-0.3000]}/{[1+0.0500]
   [1-0.1900]}
 = 48,851 millions USD

Law-1447:
If (t= T/B= Tax Rate= 30.00%), (s= Sales or Revenues= 51,294 millions USD), (A= After Tax Income= 7,899 millions USD), (v= V/S= Variable Portion=19.00%), (F= Fixed Cost= 29,750 millions USD), (s'= Sales of Past Year= 48,851 millions USD), and (i= I/S= Interest Portion= 1.00%), are known, then it's (s= Sales Growth planned) is:

$s = \{F+Si+A/[1-t]\}/\{S'[1-v]\}-1$
$= \{29,750+51,294*0.0100+7,899$
$/[1-0.3000]\}/\{48,851$
$[1-0.1900]\}-1$
$= \underline{5.00\%}$

Law-1448:
If (t= T/B= Tax Rate= 30.00%), (s= Sales or Revenues= 51,294 millions USD), (A= After Tax Income= 7,899 millions USD), (s= [S/S']-1= Sales Growth= 5.00%), (F= Fixed Cost= 29,750 millions USD), (s'= Sales of Past Year= 48,851 millions USD), and (i= I/S= Interest Portion= 1.00%), are known, then it's (v= Variable Portion planned) is:

$v = 1-\{F+Si+A/[1-t]\}/\{S'[1+s]\}$
$= 1-\{29,750+51,294*0.0100+7,899$
$/[1-0.3000]\}/\{48,851$
$[1+0.0500]\}$
$= \underline{19.00\%}$

Law-1449:

If (t= T/B= Tax Rate= 30.00%), (S= Sales or Revenues= 51,294 millions USD), (A= After Tax Income= 7,899 millions USD), (s= [S/S']-1= Sales Growth= 5.00%), (v= V/S= Variable Portion= 19.00%), (S'= Sales of Past Year= 48,851 millions USD), and (i= I/S= Interest Portion= 1.00%), are known, then it's (F= Fixed Cost planned) is:

$F = \{S'[1+s][1-v]-Si-A/[1-t]\}$
$= \{48,851[1+0.0500][1-0.1900]$
$\quad -51,294*0.0100-7,899$
$\quad /[1-0.3000]\}$
$= \underline{29,750}$ millions USD

Law-1450:

If (t= T/B= Tax Rate= 30.00%), (F= Fixed Cost= 29,750 millions USD), (A= After Tax Income= 7,899 millions USD), (s= [S/S']-1= Sales Growth= 5.00%), (v= V/S= Variable Portion= 19.00%), (S'= Sales of Past Year= 48,851 millions USD), and (i= I/S= Interest Portion= 1.00%), are known, then it's (S= Sales or Revenues planned) is:

$S = \{S'[1+s][1-v]-F-A/[1-t]\}/i$
$= \{48,851[1+0.0500][1-0.1900]-29,750$
$\quad -7,899/[1-0.3000]\}/0.0100$
$= \underline{51,294}$ millions USD

Law-1451:

If (t= T/B= Tax Rate= 30.00%), (F= Fixed Cost= 29,750 millions USD), (A= After Tax Income= 7,899 millions USD), (s= [S/S']-1= Sales Growth= 5.00%), (v= V/S= Variable Portion= 19.00%), ($\$'$= Sales of Past Year= 48,851 millions USD), and ($\$$= Sales or Revenues= 51,294 millions USD), are known, then it's (i= Interest Portion planned) is:

i = {$\$'$[1+$s$][1-$v$]-$F$-$A$/[1-$t$]}/$\$$
  = {48,851[1+0.0500][1-0.1900]-29,750
       -7,899/[1-0.3000]}/51,294
  = 1.00%

Law-1452:

If (i= I/S= Interest Portion= 1.00%), (F= Fixed Cost= 29,750 millions USD), (A= After Tax Income= 7,899 millions USD), (s= [S/S']-1= Sales Growth= 5.00%), (v= V/S= Variable Portion= 19.00%), ($\$'$= Sales of Past Year= 48,851 millions USD), and ($\$$= Sales or Revenues= 51,294 millions USD), are known, then it's (t= Tax Rate planned) is:

t = 1-A/{$\$'$[1+$s$]-$\$'v$[1+$s$]-$F$-$\$i$}
  = 1-A/{$\$'$[1+$s$][1-$v$]-$F$-$\$i$}
  = 1-7,899/{48,851[1+0.0500][1-0.1900]
       -29,750-51,294*0.0100}
  = 30.00%

Law-1453:
If (**i**= I/S= Interest Portion= 1.00%), (**F**= Fixed Cost= 29,750 millions USD), (**t**= T/B= Tax Rate= 30.00%), (**s**= [S/S']-1= Sales Growth= 5.00%), (**v**= V/S= Variable Portion= 19.00%), (**S'**= Sales of Past Year= 48,851 millions USD), and (**S**= Sales or Revenues= 51,294 millions USD), are known, then it's (**A**= After Tax Income planned) is:

$$A = [1-t]\{S'[1+s]-S'v[1+s]-F-S'i[1+s]\}$$
$$= [1-t]\{S'[1+s][1-v-i]-F\}$$
$$= [1-0.3000]\{48,851[1+0.0500][1-0.1900-0.0100]-29,750\}$$
$$= \underline{7,899} \text{ millions USD}$$

Law-1454:
If (**i**= I/S= Interest Portion= 1.00%), (**F**= Fixed Cost= 29,750 millions USD), (**t**= T/B= Tax Rate= 30.00%), (**s**= [S/S']-1= Sales Growth= 5.00%), (**v**= V/S= Variable Portion= 19.00%), (**A**= After Tax Income= 7,899 millions USD), and (**S**= Sales or Revenues= 51,294 millions USD), are known, then it's (**S'**= Sales Past) is:

$$S' = \{F+A/[1-t]\}/\{[1+s][1-v-i]\}$$
$$= \{29,750+7,899[1-0.3000]\}/[1+0.0500][1-0.1900-0.0100]\}$$
$$= \underline{48,851} \text{ millions USD}$$

Law-1455:

If (i= I/S= Interest Portion= 1.00%), (F= Fixed Cost= 29,750 millions USD), (t= T/B= Tax Rate= 30.00%), ($\$'$= Sales of Past Year= 48,851 millions USD), (v= V/S= Variable Portion= 19.00%), (A= After Tax Income= 7,899 millions USD), and ($\$$= Sales or Revenues= 51,294 millions USD), are known, then it's (s= Sales Growth planned) is:

$$s = \{F+A/[1-t]\}/\{[\$'[1-v-i]\} - 1$$
$$= \{29,750+7,899[1-0.3000]/48,851 [1-0.1900-0.0100]\} - 1$$
$$= 5.00\%$$

Law-1456:

If (i= I/S= Interest Portion= 1.00%), (F= Fixed Cost= 29,750 millions USD), (t= T/B= Tax Rate= 30.00%), ($\$'$= Sales of Past Year= 48,851 millions USD), (s= [S/S']-1= Sales Growth= 5.00%), (A= After Tax Income= 7,899 millions USD), and ($\$$= Sales or Revenues= 51,294 millions USD), are known, then it's (v= Variable Portion planned) is:

$$v = 1 - \{F+A/[1-t]\}/\{[\$'[1+s]\} - i$$
$$= 1 - \{29,750+7,899[1-0.3000]/48,851 [1+0.0500]\} - 0.0100$$
$$= 19.00\%$$

Law-1457:

If (v= V/S= Variable Portion= 19.00%), (F= Fixed Cost= 29,750 millions USD), (t= T/B= Tax Rate= 30.00%), (S'= Sales of Past Year= 48,851 millions USD), (s= [S/S']-1= Sales Growth= 5.00%), (A= After Tax Income= 7,899 millions USD), and (S= Sales or Revenues= 51,294 millions USD), are known, then it's (i= Interest Portion planned) is:

i= 1-{F+A/[1-t]}/{[S'[1+s]}-v
= 1-{29,750+7,899[1-0.3000]}/48,851
  [1+0.0500]}-0.1900
= 1.00%

Law-1458:

If (v= V/S= Variable Portion= 19.00%), (i= I/S= Interest Portion= 1.00%), (t= T/B= Tax Rate= 30.00%), (S'= Sales of Past Year= 48,851 millions USD), (s= [S/S']-1= Sales Growth= 5.00%), (A= After Tax Income= 7,899 millions USD), and (S= Sales or Revenues= 51,294 millions USD), are known, then it's (F= Fixed Cost planned) is:

F= S'[1+s][1-v-i]-A/[1-t]}
= {48,851[1+0.0500][1-0.1900-0.0100]
  -7,899/[1-0.3000]
= 29,750 millions USD

Law-1459:

If (v= V/S= Variable Portion= 19.00%), (i= I/S= Interest Portion= 1.00%), (**F**= Fixed Cost= 29,750 millions USD), ($\$'$= Sales of Past Year= 48,851 millions USD), (s= [S/S']-1= Sales Growth= 5.00%), (**A**= After Tax Income= 7,899 millions USD), and ($\$$= Sales or Revenues= 51,294 millions USD), are known, then it's (**t**= Tax Rate planned) is:

$$t = 1-A/\{\$'[1+s]-\$'v[1+s]-F-\$'i[1+s]\}$$
$$= 1-A/\{\$'[1+s][1-v-i]-F\}$$
$$= 1-7{,}899/\{48{,}851[1+0.0500][1-0.1900-0.0100]-29{,}750\}$$
$$= \underline{30.00}\%$$

Law-1460:

If (v= V/S= Variable Portion= 19.00%), (f= F/S= Fixed Portion= 58.00%), ($\$'$= Sales of Past Year= 48,851 millions USD), (s= [S/S']-1= Sales Growth= 5.00%), (**I**= Interest Expense= 513 millions USD), (**t**= T/B= Tax Rate= 30.00%), and ($\$$= Sales or Revenues= 51,294 millions USD), are known, then it's (**A**= After Tax Income planned) is:

$$A = [1-t]\{\$'[1+s]-\$'v[1+s]-\$f-I\}$$
$$= [1-t]\{\$'[1+s][1-v]-\$f-I\}$$
$$= [1-0.3000]\{48{,}851[1+0.0500][1-0.1900]-51{,}294*0.5800-513\}$$
$$= \underline{7{,}899} \text{ millions USD}$$

Law-1461:
If (v= V/S= Variable Portion= 19.00%), (f= F/S=Fixed Portion= 58.00%), (A= After Tax Income= 7,899 millions USD), (s= [S/S']-1= Sales Growth= 5.00%), (I= Interest Expense= 513 millions USD), (t= T/B= Tax Rate= 30.00%), and (S= Sales or Revenues= 51,294 millions USD), are known, then it's (S'= Sales Past) must be:

S' = {Sf+I+A/[1-t]}{[1+s][1-v]}
  = [51,294*0.5800+513+7,899
       /[1-0.3000]}/{ [1+0.0500]
       [1-0.1900]}
  = 48,851 millions USD

Law-1462:
If (v= V/S= Variable Portion= 19.00%), (f= F/S=Fixed Portion= 58.00%), (A= After Tax Income= 7,899 millions USD), (S'= Sales of Past Year= 48,851 millions USD), (I= Interest Expense= 513 millions USD), (t= T/B= Tax Rate= 30.00%), and (S= Sales or Revenues= 51,294 millions USD), are known, then it's (s= Sales Growth planned) is

s = {Sf+I+A/[1-t]}{S'[1-v]}-1
  = {[51,294*0.5800+513+7,899
       /[1-0.3000]}/{48,851
       [1-0.1900]}-1
  = 5.00%

Law-1463:
If (s= [S/S']-1= Sales Growth= 5.00%), (f= F/S=Fixed Portion= 58.00%), (A= After Tax Income= 7,899 millions USD), (S'= Sales of Past Year= 48,851 millions USD), (I= Interest Expense= 513 millions USD), (t= T/B= Tax Rate= 30.00%), and (S= Sales or Revenues= 51,294 millions USD), are known, then it's (v= Variable Portion planned) is:

v= 1-{Sf+I+A/[1-t]}{S'[1+s]}
 = 1-{[51,294*0.5800+513+7,899
   /[1-0.3000]}/{48,851
   [1+0.0500]}
 = 19.00%

Law-1464:
If (s= [S/S']-1= Sales Growth= 5.00%), (f= F/S=Fixed Portion= 58.00%), (A= After Tax Income= 7,899 millions USD), (S'= Sales of Past Year= 48,851 millions USD), (I= Interest Expense= 513 millions USD), (t= T/B= Tax Rate= 30.00%), and (v= V/S= Variable Portion= 19.00%), are known, then it's (S= Sales or Revenues planned) is:

S= {S'[1+s][1-v]-I-A/[1-t]}/f
 = {48,851[1+0.0500][1-0.1900] -513
   -7,899/[1-0.3000]}/0.5800
 = 51,294 millions USD

Law-1465:

If (s= [S/S']-1= Sales Growth= 5.00%), ($\$$= Sales or Revenues= 51,294 millions USD), (**A**= After Tax Income= 7,899 millions USD), ($\$$'= Sales of Past Year= 48,851 millions USD), (**I**= Interest Expense= 513 millions USD), (**t**= T/B= Tax Rate= 30.00%), and (**v**= V/S= Variable Portion= 19.00%), are known, then it's (**f**= Fixed Portion planned) is:

$$f = \{\$'[1+s][1-v]-I-A/[1-t]\}/\$$$
$$= \{48,851[1+0.0500][1-0.1900] -513$$
$$-7,899/[1-0.3000]\}/51,294$$
$$= \underline{58.00\%}$$

Law-1466:

If (s= [S/S']-1= Sales Growth= 5.00%), ($\$$= Sales or Revenues= 51,294 millions USD), (**A**= After Tax Income= 7,899 millions USD), ($\$$'= Sales of Past Year= 48,851 millions USD), (**f**= F/S=Fixed Portion= 58.00%), (**t**= T/B= Tax Rate= 30.00%), and (**v**= V/S= Variable Portion= 19.00%), are known, then it's (**I**= Fixed Portion planned) is:

$$I = \$'[1+s][1-v]-\$f-A/[1-t]$$
$$= 48,851[1+0.0500][1-0.1900]-51,294$$
$$+0.5800-7,899/[1-0.3000]\}$$
$$= \underline{513} \text{ millions USD}$$

Law-1467:

If (s= [S/S']-1= Sales Growth= 5.00%), (S= Sales or Revenues= 51,294 millions USD), (A= After Tax Income= 7,899 millions USD), (S'= Sales of Past Year= 48,851 millions USD), (f= F/S=Fixed Portion= 58.00%), (I= Interest Expense= 513 millions USD), and (v= V/S= Variable Portion= 19.00%), are known, then it's (t= Tax Rate planned) is:

$$\begin{aligned}
t &= 1-A/\{S'[1+s]-S'v[1+s][1-v]-Sf-I\} \\
&= 1-A/\{S'[1+s][1-v]-Sf-I\} \\
&= 1-7,899/\{48,851[1+0.0500][1-0.1900] \\
&\quad -51,294+0.5800-513\} \\
&= 30.00\%
\end{aligned}$$

Law-1468:

If (s= [S/S']-1= Sales Growth= 5.00%), (S= Sales or Revenues= 51,294 millions USD), (t= T/B= Tax Rate= 30.00%), (S'= Sales of Past Year= 48,851 millions USD), (f= F/S=Fixed Portion= 58.00%), (i= I/S= Interest Portion= 1.00%), and (v= V/S= Variable Portion= 19.00%), are known, then it's (A= After Tax Income planned) is:

$$\begin{aligned}
A &= [1-t]\{S'[1+s]-S'v[1+s][1-v]-Sf-Si\} \\
&= [1-t]\{S'[1+s][1-v]-S[f+i]\} \\
&= [1-0.3000]\{48,851[1+0.0500] \\
&\quad [1-0.1900]-51,294 \\
&\quad [0.5800+0.0100]\} \\
&= 7,899 \text{ millions USD}
\end{aligned}$$

Law-1469:
If (s= [S/S']-1= Sales Growth= 5.00%), (S= Sales or Revenues= 51,294 millions USD), (t= T/B= Tax Rate= 30.00%), (A= After Tax Income= 7,899 millions USD), (f= F/S=Fixed Portion= 58.00%), (i= I/S= Interest Portion= 1.00%), and (v= V/S= Variable Portion= 19.00%), are known, then it's (S'= Sales Past) must be:

S'= $S[f+i]+A/[1-t]\}/\{[1+s][1-v]\}$
= {51,294[0.5800+0.0100]}+7,899
/[1-0.3000]}{[1+0.0500]
[1-0.1900]}
= 48,851 millions USD

Law-1470:
If (S'= Sales of Past Year= 48,851 millions USD), (S= Sales or Revenues= 51,294 millions USD), (t= T/B= Tax Rate= 30.00%), (A= After Tax Income= 7,899 millions USD), (f= F/S=Fixed Portion= 58.00%), (i= I/S= Interest Portion= 1.00%), and (v= V/S= Variable Portion= 19.00%), are known, then it's (s= Sales Growth planned) is:

s= {$S[f+i]+A/[1-t]\}/\{S'[1-v]\}-1$
= {51,294[0.5800+0.0100]}+7,899
/[1-0.3000]}/{48,851
[1-0.1900]}-1
= 5.00%

Law-1471:
If ($'= Sales of Past Year= 48,851 millions USD), ($= Sales or Revenues= 51,294 millions USD), (t= T/B= Tax Rate= 30.00%), (A= After Tax Income= 7,899 millions USD), (f= F/S=Fixed Portion= 58.00%), (i= I/S= Interest Portion= 1.00%), and (s= [S/S']-1= Sales Growth= 5.00%), are known, then it's (v= Variable Portion planned) is:

$$v = 1 - \{\$[f+i] + A/[1-t]\}/\{\$'[1+s]\}$$
$$= 1 - \{51{,}294[0.5800+0.0100]\} + 7{,}899$$
$$/[1-0.3000]\}$$
$$/\{48{,}851[1+0.0500]\}$$
$$= \underline{19.00\%}$$

Law-1472:
If ($'= Sales of Past Year= 48,851 millions USD), (v= V/S= Variable Portion= 19.00%), (t= T/B= Tax Rate= 30.00%), (A= After Tax Income= 7,899 millions USD), (f= F/S=Fixed Portion= 58.00%), (i= I/S= Interest Portion= 1.00%), and (s= [S/S']-1= Sales Growth= 5.00%), are known, then it's ($= Sales or Revenues planned) is:

$$\$ = \{\$'[1+s][1-v] - A/[1-t]\}/[f+i]$$
$$= 48{,}851[1+0.0500][1-0.1900] - 7{,}899$$
$$/[1-0.3000]\}/[0.5800+0.0100]\}$$
$$= \underline{51{,}294} \text{ millions USD}$$

Law-1473:
If ($'$ = Sales of Past Year = 48,851 millions USD), (v = V/S = Variable Portion = 19.00%), (t = T/B = Tax Rate = 30.00%), (A = After Tax Income = 7,899 millions USD), ($\$$ = Sales or Revenues = 51,294 millions USD), (i = I/S = Interest Portion = 1.00%), and (s = [S/S']-1 = Sales Growth = 5.00%), are known, then it's (f = Fixed Portion planned) is:

$f = \{\$'[1+s][1-v] - A/[1-t]\}/\$ - i$
= 48,851[1+0.0500][1-0.1900]-7,899
/[1-0.3000]}/51,294-0.0100
= 58.00%

Law-1474:
If ($\$'$ = Sales of Past Year = 48,851 millions USD), (v = V/S = Variable Portion = 19.00%), (t = T/B = Tax Rate = 30.00%), (A = After Tax Income = 7,899 millions USD), ($\$$ = Sales or Revenues = 51,294 millions USD), (f = F/S = Fixed Portion = 58.00%), and (s = [S/S']-1 = Sales Growth = 5.00%), are known, then it's (i = Interest Portion planned) is:

$i = \{\$'[1+s][1-v] - A/[1-t]\}/\$ - f$
= 48,851[1+0.0500][1-0.1900]-7,899
/[1-0.3000]}/51,294 -0.5800
= 1.00%

Law-1475:

If ($\$'$= Sales of Past Year= 48,851 millions USD), (v= V/S= Variable Portion= 19.00%), (i= I/S= Interest Portion= 1.00%), (A= After Tax Income= 7,899 millions USD), ($\$$= Sales or Revenues= 51,294 millions USD), (f= F/S=Fixed Portion= 58.00%), and (s= [S/S']-1= Sales Growth= 5.00%), are known, then it's (t= Tax Rate planned) is:

t= 1-A/{$\$'$[1+$s$]-$\$v$[1+$s$]-$\$f$-$\$i$}
  = 1-A/{$\$'$[1+$s$][1-$v$]-$\$$[f+i]}
  = 1-7,899/{48,851[1+0.0500][1-0.1900]
       -51,294[0.5800 +0.0100]
  = 30.00%

Law-1476:

If ($\$'$= Sales of Past Year= 48,851 millions USD), (v= V/S= Variable Portion= 19.00%), (i= I/S= Interest Portion= 1.00%), (t= T/B= Tax Rate= 30.00%), ($\$$= Sales or Revenues= 51,294 millions USD), (f= F/S=Fixed Portion= 58.00%), and (s= [S/S']-1= Sales Growth= 5.00%), are known, then it's (A= After Tax Income planned) is:

A= [1-t]{$\$'$[1+$s$]-$\$v$[1+$s$]-$\$f$-$\$'i$[1+$s$]}
  = [1-t]{$\$'$[1+$s$][1-$v$-$i$]-$\$f$}
  = [1-0.3000]{48,851[1+0.0500][1-0.1900
       -0.0100] -51,294*0.5800}
  = 7,899 millions USD

Law-1477:

If (**A**= After Tax Income= 7,899 millions USD), (**v**= V/S= Variable Portion= 19.00%), (**i**= I/S= Interest Portion= 1.00%), (**t**= T/B= Tax Rate= 30.00%), (**S**= Sales or Revenues= 51,294 millions USD), (**f**= F/S=Fixed Portion= 58.00%), and (**s**= [S/S']-1= Sales Growth= 5.00%), are known, then it's (**S'**= Sales Past) must be:

$S' = \{Sf + A/[1-t]\} / \{[1+s][1-v-i]\}$
$= 51,294*0.5800 + 7,899/[1-0.3000]\}$
$/[1+0.0500][1-0.1900-0.0100]\}$
$= \underline{48,851}$ millions USD

Law-1478:

If (**A**= After Tax Income= 7,899 millions USD), (**v**= V/S= Variable Portion= 19.00%), (**i**= I/S= Interest Portion= 1.00%), (**t**= T/B= Tax Rate= 30.00%), (**S**= Sales or Revenues= 51,294 millions USD), (**f**= F/S=Fixed Portion= 58.00%), and (**S'**= Sales of Past Year= 48,851 millions USD), are known, then it's (**s**= Sales Growth planned) is:

$s = \{Sf + A/[1-t]\} / \{S'[1-v-i]\} - 1$
$= \{51,294*0.5800 + 7,899/[1-0.3000]\}$
$/\{48,851[1-0.1900-0.0100]\} - 1$
$= \underline{5.00}\%$

Law-1479:
If (A= After Tax Income= 7,899 millions USD), (s= [S/S']-1= Sales Growth= 5.00%), (i= I/S= Interest Portion= 1.00%), (t= T/B= Tax Rate= 30.00%), ($\$$= Sales or Revenues= 51,294 millions USD), (f= F/S=Fixed Portion= 58.00%), and ($\$'$= Sales of Past Year= 48,851 millions USD), are known, then it's (v= Variable Portion planned) is:

v= 1-i-{$\$f$+$A$/[1-$t$]}/{$\$'$[1+$s$]}
= 1-0.0100-{51,294*0.5800+7,899
   /[1-0.3000]}
   /{48,851[1+0.0500]}
= 19.00%

Law-1480:
If (A= After Tax Income= 7,899 millions USD), (s= [S/S']-1= Sales Growth= 5.00%), (v= V/S= Variable Portion= 19.00%), (t= T/B= Tax Rate= 30.00%), ($\$$= Sales or Revenues= 51,294 millions USD), (f= F/S=Fixed Portion= 58.00%), and ($\$'$= Sales of Past Year= 48,851 millions USD), are known, then it's (i= Interest Portion planned) is:

i= 1-v-{$\$f$+$A$/[1-$t$]}/{$\$'$[1+$s$]}
= 1-0.1900-{51,294*0.5800+7,899
   /[1-0.3000]}/ 48,851[1+0.0500]}
= 1.00%

Law-1481:
If (**A**= After Tax Income= 7,899 millions USD), (**s**= [S/S']-1= Sales Growth= 5.00%), (**v**= V/S= Variable Portion= 19.00%), (**t**= T/B= Tax Rate= 30.00%), (**i**= I/S= Interest Portion= 1.00%), (**f**= F/S=Fixed Portion= 58.00%), and (**$'**= Sales of Past Year= 48,851 millions USD), are known, then it's (**$**= Sales or Revenues planned) is:

$= {$'[1+s][1-v-i]-A/[1-t]}/f
= 48,851[1+0.0500][1-0.1900-0.0100]
    -7,899/[1-0.3000]}/0.5800
= 51,294 millions USD

Law-1482:
If (**A**= After Tax Income= 7,899 millions USD), (**s**= [S/S']-1= Sales Growth= 5.00%), (**v**= V/S= Variable Portion= 19.00%), (**t**= T/B= Tax Rate= 30.00%), (**i**= I/S= Interest Portion= 1.00%), (**$**= Sales or Revenues= 51,294 millions USD), and (**$'**= Sales of Past Year= 48,851 millions USD), are known, then it's (**f**= Fixed Portion planned) is:

f= {$'[1+s][1-v-i]-A/[1-t]}/$
= 48,851[1+0.0500][1-0.1900-0.0100]
    -7,899/[1-0.3000]}/ 51,294
= 58.00%

Law-1483:

If (A = After Tax Income = 7,899 millions USD), (s = [S/S']-1 = Sales Growth = 5.00%), (v = V/S = Variable Portion = 19.00%), (f = F/S = Fixed Portion = 58.00%), (i = I/S = Interest Portion = 1.00%), ($\$$ = Sales or Revenues = 51,294 millions USD), and ($\$'$ = Sales of Past Year = 48,851 millions USD), are known, then it's (t = Tax Rate planned) is:

t = 1-A/{ $\$'$[1+$s$]- $\$'v$[1+$s$]-$\$f$-$\$'i$[1+$s$]}
= 1-A/{$\$'$[1+$s$][1-$v$-$i$]-$\$f$}
= 1-7,899/{48,851[1+0.0500][1-0.1900
       -0.0100]- 51,294*0.5800
= **30.00%**

Law-1484:

If (t = T/B = Tax Rate = 30.00%), (s = [S/S']-1 = Sales Growth = 5.00%), (v = V/S = Variable Portion = 19.00%), (f = F/S = Fixed Portion = 58.00%), (i = I/S = Interest Portion = 1.00%), ($\$$ = Sales or Revenues = 51,294 millions USD), and ($\$'$ = Sales of Past Year = 48,851 millions USD), are known, then it's (A = After Tax Income planned) is:

A = [1-t]{$\$'$[1+$s$]- $\$'v$[1+$s$]-$\$f$[1+$s$]-$I$}
= [1-t]{$\$'$[1+$s$][1-$v$-$f$]-$I$}
= [1-0.3000]{ 48,851[1+0.0500]
       [1-0.1900-0.5800]- 513}
= **7,899** millions USD

Law-1485:
If (**t**= T/B= Tax Rate= 30.00%), (**s**= [S/S']-1= Sales Growth= 5.00%), (**v**= V/S= Variable Portion= 19.00%), (**f**= F/S=Fixed Portion= 58.00%), (**i**= I/S= Interest Portion= 1.00%), (**S**= Sales or Revenues= 51,294 millions USD), and (**A**= After Tax Income= 7,899 millions USD), are known, then it's (**S'**= Sales Past) must be:

$$S' = \{I+A/[1-t]\}/\{[1+s][1-v-f]\}$$
$$= \{513+7,899/[1-0.3000]\}/\{[1+0.0500][1-0.1900-0.5800]\}$$
$$= 48,851 \text{ millions USD}$$

Law-1486:
If (**t**= T/B= Tax Rate= 30.00%), (**S'**= Sales of Past Year= 48,851 millions USD), (**v**= V/S= Variable Portion= 19.00%), (**f**= F/S=Fixed Portion= 58.00%), (**i**= I/S= Interest Portion= 1.00%), (**S**= Sales or Revenues= 51,294 millions USD), and (**A**= After Tax Income= 7,899 millions USD), are known, then it's (**s**= Sales Growth planned) is:

$$s = \{I+A/[1-t]\}/\{S'[1-v-f]\}-1$$
$$= \{513+7,899/[1-0.3000]\}/\{48,851[1-0.1900-0.5800]\}-1$$
$$= 5.00\%$$

Law-1487:
If (**t**= T/B= Tax Rate= 30.00%), (**$'**= Sales of Past Year= 48,851 millions USD), (**s**= [S/S']-1= Sales Growth= 5.00%), (**f**= F/S=Fixed Portion= 58.00%), (**i**= I/S= Interest Portion= 1.00%), (**$**= Sales or Revenues= 51,294 millions USD), and (**A**= After Tax Income= 7,899 millions USD), are known, then it's (**v**= Variable Portion planned) is:

$$v = 1-f-\{I+A/[1-t]\}/\{\$'[1+s]\}$$
$$= 1-0.5800-\{513+7{,}899/[1-0.3000]\}/\{48{,}851[1+0.0500]\}$$
$$= 19.00\%$$

Law-1488:
If (**t**= T/B= Tax Rate= 30.00%), (**$'**= Sales of Past Year= 48,851 millions USD), (**s**= [S/S']-1= Sales Growth= 5.00%), (**v**= V/S= Variable Portion= 19.00%), (**i**= I/S= Interest Portion= 1.00%), (**$**= Sales or Revenues= 51,294 millions USD), and (**A**= After Tax Income= 7,899 millions USD), are known, then it's (**f**= Fixed Portion planned) is:

$$f = 1-v-\{I+A/[1-t]\}/\{\$'[1+s]\}$$
$$= 1-0.1900-\{513+7{,}899/[1-0.3000]\}/\{48{,}851[1+0.0500]\}$$
$$= 58.00\%$$

Law-1489:

If (**t**= T/B= Tax Rate= 30.00%), (**$'**= Sales of Past Year= 48,851 millions USD), (**s**= [S/S']-1= Sales Growth= 5.00%), (**v**= V/S= Variable Portion= 19.00%), (**f**= F/S=Fixed Portion= 58.00%), (**$**= Sales or Revenues= 51,294 millions USD), and (**A**= After Tax Income= 7,899 millions USD), are known, then it's (**I**= Interest Expense planned) is:

$$I = \{\$'[1+s][1-v-f]-A/[1-t]\}$$
$$= \{48,851[1+0.0500][1-0.1900-0.5800]$$
$$-7,899/[1-0.3000]\}$$
$$= \underline{513} \text{ millions USD}$$

Law-1490:

If (**I**= Interest Expense= 513 millions USD), (**$'**= Sales of Past Year= 48,851 millions USD), (**s**= [S/S']-1= Sales Growth= 5.00%), (**v**= V/S= Variable Portion= 19.00%), (**f**= F/S=Fixed Portion= 58.00%), (**$**= Sales or Revenues= 51,294 millions USD), and (**A**= After Tax Income= 7,899 millions USD), are known, then it's (**t**= Tax Rate planned) is:

$$t = 1-A/\{\$'[1+s]-\$'v[1+s]-\$'f[1+s]-I\}$$
$$= 1-A/\{\$'[1+s][1-v-f]-I\}$$
$$= 1-7,899/\{48,851[1+0.0500][1-0.1900$$
$$-0.5800]-513\}$$
$$= \underline{30.00\%}$$

Law-1491:
If (i= I/S= Interest Portion= 1.00%), (S'= Sales of Past Year= 48,851 millions USD), (s= [S/S']-1= Sales Growth= 5.00%), (v= V/S= Variable Portion= 19.00%), (f= F/S=Fixed Portion= 58.00%), (S= Sales or Revenues= 51,294 millions USD), and (t= T/B= Tax Rate=30.00%), are known, then it's (A= After Tax Income planned) is:

A = [1-t]{S'[1+s]-$S'v$[1+s]-$S'f$[1+s]-Si}
= [1-t]{S'[1+s][1-v-f]-Si}
= [1-0.3000]{48,851[1+0.0500]
[1-0.1900-0.5800]
-51,294*0.0100}
= 7,899 millions USD

Law-1492:
If (i= I/S= Interest Portion= 1.00%), (A= After Tax Income= 7,899 millions USD), (s= [S/S']-1= Sales Growth= 5.00%), (v= V/S= Variable Portion= 19.00%), (f= F/S=Fixed Portion= 58.00%), (S= Sales or Revenues= 51,294 millions USD), and (t= T/B= Tax Rate= 30.00%), are known, then it's (S'= Sales Past) must be:

S' = {Si+A/[1-t]/{[1+s][1-v-f]}
= {51,294*0.0100 +7,899/[1-0.3000]}
/{[1+0.0500][1-0.1900-0.5800]}
= 48,851 millions USD

Law-1493:

If (i= I/S= Interest Portion= 1.00%), (A= After Tax Income= 7,899 millions USD), (S'= Sales of Past Year= 48,851 millions USD), (v= V/S= Variable Portion= 19.00%), (f= F/S=Fixed Portion= 58.00%), (S= Sales or Revenues= 51,294 millions USD), and (t= T/B= Tax Rate= 30.00%), are known, then it's (s= Sales Growth planned) is:

$$s = \{Si + A/[1-t]\}/\{S'[1-v-f]\} - 1$$
$$= \{51,294*0.0100 + 7,899/[1-0.3000]\}$$
$$/\{48,851[1-0.1900-0.5800]\} - 1$$
$$= 5.00\%$$

Law-1494:

If (i= I/S= Interest Portion= 1.00%), (A= After Tax Income= 7,899 millions USD), (S'= Sales of Past Year= 48,851 millions USD), (s= [S/S']-1= Sales Growth= 5.00%), (f= F/S=Fixed Portion= 58.00%), (S= Sales or Revenues= 51,294 millions USD), and (t= T/B= Tax Rate= 30.00%), are known, then it's (v= Variable Portion planned) is:

$$v = 1 - f - \{Si + A/[1-t]\}/\{S'[1+s]\}$$
$$= 1 - 0.5800 - \{51,294*0.0100 + 7,899$$
$$/[1-0.3000]\}$$
$$/\{48,851[1+0.0500]\}$$
$$= 19.00\%$$

Law-1495:
If (i= I/S= Interest Portion= 1.00%), (A= After Tax Income= 7,899 millions USD), ($\$'$= Sales of Past Year= 48,851 millions USD), (s= [S/S']-1= Sales Growth= 5.00%), (v= V/S= Variable Portion= 19.00%), ($\$$= Sales or Revenues= 51,294 millions USD), and (t= T/B= Tax Rate= 30.00%), are known, then it's (f= Fixed Portion planned) is:

f= 1-v-{$\$i$+$A$/[1-$t$]/{$\$'$[1+$s$]}
    = 1-0.1900-{51,294*0.0100 +7,899
        /[1-0.3000]}
        /{48,851[1+0.0500]}
    = 58.00%

Law-1496:
If (i= I/S= Interest Portion= 1.00%), (A= After Tax Income= 7,899 millions USD), ($\$'$= Sales of Past Year= 48,851 millions USD), (s= [S/S']-1= Sales Growth= 5.00%), (v= V/S= Variable Portion= 19.00%), (f= F/S=Fixed Portion= 58.00%), and (t= T/B= Tax Rate= 30.00%), are known, then it's ($\$$= Sales or Revenues planned) is:

$\$$= {$\$'$[1+$s$][1-$v$-$f$]-$A$/[1-$t$]/$i$
    = {48,851[1+0.0500][1-0.1900-0.5800]
        -7,899/[1-0.3000]}/0.0100
    = 51,294 millions USD

Law-1497:

If ($= Sales or Revenues= 51,294 millions USD), (A= After Tax Income= 7,899 millions USD), ($'= Sales of Past Year= 48,851 millions USD), (s= [S/S']-1= Sales Growth= 5.00%), (v= V/S= Variable Portion= 19.00%), (f= F/S=Fixed Portion= 58.00%), and (t= T/B= Tax Rate= 30.00%), are known, then it's (i= Interest Portion planned) is:

$$i = \{\$'[1+s][1-v-f] - A/[1-t]\}/\$$$
$$= \{48,851[1+0.0500][1-0.1900-0.5800]$$
$$\quad -7,899/[1-0.3000]\}/51,294$$
$$= 1.00\%$$

Law-1498:

If ($= Sales or Revenues= 51,294 millions USD), (A= After Tax Income= 7,899 millions USD), ($'= Sales of Past Year= 48,851 millions USD), (s= [S/S']-1= Sales Growth= 5.00%), (v= V/S= Variable Portion= 19.00%), (i= I/S= Interest Portion= 1.00%), and (f= F/S=Fixed Portion= 58.00%), are known, then it's (t= Tax Rate planned) is:

$$t = 1 - A/\{\$'[1+s] - \$'v[1+s] - \$'f[1+s] - \$i\}$$
$$= 1 - A/\{\$'[1+s][1-v-f] - \$i\}$$
$$= 1 - 7,899/\{48,851[1+0.0500][1-0.1900$$
$$\quad -0.5800] - 51,294*0.0100\}$$
$$= 30.00\%$$

Law-1499:
If ($ = Sales or Revenues= 51,294 millions USD), (t= T/B= Tax Rate= 30.00%), ($'= Sales of Past Year= 48,851 millions USD), (s= [S/S']-1= Sales Growth= 5.00%), (v= V/S= Variable Portion= 19.00%), (i= I/S= Interest Portion= 1.00%), and (f= F/S=Fixed Portion= 58.00%), are known, then it's (A= After Tax Income planned) is:

$$A = [1-t]\{\$'[1+s] - \$'v[1+s] - \$'f[1+s] - \$'i[1+s]\}$$
$$= [1-t]\{\$'[1+s][1-v-f-i]\}$$
$$= [1-0.3000]\{48,851[1+0.0500][1-0.1900-0.5800-0.0100]\}$$
$$= 7,899 \text{ millions USD}$$

Law-1500:
If ($ = Sales or Revenues= 51,294 millions USD), (t= T/B= Tax Rate= 30.00%), (A= After Tax Income= 7,899 millions USD), (s= [S/S']-1= Sales Growth= 5.00%), (v= V/S= Variable Portion= 19.00%), (i= I/S= Interest Portion= 1.00%), and (f= F/S= Fixed Portion= 58.00%), are known, then it's ($'= Sales Past) must be:

$$\$' = \{A/[1-t]\} / \{[1+s][1-v-f-i]\}$$
$$= \{7,899/[1-0.3000]\} / \{[1+0.0500][1-0.1900-0.5800-0.0100]\}$$
$$= 48,851 \text{ millions USD}$$

Law-1501:

If ($= Sales or Revenues= 51,294 millions USD), (t= T/B= Tax Rate= 30.00%), (A= After Tax Income= 7,899 millions USD), ($'= Sales of Past Year= 48,851 millions USD), (v= V/S= Variable Portion= 19.00%), (i= I/S= Interest Portion= 1.00%), and (f= F/S= Fixed Portion= 58.00%), are known, then it's (s= Sales Growth planned) is:

$$s = \{A/[1-t]\}/\{\$'[1-v-f-i]\} - 1$$
$$= \{7,899/[1-0.3000]\}/\{48,851[1-0.1900 - 0.5800 - 0.0100]\} - 1$$
$$= 5.00\%$$

Law-1502:

If ($= Sales or Revenues= 51,294 millions USD), (t= T/B= Tax Rate= 30.00%), (A= After Tax Income= 7,899 millions USD), ($'= Sales of Past Year= 48,851 millions USD), (s= [S/S']-1= Sales Growth= 5.00%), (i= I/S= Interest Portion= 1.00%), and (f= F/S= Fixed Portion= 58.00%), are known, then it's (v= Variable Portion planned) is:

$$v = 1 - f - i - \{A/[1-t]\}\{\$'[1+s]\}$$
$$= 1 - 0.5800 - 0.0100 - \{7,899/[1-0.3000]\} /\{48,851[1+0.0500]\}$$
$$= 19.00\%$$

Law-1503:

If ($= Sales or Revenues= 51,294 millions USD), (t= T/B= Tax Rate= 30.00%), (A= After Tax Income= 7,899 millions USD), ($'= Sales of Past Year= 48,851 millions USD), (s= [S/S']-1= Sales Growth= 5.00%), (i= I/S= Interest Portion= 1.00%), and (v= V/S= Variable Portion= 19.00%), are known, then it's (f= Fixed Portion planned) is:

$$f = 1 - v - i - \{A/[1-t]\}\{S'[1+s]\}$$
$$= 1 - 0.1900 - 0.0100 - \{7,899/[1-0.3000]\}/\{48,851[1+0.0500]\}$$
$$= 58.00\%$$

Law-1504:

If ($= Sales or Revenues= 51,294 millions USD), (t= T/B= Tax Rate= 30.00%), (A= After Tax Income= 7,899 millions USD), ($'= Sales of Past Year= 48,851 millions USD), (s= [S/S']-1= Sales Growth= 5.00%), (f= F/S= Fixed Portion= 58.00%), and (v= V/S= Variable Portion= 19.00%), are known, then it's (i= Interest Portion planned) is:

$$i = 1 - f - v - \{A/[1-t]\}\{S'[1+s]\}$$
$$= 1 - 0.5800 - 0.1900 - \{7,899/[1-0.3000]\}/\{48,851[1+0.0500]\}$$
$$= 1.00\%$$

Law-1505:

If ($\$$= Sales or Revenues= 51,294 millions USD), (i= I/S= Interest Portion= 1.00%), (A= After Tax Income= 7,899 millions USD), ($\$'$= Sales of Past Year= 48,851 millions USD), (s= [S/S']-1= Sales Growth= 5.00%), (f= F/S=Fixed Portion= 58.00%), and (v= V/S= Variable Portion= 19.00%), are known, then it's (t= Tax Rate planned) is:

t = 1-A/{$\$'$[1+$s$]-$\$'v$[1+$s$]-$\$'f$[1+$s$]-$\$'i$[1+$s$]}
= 1-A/{$\$'$[1+$s$][1-$f$-$v$-$i$]}
= 1-7,899/{48,851[1+0.0500][1-0.5800-0.1900-0.0100}
= __30.00%__

CHAPTER-10:
Manageable D= Dividend Optimization

Based on 12/31/15 Pfizer Annual Report
(http://www.nasdaq.com/symbol/pfe/financials?query=income-statement)

Law-1506:
　If (A= After Tax Income= 7,899 millions USD), and (d= D/A=Dividend Portion or Payout= 30.00%), are known, then it's (D= Dividend Paid planned) is:
　$D = Ad$
　　$= 7,899 * 0.3000$
　　$= 2,370$ millions USD

Law-1507:
　If (D= Dividend Paid= 2,370 millions USD), and (d= D/A= Dividend Portion or Payout= 30.00%), are known, then it's (A= After Tax Income planned) is:
　$A = D/d$
　　$= 2,370 / 0.3000$
　　$= 7,899$ millions USD

Law-1508:
   If (**A**= After Tax Income= 7,899 millions USD), and (**D**= Dividend Paid= 2,370 millions USD), are known, then it's (**d**= Dividend Portion planned) is:
   d= D/A
   $= 2,370/7,899$
   $= \underline{30.00}\%$

Law-1509:
   If (**B**= Before Tax Income= 11,285 millions USD), (**T**= Tax Paid= 3,385 millions USD), and (**d**= D/A= Dividend Portion or Payout= 30.00%), are known, then it's (**D**= Dividend Paid planned) is:
   D= d[B-T]
   $= 0.3000[11,285-3,385]$
   $= \underline{2,370}$ millions USD

Law-1510:
   If (**D**= Dividend Paid= 2,370 millions USD), (**T**= Tax Paid= 3,385 millions USD), and (**d**= D/A= Dividend Portion or Payout= 30.00%), are known, then it's (**B**= Before Tax Income planned) is:
   B= T+[D/d]
   $= 3,385+[2,370/0.3000]$
   $= \underline{11,285}$ millions USD

Law-1511:

If (**D**= Dividend Paid= 2,370 millions USD), (**B**= Before Tax Income= 11,285 millions USD), and (**d**=D/A=Dividend Portion or Payout= 30.00%), are known, then it's (**T**= Tax Paid planned) is:

T= **B**-[**D**/**d**]
$$= 11,285-[2,370/0.3000]$$
$$= 3,385 \text{ millions USD}$$

Law-1512:

If (**D**= Dividend Paid= 2,370 millions USD, (**B**= Before Tax Income= 11,285 millions USD), and (**T**= Tax Paid= 3,385 millions USD), are known, then it's (**d**= Dividend Portion planned) is:

d= **D**/[**B**-**T**]
$$= 2,370/[11,285-3,385]$$
$$= 30.00\%$$

Law-1513:

If (**t**= T/B= Tax Rate= 30.00%), (**d**= D/A=Dividend Portion or Payout= 30.00%), and (**B**= Before Tax Income= 11,285 millions USD), are known, then it's (**D**= Dividend Paid planned) is:

D= **d**[**B**-**Bt**]= **dB**[1-**t**]
$$= 0.3000*11,285[1-0.3000]$$
$$= 2,370 \text{ millions USD}$$

Law-1514:
If (t= T/B= Tax Rate= 30.00%), (d= D/A=Dividend Portion or Payout= 30.00%), and (D= Dividend Paid= 2,370 millions USD), are known, then it's (B= Before Tax Income planned) is:
B= [D/d]/[1-t]
   = [2,370/0.3000]/[1-0.3000]
   = 11,285 millions USD

Law-1515:
If (B= Before Tax Income= 11,285 millions USD), (d= D/A= Dividend Portion or Payout= 30.00%), and (D= Dividend Paid= 2,370 millions USD), are known, then it's (t= Tax Rate planned) is:
t= 1-[D/d]/B
   = 1-[2,370/0.3000]/11,285
   = 30.00%

Law-1516:
If (B= Before Tax Income= 11,285 millions USD), (t= T/B= Tax Rate= 30.00%), and (D= Dividend Paid= 2,370 millions USD), are known, then it's (d=vDividend Portion planned) is:
d= D/[B-Bt]= D/B[1-t]
   = 2,370/11,285[1-0.3000]
   = 30.00%

Law-1517:

If (O= Operational Surplus= 11,798 millions USD), (I= Interest Expense= 513 millions USD), (t= T/B= Tax Rate= 30.00%), and (d= D/A=Dividend Portion or Payout= 30.00%), are known, then it's (D= Dividend Paid planned) is:

$D = d[1-t][O-I]$
$= 0.3000[1-0.3000][11,798-513]$
$= 2,370$ millions USD

Law-1518:

If (O= Operational Surplus= 11,798 millions USD), (I= Interest Expense= 513 millions USD), (D= Dividend Paid= 2,370 millions USD), and (d= D/A=Dividend Portion or Payout= 30.00%), are known, then it's (t= Tax Rate planned) is:

$t = 1-D/\{d/[O-I]\}$
$= 1- 2,370/\{0.3000\,[11,798-513]\}$
$= 30.00\%$

Law-1519:

If (t= T/B= Tax Rate= 30.00%), (I= Interest Expense= 513 millions USD), (D= Dividend Paid= 2,370 millions USD), and (d= D/A=Dividend Portion or Payout= 30.00%), are known, then it's (O=Operational Surplus planned) is:

$O = I+D/\{d/[1-t]\}$
$= 513+2,370/\{0.3000[1- 0.3000]\}$
$= 11,798$ millions USD

Law-1520:

If (t= T/B= Tax Rate= 30.00%), (O= Operational Surplus= 11,798 millions USD), (D= Dividend Paid= 2,370 millions USD), and (d= D/A=Dividend Portion or Payout= 30.00%), are known, then it's (I= Interest Expense planned) is:

I= O-D/{d/[1-t]}
 = 11,798-2,370/{0.3000[1- 0.3000]}
 = 513 millions USD

Law-1521:

If (t= T/B= Tax Rate= 30.00%), (O= Operational Surplus= 11,798 millions USD), (D= Dividend Paid= 2,370 millions USD), and (I= Interest Expense= 513 millions USD), are known, then it's (d= Dividend Portion planned) is:

d= D/{[1-t][O-I]}
 = 2,370/{[1-0.3000][11,798- 513]}
 = 30.00%

Law-1522:

If (t= T/B= Tax Rate= 30.00%), (S= Sales or Revenues= 51,294 millions USD), (O= Operational Surplus= 11,798 millions USD), (d= D/A= Dividend Portion or Payout= 30.00%), and (i= I/S= Interest Portion= 1.00%), are known, then it's (D= Dividend Paid planned) is:

$$D = d[1-t][O-Si]$$
$$= 0.3000[1-0.3000][11,798 - 51,294 *0.0100]$$
$$= \underline{2,370} \text{ millions USD}$$

Law-1523:

If (D= Dividend Paid= 2,370 millions USD), (S= Sales or Revenues= 51,294 millions USD), (O= Operational Surplus= 11,798 millions USD), (d= D/A=Dividend Portion or Payout= 30.00%), and (i= I/S= Interest Portion= 1.00%), are known, then it's (t= Tax Rate planned) is:

$$t = 1 - D\{d[O-Si]\}$$
$$= 1 - 2,370\{0.3000[11,798 - 51,294 *0.0100]\}$$
$$= \underline{30.00}\%$$

Law-1524:

If (**D**= Dividend Paid= 2,370 millions USD), (**$**= Sales or Revenues= 51,294 millions USD), (**t**= T/B= Tax Rate= 30.00%), (**d**= D/A=Dividend Portion or Payout= 30.00%), and (**i**= I/S= Interest Portion= 1.00%), are known, then it's (**O**= Operational Surplus planned) is:

$$\mathbf{O} = \$i + D/\{d[1-t]\}$$
$$= 51,294*0.0100 + 2,370/\{0.3000[1-0.3000]\}$$
$$= 11,798 \text{ millions USD}$$

Law-1525:

If (**D**= Dividend Paid= 2,370 millions USD), (**O**= Operational Surplus= 11,798 millions USD), (**t**= T/B= Tax Rate= 30.00%), (**d**= D/A= Dividend Portion or Payout= 30.00%), and (**i**= I/S= Interest Portion= 1.00%), are known, then it's (**$**= Sales or Revenues planned) is:

$$\$ = (\mathbf{O} - D/\{d[1-t]\})/i$$
$$= (11,798 - 2,370/\{0.3000[1-0.3000]\})/0.0100$$
$$= 51,294 \text{ millions USD}$$

Law-1526:

If (**D**= Dividend Paid= 2,370 millions USD), (**O**= Operational Surplus= 11,798 millions USD), (**t**= T/B= Tax Rate= 30.00%), (**d**= D/A= Dividend Portion or Payout= 30.00%), and (**$**= Sales or Revenues= 51,294 millions USD), are known, then it's (**i**= Interest Portion planned) is:

$$i = (O-D/\{d[1-t]\})/\$$$
$$= (11{,}798 - 2{,}370/\{0.3000[1-0.3000]\})/51{,}294$$
$$= \underline{1.00\%}$$

Law-1527:

If (**D**= Dividend Paid= 2,370 millions USD), (**O**= Operational Surplus= 11,798 millions USD), (**t**= T/B= Tax Rate= 30.00%), (**i**= I/S= Interest Portion= 1.00%), and (**$**= Sales or Revenues= 51,294 millions USD), are known, then it's (**d**= Dividend Portion planned) is:

$$d = D/\{[1-t][O-\$i]\}$$
$$= 2{,}370/\{[1-0.3000][11{,}798 - 51{,}294 * 0.0100]\}$$
$$= \underline{30.00\%}$$

Law-1528:

If (**d**= D/A= Dividend Portion or Payout= 30.00%), (**O**= Operational Surplus= 11,798 millions USD), (**t**= T/B= Tax Rate= 30.00%), (**s**= [S/S']-1= Sales Growth= 5.00%), (**i**= I/S= Interest Portion= 1.00%), and (**$**= Sales or Revenues= 51,294 millions USD), are known, then it's (**D**= Dividend Paid planned) is:

$$D = d[1-t]\{O - \$'i[1+s]\}$$
$$= 0.3000[1-0.3000]\{11,798 - 48,851$$
$$\qquad *0.0100[1+0.0500]\}$$
$$= \underline{2,370} \text{ millions USD}$$

Law-1529:

If (**d**= D/A=Dividend Portion or Payout= 30.00%), (**D**= Dividend Paid= 2,370 millions USD), (**t**= T/B= Tax Rate= 30.00%), (**i**= I/S= Interest Portion= 1.00%), (**$'**= Sales of Past Year= 48,851 millions USD), (**s**= [S/S']-1= Sales Growth= 5.00%), and are known, then it's (**O**= Operational Surplus planned) is:

$$O = \{\$'i[1+s] + D / \{d[1-t]\}$$
$$= \{48,851*0.0100[1+0.05] + 2,370$$
$$\qquad / \{0.3000[1-0.3000]\}$$
$$= \underline{11,798} \text{ millions USD}$$

Law-1530:

If (d= D/A=Dividend Portion or Payout= 30.00%), (D= Dividend Paid= 2,370 millions USD), (t= T/B= Tax Rate= 30.00%), (O= Operational Surplus= 11,798 millions USD), (s= [S/S']-1= Sales Growth= 5.00%), (i= I/S= Interest Portion= 1.00%), and are known, then it's (S'= Sales Past) must be:

$$S' = (O - D/\{d[1-t]\})/\{i[1+s]\}$$
$$= (11,798 - 2,370/\{0.3000[1-0.3000]\}) / 0.0100[1+0.0500]\}$$
$$= \underline{48,851} \text{ millions USD}$$

Law-1531:

If (d= D/A=Dividend Portion or Payout= 30.00%), (D= Dividend Paid= 2,370 millions USD), (t= T/B= Tax Rate= 30.00%), (O= Operational Surplus= 11,798 millions USD), (s= [S/S']-1= Sales Growth= 5.00%), (S'= Sales of Past Year= 48,851 millions USD), and are known, then it's (i= Interest Portion planned) is:

$$i = (O - D/\{d[1-t]\})/\{S[1+s]\}$$
$$= (11,798 - 2,370/\{0.3000[1-0.3000]\}) / 51,294[1+0.0500]\}$$
$$= \underline{1.00\%}$$

Law-1532:

If (**d**= D/A=Dividend Portion or Payout= 30.00%), (**D**= Dividend Paid= 2,370 millions USD), (**t**= T/B= Tax Rate= 30.00%), (**O**= Operational Surplus= 11,798 millions USD), (**i**= I/S= Interest Portion= 1.00%), (**S'**= Sales of Past Year= 48,851 millions USD), and are known, then it's (**s**= Sales Growth planned) is:

$$s = (O-D/\{d[1-t]\})/[S'i]-1$$
$$= (11,798-2,370/\{0.3000[1-0.3000]\})/[48,851*0.0100]-1$$
$$= 5.00\%$$

Law-1533:

If (**d**= D/A= Dividend Portion or Payout= 30.00%), (**D**= Dividend Paid= 2,370 millions USD), (**s**= [S/S']-1= Sales Growth= 5.00%), (**O**= Operational Surplus= 11,798 millions USD), (**i**= I/S= Interest Portion= 1.00%), and (**S'**= Sales of Past Year= 48,851 millions USD), are known, then it's (**t**= Tax Rate planned) is:

$$t = 1-D/(d\{O-S'i[1+s]\})$$
$$= 1- 2,370/(0.3000\{11,798-48,851*0.0100[1+0.0500]\})$$
$$= 30.00\%$$

Law-1534:

If (**t**= T/B= Tax Rate= 30.00%), (**D**= Dividend Paid= 2,370 millions USD), (**s**= [S/S']-1= Sales Growth= 5.00%), (**O**= Operational Surplus= 11,798 millions USD), (**i**= I/S= Interest Portion= 1.00%), and (**S'**= Sales of Past Year= 48,851 millions USD), are known, then it's (**d**=Dividend Portion planned) is:

$$d = D/([1-t]\{O-S'i[1+s]\})$$
$$= 2,370/([1-0.3000]\{11,798-48,851 \\ *0.0100[1+0.0500]\})$$
$$= 30.00\%$$

Law-1535:

If (**t**= T/B= Tax Rate= 30.00%), (**d**= D/A= Dividend Portion or Payout= 30.00%), (**F**= Fixed Cost= 29,750 millions USD), (**M**= Margin of Contribution= 41,548 millions USD), and (**I**= Interest Expense= 513 millions USD), are known, then it's (**D**= Dividend Paid planned) is:

$$D = d[1-t][M-F-I]$$
$$= 0.3000[1-0.3000][41,548-29,750-513]$$
$$= 2,370 \text{ millions USD}$$

Law-1536:
If (**t**= T/B= Tax Rate= 30.00%), (**d**= D/A= Dividend Portion or Payout= 30.00%), (**F**= Fixed Cost= 29,750 millions USD), (**D**= Dividend Paid= 2,370 millions USD), and (**I**= Interest Expense= 513 millions USD), are known, then it's (**M**= Margin of Contribution planned) is:

M= **F**+**I**+**D**/{**d**[1-**t**]}
= 29,750+513+2,370/{0.3000[1-0.3000]}
= 41,548 millions USD

Law-1537:
If (**t**= T/B= Tax Rate= 30.00%), (**d**= D/A= Dividend Portion or Payout= 30.00%), (**M**= Margin of Contribution= 41,548 millions USD), (**D**= Dividend Paid= 2,370 millions USD), and (**I**= Interest Expense= 513 millions USD), are known, then it's (**F**=Fixed Cost planned) is:

F= **M**-**I**-**D**/{**d**[1-**t**]})
= 41,548-513-2,370/{0.3000
[1-0.3000]}
= 29,750 millions USD

Law-1538:

If (**t**= T/B= Tax Rate= 30.00%), (**d**= D/A=Dividend Portion or Payout= 30.00%), (**M**= Margin of Contribution= 41,548 millions USD), (**D**= Dividend Paid= 2,370 millions USD), and (**F**= Fixed Cost= 29,750 millions USD), are known, then it's (**I**=Interest Expense planned) is:

$$I = M-F-D/\{d[1-t]\}$$
$$= 41,548-29,750-2,370/\{0.3000[1-0.3000]\}$$
$$= 513 \text{ millions USD}$$

Law-1539:

If (**I**= Interest Expense= 513 millions USD), (**d**= D/A=Dividend Portion or Payout= 30.00%), (**M**= Margin of Contribution= 41,548 millions USD), (**D**= Dividend Paid= 2,370 millions USD), and (**F**= Fixed Cost= 29,750 millions USD), are known, then it's (**t**=Tax Rate planned) is:

$$t = 1-D/\{d[M-F-I]\}$$
$$= 1-2,370/\{0.3000[41,548-29,750-513]\}$$
$$= 30.00\%$$

Law-1540:
If (**I**= Interest Expense= 513 millions USD), (**t**= T/B= Tax Rate= 30.00%), (**M**= Margin of Contribution= 41,548 millions USD), (**D**= Dividend Paid= 2,370 millions USD), and (**F**= Fixed Cost= 29,750 millions USD), are known, then it's (**d**=Dividend Portion planned) is:

$$d = D/\{[1-t][M-F-I]\}$$
$$= 2,370/\{[1-0.3000][41,548-29,750 -513]\}$$
$$= 30.00\%$$

Law-1541:
If (**i**= I/S= Interest Portion= 1.00%), (**t**= T/B= Tax Rate= 30.00%), (**M**= Margin of Contribution= 41,548 millions USD), (**d**= D/A= Dividend Portion or Payout= 30.00%), and (**F**= Fixed Cost= 29,750 millions USD), are known, then it's (**D**= Dividend Paid planned) is:

$$D = d[1-t][M-F-Si]\}$$
$$= 0.3000[1-0.3000][41,548-29,750 -51,294*0.0100]\}$$
$$= 2,370 \text{ millions USD}$$

Law-1542:

If (i= I/S= Interest Portion= 1.00%), (t= T/B= Tax Rate= 30.00%), (D= Dividend Paid= 2,370 millions USD), (d= D/A= Dividend Portion or Payout= 30.00%), and (F= Fixed Cost= 29,750 millions USD), are known, then it's (M= Margin of Contribution planned) is:

$$M = [F+Si+D]/\{d[1-t]\}$$
$$= [29,750+51,294*0.0100]+2,370 /\{0.3000[1-0.3000]\}$$
$$= 2,370 \text{ millions USD}$$

Law-1543:

If (i= I/S= Interest Portion= 1.00%), (t= T/B= Tax Rate= 30.00%), (D= Dividend Paid= 2,370 millions USD), (S= Sales or Revenues= 51,294 millions USD), (d= D/A= Dividend Portion or Payout= 30.00%), and (M= Margin of Contribution= 41,548 millions USD), are known, then it's (F= Fixed Cost planned) is:

$$F = M-Si-D/\{d[1-t]\}$$
$$= [41,548-51,294*0.0100-2,370] /\{ 0.3000[1-0.3000]\}$$
$$= 29,750 \text{ millions USD}$$

Law-1544:
If (i= I/S= Interest Portion= 1.00%), (t= T/B= Tax Rate= 30.00%), (D= Dividend Paid= 2,370 millions USD), (F= Fixed Cost= 29,750 millions USD), (d= D/A= Dividend Portion or Payout= 30.00%), and (M= Margin of Contribution= 41,548 millions USD), are known, then it's (S= Sales or Revenues planned) is:

$S = (M-F-D/\{d[1-t]\})/i$
  = ([41,548-29,750-2,370]/{ 0.3000 [1-0.3000]})/0.0100
  = 51,294 millions USD

Law-1545:
If (S= Sales or Revenues= 51,294 millions USD), (t= T/B= Tax Rate= 30.00%), (D= Dividend Paid= 2,370 millions USD), (F= Fixed Cost= 29,750 millions USD), (d= D/A= Dividend Portion or Payout= 30.00%), and (M= Margin of Contribution= 41,548 millions USD), are known, then it's (i= Interest Portion planned) is:

$i = (M-F-D/\{d[1-t]\})/S$
  = ([41,548-29,750-2,370]/{ 0.3000 [1-0.3000]})/51,294
  = 1.00%

Law-1546:
If ($= Sales or Revenues= 51,294 millions USD), (i= I/S= Interest Portion= 1.00%), (D= Dividend Paid= 2,370 millions USD), (F= Fixed Cost= 29,750 millions USD), (d= D/A= Dividend Portion or Payout= 30.00%), and (M= Margin of Contribution= 41,548 millions USD), are known, then it's (t= Tax Rate planned) is:

$$t = 1-D/\{d[M-F-Si]\}$$
$$= (1-2,370 \{[41,548-29,750-51,294 *0.0100]\})$$
$$= 30.00\%.$$

Law-1547:
If ($= Sales or Revenues= 51,294 millions USD), (i= I/S= Interest Portion= 1.00%), (D= Dividend Paid= 2,370 millions USD), (F= Fixed Cost= 29,750 millions USD), (t= T/B= Tax Rate= 30.00%), and (M= Margin of Contribution= 41,548 millions USD), are known, then it's (d= Dividend Portion or Payout planned) is:

$$d = D/\{[1-t][M-F-Si]\}$$
$$= 2,370/\{[1-0.3000][41,548-29,750 -51,294*0.0100]\})$$
$$= 30.00\%.$$

Law-1548:
   If ($= Sales or Revenues= 51,294 millions USD), (i= I/S= Interest Portion= 1.00%), (d= D/A=Dividend Portion or Payout= 30.00%), (F= Fixed Cost= 29,750 millions USD), (s= [S/S']-1= Sales Growth= 5.00%), (t= T/B= Tax Rate= 30.00%), and (M= Margin of Contribution= 41,548 millions USD), are known, then it's (D= Dividend Paid planned) is:
   D= d[1-t]{M-F-$'i[1+s]})
    = 0.3000 [1-0.3000]{41,548-29,750
    -48,851*0.0100[1+0.0500]}
    = 2,370 millions USD

Law-1549:
   If ($= Sales or Revenues= 51,294 millions USD), (i= I/S= Interest Portion= 1.00%), (d= D/A=Dividend Portion or Payout= 30.00%), (F= Fixed Cost= 29,750 millions USD), (s= [S/S']-1= Sales Growth= 5.00%), (t= T/B= Tax Rate= 30.00%), and (D= Dividend Paid= 2,370 millions USD), are known, then it's (M= Margin of Contribution planned) is:
   M= F-$'i[1+s]+D/{d[1-t]}
    = 29,750+48,851*0.0100[1+0.0500]
    +2,370/{0.3000[1-0.3000]}
    = 41,548 millions USD

Law-1550:

If ($= Sales or Revenues= 51,294 millions USD), (i= I/S= Interest Portion= 1.00%), (d= D/A=Dividend Portion or Payout= 30.00%), (M= Margin of Contribution= 41,548 millions USD), (t= T/B= Tax Rate= 30.00%), (s= [S/S']-1= Sales Growth= 5.00%), and (D= Dividend Paid= 2,370 millions USD), are known, then it's (F= Fixed Cost planned) is:

$$F = M - S'i[1+s] - D/\{d[1-t]\}$$
$$= 41{,}548 - 48{,}851 * 0.0100[1+0.0500]$$
$$\quad - 2{,}370/\{0.3000[1-0.3000]\}$$
$$= \underline{29{,}750} \text{ millions USD}$$

Law-1551:

If ($= Sales or Revenues= 51,294 millions USD), (i= I/S= Interest Portion= 1.00%), (d= D/A=Dividend Portion or Payout= 30.00%), (M= Margin of Contribution= 41,548 millions USD), (t= T/B= Tax Rate= 30.00%), (s= [S/S']-1= Sales Growth= 5.00%), and (D= Dividend Paid= 2,370 millions USD), are known, then it's ($'= Sales Past) must be:

$$S' = \{M - F - [D/d]/[1-t]\} / \{i[1+s]\}$$
$$= \{41{,}548 - 29{,}750 - [2{,}370/0.3000]$$
$$\quad /[1-0.3000]\}$$
$$\quad /\{0.0100[1+0.0500]\}$$
$$= \underline{48{,}851} \text{ millions USD}$$

Finance Construction-2, *Tim Asikin, Steve Asikin, Indra Senihardja*

Law-1552:

If ($= Sales or Revenues= 51,294 millions USD), ($'= Sales of Past Year= 48,851 millions USD), (d= D/A=Dividend Portion or Payout= 30.00%), (M= Margin of Contribution= 41,548 millions USD), (t= T/B= Tax Rate= 30.00%), (s= [S/S']-1= Sales Growth= 5.00%), and (D= Dividend Paid= 2,370 millions USD), are known, then it's (i= Interest Portion planned) is:

$$i = (\{M-F-D/\{d[1-t]\}/\{\$'[1+s]\} $$
$$= (\{41,548-29,750- 2,370/\{0.3000]$$
$$/[1-0.3000]\}$$
$$/\{48,851[1+0.0500]\}$$
$$= 1.00\%$$

Law-1553:

If ($= Sales or Revenues= 51,294 millions USD), ($'= Sales of Past Year= 48,851 millions USD), (d= D/A=Dividend Portion or Payout= 30.00%), (M= Margin of Contribution= 41,548 millions USD), (s= [S/S']-1= Sales Growth= 5.00%), (t= T/B= Tax Rate= 30.00%), and (D= Dividend Paid= 2,370 millions USD), are known, then it's (i= Interest Portion planned) is:

$$i = (\{M-F-D/\{d[1-t]\}/\{\$'[1+s]\} $$
$$= (\{41,548-29,750- 2,370/\{0.3000]$$
$$/[1-0.3000]\}$$
$$/\{48,851[1+0.0500]\}$$
$$= 1.00\%$$

Law-1554:
If ($= Sales or Revenues= 51,294 millions USD), ($'= Sales of Past Year= 48,851 millions USD), (d= D/A=Dividend Portion or Payout= 30.00%), (M= Margin of Contribution= 41,548 millions USD), (s= [S/S']-1= Sales Growth= 5.00%), (i= I/S= Interest Portion= 1.00%), and (D= Dividend Paid= 2,370 millions USD), are known, then it's (t= Tax Rate planned) is:

$$t = 1 - D/(d\{[M-F-S'[1+s]]\})$$
$$= 1 - 2{,}370/(0.3000]\{41{,}548 - 29{,}750 - 48{,}851*0.0100[1+0.0500]\})$$
$$= 30.00\%$$

Law-1555:
If ($= Sales or Revenues= 51,294 millions USD), ($'= Sales of Past Year= 48,851 millions USD), (t= T/B= Tax Rate= 30.00%), (M= Margin of Contribution= 41,548 millions USD), (s= [S/S']-1= Sales Growth= 5.00%), (i= I/S= Interest Portion= 1.00%), and (D= Dividend Paid= 2,370 millions USD), are known, then it's (d= Dividend Portion planned) is:

$$d = D/([1-t]\{[M-F-S'[1+s]]\})$$
$$= 2{,}370/([1-0.3000]\{[41{,}548 - 29{,}750 - 48{,}851*0.0100[1+0.0500]\})$$
$$= 30.00\%$$

Law-1556:
   If ($= Sales or Revenues= 51,294 millions USD), (f= F/S=Fixed Portion= 58.00%), (t= T/B= Tax Rate= 30.00%), (M= Margin of Contribution= 41,548 millions USD), (I= Interest Expense= 513 millions USD), and (d= D/A=Dividend Portion or Payout= 30.00%), are known, then it's (D= Dividend Paid planned) is:
   $$D = d[1-t][[M-Sf-I]]$$
   $$= 0.3000[1-0.3000]\{41,548-51,294$$
   $$*0.5800-513]$$
   $$= 2,370 \text{ millions USD}$$

Law-1557:
   If ($= Sales or Revenues= 51,294 millions USD), (f= F/S=Fixed Portion= 58.00%), (t= T/B= Tax Rate= 30.00%), (D= Dividend Paid= 2,370 millions USD), (I= Interest Expense= 513 millions USD), and (d= D/A= Dividend Portion or Payout= 30.00%), are known, then it's (M= Margin of Contribution planned) is:
   $$M = Sf+I+D/\{d[1-t]\}$$
   $$= 51,294*0.5800+513+2,370/\{0.3000$$
   $$[1-0.3000]\}$$
   $$= 41,548 \text{ millions USD}$$

Law-1558:
If (**M**= Margin of Contribution= 41,548 millions USD), (**f**= F/S=Fixed Portion= 58.00%), (**t**= T/B= Tax Rate= 30.00%), (**D**= Dividend Paid= 2,370 millions USD), (**I**= Interest Expense= 513 millions USD), and (**d**= D/A= Dividend Portion or Payout= 30.00%), are known, then it's (**$**= Sales or Revenues planned) is:
$$\$ = (M-I-D/\{d[1-t]\})/f$$
$$= (41,548-513-2,370/\{0.3000 [1-0.3000]\})/0.5800$$
$$= 51,294 \text{ millions USD}$$

Law-1559:
If (**M**= Margin of Contribution= 41,548 millions USD), (**$**= Sales or Revenues= 51,294 millions USD), (**t**= T/B= Tax Rate= 30.00%), (**D**= Dividend Paid= 2,370 millions USD), (**I**= Interest Expense= 513 millions USD), and (**d**= D/A= Dividend Portion or Payout= 30.00%), are known, then it's (**f**= Fixed Portion planned) is:
$$f = (M-I-D/\{d[1-t]\})/\$$$
$$= (41,548-513-2,370/\{0.3000 [1-0.3000]\})/51,294$$
$$= 58.00\%$$

Law-1560:
If (**M**= Margin of Contribution= 41,548 millions USD), (**S**= Sales or Revenues= 51,294 millions USD), (**t**= T/B= Tax Rate= 30.00%), (**D**= Dividend Paid= 2,370 millions USD), (**f**= F/S= Fixed Portion= 58.00%), and (**d**= D/A= Dividend Portion or Payout= 30.00%), are known, then it's (**I**= Interest Expense planned) is:

$$I = (M\text{-}Sf\text{-}D/\{d[1\text{-}t]\})$$
$$= (41{,}548\text{-}51{,}294*0.5800\text{-}2{,}370/\{0.3000 [1\text{-}0.3000]\})$$
$$= \underline{513} \text{ millions USD}$$

Law-1561:
If (**M**= Margin of Contribution= 41,548 millions USD), (**S**= Sales or Revenues= 51,294 millions USD), (**I**= Interest Expense= 513 millions USD), (**D**= Dividend Paid= 2,370 millions USD), (**f**= F/S= Fixed Portion= 58.00%), and (**d**= D/A= Dividend Portion or Payout= 30.00%), are known, then it's (**t**= Tax Rate planned) is:

$$t = 1\text{-}D/\{d[M\text{-}Sf\text{-}I]\}$$
$$= 1\text{-}2{,}370/\{0.3000 [41{,}548\text{-}51{,}294 *0.5800\text{-}513\}$$
$$= \underline{30.00\%}$$

Law-1562:
If (**M**= Margin of Contribution= 41,548 millions USD), (**$**= Sales or Revenues= 51,294 millions USD), (**I**= Interest Expense= 513 millions USD), (**D**= Dividend Paid= 2,370 millions USD), (**f**= F/S= Fixed Portion= 58.00%), and (**t**= T/B= Tax Rate= 30.00%), are known, then it's (**d**= Dividend Portion planned) is:
$$d = D/\{[1-t][M-\$f-I]\}$$
$$= 2,370/\{[1-0.3000][41,548-51,294*0.5800-513]\}$$
$$= 30.00\%$$

Law-1563:
If (**M**= Margin of Contribution= 41,548 millions USD), (**$**= Sales or Revenues= 51,294 millions USD), (**i**= I/S= Interest Portion= 1.00%), (**d**= D/A= Dividend Portion or Payout= 30.00%), (**f**= F/S= Fixed Portion= 58.00%), and (**t**= T/B= Tax Rate= 30.00%), are known, then it's (**D**= Dividend Paid planned) is:
$$D = d[1-t][M-\$f-\$i] = d[1-t]\{M-\$[f+i]\}$$
$$= 0.3000[1-0.3000][41,548-51,294[0.5800+0.0100]]$$
$$= 2,370 \text{ millions USD}$$

Law-1564:

If (**D**= Dividend Paid= 2,370 millions USD), (**$**= Sales or Revenues= 51,294 millions USD), (**i**= I/S= Interest Portion= 1.00%), (**d**= D/A= Dividend Portion or Payout= 30.00%), (**f**= F/S= Fixed Portion= 58.00%), and (**t**= T/B= Tax Rate= 30.00%), are known, then it's (**M**= Margin of Contribution planned) is:

$M = \$[f+i] + D/\{d[1-t]\}$
  = 51,294[0.5800+0.0100] +2,370
    /{0.3000[1-0.3000]}
  = 41,548 millions USD

Law-1565:

If (**D**= Dividend Paid= 2,370 millions USD), (**M**= Margin of Contribution= 41,548 millions USD), (**i**= I/S= Interest Portion= 1.00%), (**d**=D/A= Dividend Portion or Payout= 30.00%), (**f**= F/S= Fixed Portion= 58.00%), and (**t**= T/B= Tax Rate= 30.00%), are known, then it's (**$**= Sales or Revenues planned) is:

$\$ = (M-D/\{d[1-t]\})/[f+i]$
  = (41,548-2,370/ {0.3000[1-0.3000]}
    /[0.5800+0.0100]
  = 51,294 millions USD

Law-1566:

If (D= Dividend Paid= 2,370 millions USD), (M= Margin of Contribution= 41,548 millions USD), (i= I/S= Interest Portion= 1.00%), (d=D/A= Dividend Portion or Payout= 30.00%), (S= Sales or Revenues= 51,294 millions USD), and (t= T/B= Tax Rate= 30.00%), are known, then it's (f= Fixed Portion planned) is:

$f = (M-D/\{d[1-t]\})/S-i$
= (41,548-2,370/ {0.3000[1-0.3000]})
/51,294 - 0.0100
= 58.00%

Law-1567:

If (D= Dividend Paid= 2,370 millions USD), (M= Margin of Contribution= 41,548 millions USD), (f= F/S=Fixed Portion= 58.00%), (d=D/A= Dividend Portion or Payout= 30.00%), (S= Sales or Revenues= 51,294 millions USD), and (t= T/B= Tax Rate= 30.00%), are known, then it's (i= Interest Portion planned) is:

$i = (M-D/\{d[1-t]\})/S-f$
= (41,548-2,370/{0.3000[1-0.3000]})
/51,294 - 0.5800
= 1.00%

Law-1568:
If (**D**= Dividend Paid= 2,370 millions USD), (**M**= Margin of Contribution= 41,548 millions USD), (**f**= F/S=Fixed Portion= 58.00%), (**d**=D/A= Dividend Portion or Payout= 30.00%), (**$**= Sales or Revenues= 51,294 millions USD), and (**i**= I/S= Interest Portion= 1.00%), are known, then it's (**t**= Tax Rate planned) is:
$$t = 1 - D/\{d[M-\$[f+i]]\}$$
$$= 1 - 2,370/(0.3000\{41,548 - 51,294 [0.5800+0.0100]\}$$
$$= 30.00\%$$

Law-1569:
If (**D**= Dividend Paid= 2,370 millions USD), (**M**= Margin of Contribution= 41,548 millions USD), (**f**= F/S=Fixed Portion= 58.00%), (**t**= T/B= Tax Rate= 30.00%), (**$**= Sales or Revenues= 51,294 millions USD), and (**i**= I/S= Interest Portion= 1.00%), are known, then it's (**d**= Dividend Portion or Payout planned) is:
$$d = D/\{[1-t][M-\$f-\$i]\} = D/([1-t]\{[M-\$[f+i]]\})$$
$$= 2,370/([1-0.3000]\{41,548 - 51,294 [0.5800+0.0100]\})$$
$$= 30.00\%$$

Law-1570:
If (d= D/A=Dividend Portion or Payout= 30.00%), (M= Margin of Contribution= 41,548 millions USD), (f= F/S=Fixed Portion= 58.00%), (t= T/B= Tax Rate= 30.00%), (S'= Sales of Past Year= 48,851 millions USD), (s= [S/S']-1= Sales Growth= 5.00%), (S= Sales or Revenues= 51,294 millions USD), and (i= I/S= Interest Portion= 1.00%), are known, then it's (D= Dividend Paid planned) is:

$D = d[1-t]\{[M-Sf-S'i[1+s]]\}$
= 0.3000[1-0.3000]{41,548-51,294
*0.5800 -48,851*0.0100
[1+0.0500]}
= 2,370 millions USD

Law-1571:
If (d= D/A=Dividend Portion or Payout= 30.00%), (D= Dividend Paid= 2,370 millions USD), (f= F/S=Fixed Portion= 58.00%), (t= T/B= Tax Rate= 30.00%),(S'= Sales of Past Year= 48,851 millions USD), (s= [S/S']-1= Sales Growth= 5.00%), (S= Sales or Revenues= 51,294 millions USD), and (i= I/S= Interest Portion= 1.00%), are known, then it's (M= Margin of Contribution planned) is:

$M = Sf+S'i[1+s]+D/\{d[1-t]\}$
= 51,294*0.5800 +48,851*0.0100
[1+0.0500]+ 2,370
/{0.3000[1-0.3000}
= 41,548 millions USD

Law-1572:

If (d= D/A=Dividend Portion or Payout= 30.00%), (D= Dividend Paid= 2,370 millions USD), (f= F/S= Fixed Portion= 58.00%), (t= T/B= Tax Rate= 30.00%), (S'= Sales of Past Year= 48,851 millions USD), (s= [S/S']-1= Sales Growth= 5.00%), (M= Margin of Contribution= 41,548 millions USD), and (i= I/S= Interest Portion= 1.00%), are known, then it's (S= Sales or Revenues planned) is:

$$S = (M-S'i[1+s]-D/\{d[1-t]\})/f$$
$$= (41,548 - 48,851*0.0100[1+0.0500]$$
$$-2,370/\{0.3000[1-0.3000]\})$$
$$/0.5800$$
$$= 51,294 \text{ millions USD}$$

Law-1573:

If (d= D/A= Dividend Portion or Payout= 30.00%), (D= Dividend Paid= 2,370 millions USD), (S= Sales or Revenues= 51,294 millions USD), (t= T/B= Tax Rate= 30.00%), (S'= Sales of Past Year= 48,851 millions USD), (s= [S/S']-1= Sales Growth= 5.00%), (M= Margin of Contribution= 41,548 millions USD), and (i= I/S= Interest Portion= 1.00%), are known, then it's (f= Fixed Portion planned) is:

$$f = (M-S'i[1+s]-D/\{d[1-t]\})/S$$
$$= (41,548 - 48,851*0.0100[1+0.0500]$$
$$-2,370/\{0.3000[1-0.3000]\})$$
$$/51,294$$
$$= 58.00\%$$

Law-1574:

If (d= D/A=Dividend Portion or Payout= 30.00%), (D= Dividend Paid= 2,370 millions USD), (S= Sales or Revenues= 51,294 millions USD), (t= T/B= Tax Rate= 30.00%), (f= F/S=Fixed Portion= 58.00%), (s= [S/S']-1= Sales Growth= 5.00%), (M= Margin of Contribution= 41,548 millions USD), and (i= I/S= Interest Portion= 1.00%), are known, then it's (S'= Sales Past) must be:

$$S' = (M-Sf-D/\{d[1-t]\})/\{i[1+s]\}$$
$$= (41,548 - 51,294*0.5800 - 2,370$$
$$/\{0.3000[1-0.3000]\})$$
$$/\{0.0100[1+0.0500]\}$$
$$= 48,851 \text{ millions USD}$$

Law-1575:

If (d= D/A=Dividend Portion or Payout= 30.00%), (D= Dividend Paid= 2,370 millions USD), (S= Sales or Revenues= 51,294 millions USD), (t= T/B= Tax Rate= 30.00%), (f= F/S= Fixed Portion= 58.00%), (s= [S/S']-1= Sales Growth= 5.00%), (M= Margin of Contribution= 41,548 millions USD), and (S'= Sales of Past Year= 48,851 millions USD), are known, then it's (i= Interest Portion planned) is:

$$i = (M-Sf-D/\{d[1-t]\})/\{S'[1+s]\}$$
$$= (41,548 - 51,294*0.5800 - 2,370$$
$$/\{0.3000[1-0.3000]\})$$
$$/\{48,851[1+0.0500]\}$$
$$= 1.00\%$$

Law-1576:
If (**d**= D/A=Dividend Portion or Payout= 30.00%), (**D**= Dividend Paid= 2,370 millions USD), (**S**= Sales or Revenues= 51,294 millions USD), (**t**= T/B= Tax Rate= 30.00%), (**f**= F/S= Fixed Portion= 58.00%), (**i**= I/S= Interest Portion= 1.00%), (**M**= Margin of Contribution= 41,548 millions USD), and (**S'**= Sales of Past Year= 48,851 millions USD), are known, then it's (**s**= Sales Growth planned) is:

s= (**M-Sf-D**/{**d**[1-**t**]})/[**S'i**]-1
 = (41,548- 51,294*0.5800 -2,370
  /{0.3000[1-0.3000]})
  /[48,851*0.0100]-1
 = 5.00%

Law-1577:
If (**d**= D/A=Dividend Portion or Payout= 30.00%), (**D**= Dividend Paid= 2,370 millions USD), (**S**= Sales or Revenues= 51,294 millions USD), (**s**= [S/S']-1= Sales Growth= 5.00%), (**f**= F/S= Fixed Portion= 58.00%), (**i**= I/S= Interest Portion= 1.00%), (**M**= Margin of Contribution= 41,548 millions USD), and (**S'**= Sales of Past Year= 48,851 millions USD), are known, then it's (**t**= Tax Rate planned) is:

t= 1-**D**/(**d**{**M-Sf-S'i**[1+**s**]})
 = 1-2,370/(0.3000{41,548-51,294
  *0.5800-48,851*0.0100]
  [1+0.0500]})
 = 30.00%

Law-1578:

If (**D**= Dividend Paid= 2,370 millions USD), (**S**= Sales or Revenues= 51,294 millions USD), (**s**= [S/S']-1= Sales Growth= 5.00%), (**f**= F/S= Fixed Portion= 58.00%), (**i**= I/S= Interest Portion= 1.00%), (**t**= T/B= Tax Rate= 30.00%), (**M**= Margin of Contribution= 41,548 millions USD), and (**S'**= Sales of Past Year= 48,851 millions USD), are known, then it's (**d**= Dividend Portion or Payout planned) is:

$$d = D/([1-t]\{M-Sf-S'i[1+s]\})$$
$$= 2,370/([1-0.3000]\{41,548-51,294$$
$$*0.5800-48,851*0.0100]$$
$$[1+0.0500]\})$$
$$= 30.00\%$$

Law-1579:

If (**d**= D/A=Dividend Portion or Payout= 30.00%), (**s**= [S/S']-1= Sales Growth= 5.00%), (**f**= F/S= Fixed Portion= 58.00%), (**I**= Interest Expense= 513 millions USD), (**t**= T/B= Tax Rate= 30.00%), (**M**= Margin of Contribution= 41,548 millions USD), and (**S'**= Sales of Past Year= 48,851 millions USD), are known, then it's (**D**= Dividend Paid planned) is:

$$D = d([1-t]\{M-I-S'f[1+s]\})$$
$$= 0.3000([1-0.3000]\{41,548- 513$$
$$-48,851*0.5800][1+0.0500]\})$$
$$= 2,370 \text{ millions USD}$$

Law-1580:
If (**d**= D/A=Dividend Portion or Payout= 30.00%), (**s**= [S/S']-1= Sales Growth= 5.00%), (**f**= F/S= Fixed Portion= 58.00%), (**I**= Interest Expense= 513 millions USD), (**t**= T/B= Tax Rate= 30.00%), (**D**= Dividend Paid= 2,370 millions USD), and (**$'**= Sales of Past Year= 48,851 millions USD), are known, then it's (**M**= Margin of Contribution planned) is:

$$M = \$'f[1+s]+I+D/\{d([1-t]\}$$
$$= 48,851*0.5800\ [1+0.0500]+513$$
$$+2,370/\{\ 0.3000([1-0.3000]\}$$
$$= 41,548 \text{ millions USD}$$

Law-1581:
If (**d**= D/A=Dividend Portion or Payout= 30.00%), (**s**= [S/S']-1= Sales Growth= 5.00%), (**f**= F/S= Fixed Portion= 58.00%), (**I**= Interest Expense= 513 millions USD), (**t**= T/B= Tax Rate= 30.00%), (**D**= Dividend Paid= 2,370 millions USD), and (**M**= Margin of Contribution= 41,548 millions USD), are known, then it's (**$'**= Sales Past) is:

$$\$' = (M-I-D/\{d([1-t]\})/\{f[1+s]\}$$
$$= (41,548-513-2,370/\{0.3000[1-0.3000]\}$$
$$/\{0.5800\ [1+0.0500]\}$$
$$= 48,851 \text{ millions USD}$$

Law-1582:
If (**d**= D/A=Dividend Portion or Payout= 30.00%), (**s**= [S/S']-1= Sales Growth= 5.00%), (**S'**= Sales of Past Year= 48,851 millions USD), (**I**= Interest Expense= 513 millions USD), (**t**= T/B= Tax Rate= 30.00%), (**D**= Dividend Paid= 2,370 millions USD), and (**M**= Margin of Contribution= 41,548 millions USD), are known, then it's (**f**= Fixed Portion planned) is:

$$f = (M-I-D/\{d[1-t]\})/\{S'[1+s]\}$$
$$= (41,548-513-2,370$$
$$/\{0.3000[1-0.3000]\})$$
$$/\{48,851\,[1+0.0500]\}$$
$$= \underline{58.00\%}$$

Law-1583:
If (**d**= D/A=Dividend Portion or Payout= 30.00%), (**f**= F/S=Fixed Portion= 58.00%), (**S'**= Sales of Past Year= 48,851 millions USD), (**I**= Interest Expense= 513 millions USD), (**t**= T/B= Tax Rate= 30.00%), (**D**= Dividend Paid= 2,370 millions USD), and (**M**= Margin of Contribution= 41,548 millions USD), are known, then it's (**s**= Sales Growth planned) is:

$$s = (M-I-D/\{d([1-t]\})/[S'f]-1$$
$$= (41,548-513-2,370/\{0.3000$$
$$[1-0.3000]\})/[48,851*0.5800]-1$$
$$= \underline{5.00\%}$$

Law-1584:

If (**d**= D/A=Dividend Portion or Payout= 30.00%), (**f**= F/S=Fixed Portion= 58.00%), (**S'**= Sales of Past Year= 48,851 millions USD), (**s**= [S/S']-1= Sales Growth= 5.00%), (**t**= T/B= Tax Rate= 30.00%), (**D**= Dividend Paid= 2,370 millions USD), and (**M**= Margin of Contribution= 41,548 millions USD), are known, then it's (**I**= Interest Expense planned) is:

$$I = M - S'f[1+s] - D/\{d([1-t])\}$$
$$= (41,548 - 51,294 * 0.5800 - 2,370$$
$$/\{0.3000([1-0.3000])\}$$
$$= \underline{513} \text{ millions USD}$$

Law-1585:

If (**d**= D/A=Dividend Portion or Payout= 30.00%), (**f**= F/S=Fixed Portion= 58.00%), (**S'**= Sales of Past Year= 48,851 millions USD), (**s**= [S/S']-1= Sales Growth= 5.00%), (**I**= Interest Expense= 513 millions USD), (**D**= Dividend Paid= 2,370 millions USD), and (**M**= Margin of Contribution= 41,548 millions USD), are known, then it's (**t**= Tax rate planned) is:

$$t = 1 - D/(d\{M - I - S'f[1+s]\})$$
$$= 1 - 2,370/0.3000\{41,548 - 513 - 48,851$$
$$* 0.5800[1+0.0500]\}$$
$$= \underline{30.00\%}$$

Law-1586:

If (t= T/B= Tax Rate= 30.00%), (f= F/S=Fixed Portion= 58.00%), (S'= Sales of Past Year= 48,851 millions USD), (s= [S/S']-1= Sales Growth= 5.00%), (I= Interest Expense= 513 millions USD), (D= Dividend Paid= 2,370 millions USD), and (M= Margin of Contribution= 41,548 millions USD), are known, then it's (d= Dividend Portion planned) is:

d= D/([1-t]{M-I-$S'f$[1+s]})
= 2,370/([1-0.3000]{41,548-513
-48,851*0.5800[1+0.0500]})
= 30.00%

Law-1587:

If (t= T/B= Tax Rate= 30.00%), (f= F/S= Fixed Portion= 58.00%), (S'= Sales of Past Year= 48,851 millions USD), (s= [S/S']-1= Sales Growth= 5.00%), (i= I/S= Interest Portion= 1.00%), (I= Interest Expense= 513 millions USD), (d= D/A= Dividend Portion or Payout= 30.00%), and (M= Margin of Contribution= 41,548 millions USD), are known, then it's (D= Dividend Paid planned) is:

D= d([1-t]{M-I-$S'f$[1+s]})
= 0.3000([1-0.3000]{41,548-513
-48,851*0.5800[1+0.0500]})
= 2,370 millions USD

Law-1588:
If (**t**= T/B= Tax Rate= 30.00%), (**f**= F/S=Fixed Portion= 58.00%), (**S'**= Sales of Past Year= 48,851 millions USD), (**s**= [S/S']-1= Sales Growth= 5.00%), (**I**= Interest Expense= 513 millions USD), (**D**= Dividend Paid= 2,370 millions USD), and (**d**= D/A= Dividend Portion or Payout= 30.00%), are known, then it's (**M**= Margin of Contribution planned) is:

$$M = S'f[1+s]+I+D/\{d([1-t])\}$$
$$= 48{,}851 * 0.5800[1+0.0500]+513$$
$$+2{,}370/\{0.3000[1-0.3000]\}$$
$$= 41{,}548 \text{ millions USD}$$

Law-1589:
If (**t**= T/B= Tax Rate= 30.00%), (**f**= F/S=Fixed Portion= 58.00%), (**M**= Margin of Contribution= 41,548 millions USD), (**s**= [S/S']-1= Sales Growth= 5.00%), (**I**= Interest Expense= 513 millions USD), (**D**= Dividend Paid= 2,370 millions USD), and (**d**= D/A=Dividend Portion or Payout= 30.00%), are known, then it's (**S'**= Sales Past) must be:

$$S' = (M-I-D/\{d[1-t]\})/\{f[1+s]\}$$
$$= (41{,}548 - 513 - 2{,}370/\{0.3000$$
$$[1-0.3000]\})$$
$$/\{0.5800[1+0.0500]\}$$
$$= 48{,}851 \text{ millions USD}$$

Law-1590:
   If (**t**= T/B= Tax Rate= 30.00%), (**$'**= Sales of Past Year= 48,851 millions USD), (**M**= Margin of Contribution= 41,548 millions USD), (**s**= [S/S']-1= Sales Growth= 5.00%), (**I**= Interest Expense= 513 millions USD), (**D**= Dividend Paid= 2,370 millions USD), and (**d**= D/A=Dividend Portion or Payout= 30.00%), are known, then it's (**f**= Fixed Portion planned) is:
   f= (M-I-D/{d[1-t]})/{$'[1+s]}
   $\quad$ = (41,548- 513-2,370/{0.3000
   $\quad\quad$ [1-0.3000]})
   $\quad\quad$ /{48,851[1+0.0500]}
   $\quad$ = 58.00%

Law-1591:
   If (**t**= T/B= Tax Rate= 30.00%), (**$'**= Sales of Past Year= 48,851 millions USD), (**$**= Sales or Revenues= 51,294 millions USD), (**M**= Margin of Contribution= 41,548 millions USD), (**f**= F/S=Fixed Portion= 58.00%), (**i**= I/S= Interest Portion= 1.00%), (**D**= Dividend Paid= 2,370 millions USD), and (**d**= D/A=Dividend Portion or Payout= 30.00%), are known, then it's (**s**= Sales Growth planned) is:
   s= (M-$i-D/{d[1-t]})/[$'f] -1
   $\quad$ = (41,548- 51,294*0.0100-2,370/{0.3000
   $\quad\quad$ [1-0.3000]})
   $\quad\quad$ /[48,851*0.58] - 1
   $\quad$ = 5.00%

Law-1592:
If (**t**= T/B= Tax Rate= 30.00%), (**$'**= Sales of Past Year= 48,851 millions USD), (**s**= [S/S']-1= Sales Growth= 5.00%), (**M**= Margin of Contribution= 41,548 millions USD), (**f**= F/S=Fixed Portion= 58.00%), (**i**= I/S= Interest Portion= 1.00%), (**D**= Dividend Paid= 2,370 millions USD), and (**d**= D/A=Dividend Portion or Payout= 30.00%), are known, then it's (**$**= Sales or Revenues planned) is:
$$\$= (M-\$'f[1+s]-D/\{d[1-t]\})/i$$
$$= (41,548-48,851*0.5800 [1+0.0500]$$
$$-2,370/\{0.3000$$
$$[1-0.3000]\})/0.0100$$
$$= \underline{51,294} \text{ millions USD}$$

Law-1593:
If (**t**= T/B= Tax Rate= 30.00%), (**$'**= Sales of Past Year= 48,851 millions USD), (**s**= [S/S']-1= Sales Growth= 5.00%), (**M**= Margin of Contribution= 41,548 millions USD), (**f**= F/S=Fixed Portion= 58.00%), (**$**= Sales or Revenues= 51,294 millions USD), (**D**= Dividend Paid= 2,370 millions USD), and (**d**= D/A= Dividend Portion or Payout= 30.00%), are known, then it's (**i**= Interest Portion planned) is:
$$i= (M-\$'f[1+s]-D/\{d[1-t]\})/\$$$
$$= (41,548- 48,851*0.5800[1+0.0500]$$
$$-2,370/\{0.3000[1-0.3000]\})$$
$$/51,294$$
$$= \underline{1.00}\%$$

Law-1594:

If (i= I/S= Interest Portion= 1.00%), ($\$'$= Sales of Past Year= 48,851 millions USD), (s= [S/S']-1= Sales Growth= 5.00%), (M= Margin of Contribution= 41,548 millions USD), (f= F/S=Fixed Portion= 58.00%), ($\$$= Sales or Revenues= 51,294 millions USD), (D= Dividend Paid= 2,370 millions USD), and (d= D/A= Dividend Portion or Payout= 30.00%), are known, then it's (t= Tax Rate planned) is:

t= 1-D/(d{M-$i-$'f[1+s]})
 = 1-2,370/(0.3000{41,548- 51,294
     *0.0100-48,851*0.5800
     [1+0.0500]})
 = 30.00%

Law-1595:

If (i= I/S= Interest Portion= 1.00%), ($\$'$= Sales of Past Year= 48,851 millions USD), (s= [S/S']-1= Sales Growth= 5.00%), (M= Margin of Contribution= 41,548 millions USD), (f= F/S=Fixed Portion= 58.00%), ($\$$= Sales or Revenues= 51,294 millions USD), (D= Dividend Paid= 2,370 millions USD), and (t= T/B= Tax Rate= 30.00%), are known, then it's (d= Dividend Portion planned) is:

d= D/([1-t]{M-$i-$'f[1+s]})
 = 2,370/([1-0.3000]{41,548- 51,294
     *0.0100-48,851*0.5800
     [1+0.0500]})
 = 30.00%

Law-1596:
If (i = I/S = Interest Portion = 1.00%), (S' = Sales of Past Year = 48,851 millions USD), (s = [S/S']-1 = Sales Growth = 5.00%), (M = Margin of Contribution = 41,548 millions USD), (f = F/S = Fixed Portion = 58.00%), (S = Sales or Revenues = 51,294 millions USD), (d = D/A = Dividend Portion or Payout = 30.00%), and (t = T/B = Tax Rate = 30.00%), are known, then it's (D = Dividend Paid planned) is:

$D = d[1-t]\{M-S'f[1+s]-S'i[1+s]\}$
$ = d[1-t]\{M-S'[1+s][f+i]\}$
$ = 0.3000[1-0.3000]\{41,548-48,851$
$\phantom{D = 0.3000[1-0.3000]\{}[1+0.0500][0.5800+0.0100]\}$
$ = 2,370$ millions USD

Law-1597:
If (i = I/S = Interest Portion = 1.00%), (S' = Sales of Past Year = 48,851 millions USD), (s = [S/S']-1 = Sales Growth = 5.00%), (D = Dividend Paid = 2,370 millions USD), (f = F/S = Fixed Portion = 58.00%), (S = Sales or Revenues = 51,294 millions USD), (d = D/A = Dividend Portion or Payout = 30.00%), and (t = T/B = Tax Rate = 30.00%), are known, then it's (M = Margin of Contribution planned) is:

$M = S'[1+s][f+i]+D/\{d[1-t]\}$
$ = 48,851[1+0.0500][0.5800+0.0100]$
$ +2,370/\{0.3000[1-0.3000]\}$
$ = 41,548$ millions USD

Law-1598:
If (**i**= I/S= Interest Portion= 1.00%), (**M**= Margin of Contribution= 41,548 millions USD), (**s**= [S/S']-1= Sales Growth= 5.00%), (**D**= Dividend Paid= 2,370 millions USD), (**f**= F/S=Fixed Portion= 58.00%), (**S**= Sales or Revenues= 51,294 millions USD), (**d**= D/A= Dividend Portion or Payout= 30.00%), and (**t**= T/B= Tax Rate= 30.00%), are known, then it's (**S'**= Sales Past) must be:

$S' = (M-D/\{d[1-t]\})/\{[1+s][f+i]\}$
$= (41,548 - 2,370/\{0.3000[1-0.3000]\})$
$/\{[1+0.0500][0.5800+0.0100]\}$
$= 48,851$ millions USD

Law-1599:
If (**i**= I/S= Interest Portion= 1.00%), (**M**= Margin of Contribution= 41,548 millions USD), (**S'**= Sales of Past Year= 48,851 millions USD), (**D**= Dividend Paid= 2,370 millions USD), (**f**= F/S= Fixed Portion= 58.00%), (**S**= Sales or Revenues= 51,294 millions USD), (**d**= D/A=Dividend Portion or Payout= 30.00%), and (**t**= T/B= Tax Rate= 30.00%), are known, then it's (**s**= Sales Growth planned) is:

$s = (M-D/\{d[1-t]\})/\{S'[f+i]\} - 1$
$= (41,548 - 2,370/\{0.3000[1-0.3000]\})$
$/\{48,851[0.5800+0.0100]\} - 1$
$= 5.00\%$

Law-1600:
If (i= I/S= Interest Portion= 1.00%), (M= Margin of Contribution= 41,548 millions USD), ($\$'$= Sales of Past Year= 48,851 millions USD), (D= Dividend Paid= 2,370 millions USD), (s= [S/S']-1= Sales Growth= 5.00%), ($\$$= Sales or Revenues= 51,294 millions USD), (d= D/A=Dividend Portion or Payout= 30.00%), and (t= T/B= Tax Rate= 30.00%), are known, then it's (f= Fixed Portion planned) is:

f= (M-D/{d[1-t]})/{$\$'$[1+$s$]}-$i$
 = (41,548- 2,370/{0.3000[1-0.3000]})
   /{48,851[1+0.0500]}-0.0100
 = 58.00%

Law-1601:
If (f= F/S=Fixed Portion= 58.00%), (M= Margin of Contribution= 41,548 millions USD), ($\$'$= Sales of Past Year= 48,851 millions USD), (D= Dividend Paid= 2,370 millions USD), (s= [S/S']-1= Sales Growth= 5.00%), ($\$$= Sales or Revenues= 51,294 millions USD), (d= D/A= Dividend Portion or Payout= 30.00%), and (t= T/B= Tax Rate= 30.00%), are known, then it's (i= Interest Portion planned) is:

i= (M-D/{d[1-t]})/{$\$'$[1+$s$]}-$f$
 = (41,548- 2,370/{0.3000[1-0.3000]})
   /{48,851[1+0.0500]}-0.5800
 = 1.00%

Law-1602:
If (**f**= F/S=Fixed Portion= 58.00%), (**M**= Margin of Contribution= 41,548 millions USD), (**$'**= Sales of Past Year= 48,851 millions USD), (**D**= Dividend Paid= 2,370 millions USD), (**s**= [S/S']-1= Sales Growth= 5.00%), (**$**= Sales or Revenues= 51,294 millions USD), (**d**= D/A=Dividend Portion or Payout= 30.00%), and (**i**= I/S= Interest Portion= 1.00%), are known, then it's (**t**= Tax Rate planned) is:
$$t = 1 - D/(d\{M - \$'f[1+s]\} - \$'i[1+s]\})$$
$$= 1 - D/(d\{M - \$'[1+s][f+i]\})$$
$$= 1 - 2{,}370/\,(0.3000\{41{,}548 - 48{,}851$$
$$[1+0.0500][0.5800+0.0100]\})$$
$$= 30.00\%$$

Law-1603:
If (**f**= F/S=Fixed Portion= 58.00%), (**M**= Margin of Contribution= 41,548 millions USD), (**$'**= Sales of Past Year= 48,851 millions USD), (**D**= Dividend Paid= 2,370 millions USD), (**s**= [S/S']-1= Sales Growth= 5.00%), (**$**= Sales or Revenues= 51,294 millions USD), (**t**= T/B= Tax Rate= 30.00%), and (**i**= I/S= Interest Portion= 1.00%), are known, then it's (**d**= Dividend Portion planned) is:
$$d = D/([1-t]\{M - \$'f[1+s]\} - \$'i[1+s]\})$$
$$= D/([1-t]\{M - \$'[1+s][f+i]\})$$
$$= 2{,}370/\,([1-0.3000]\{41{,}548 - 48{,}851$$
$$[1+0.0500][0.5800+0.0100]\})$$
$$= 30.00\%$$

Law-1604:
If (**F**= Fixed Cost= 29,750 millions USD), (**V**= Variable Cost= 9,746 millions USD), (**d**= D/A=Dividend Portion or Payout= 30.00%), (**$**= Sales or Revenues= 51,294 millions USD), (**t**= T/B= Tax Rate= 30.00%), and (**I**= Interest Expense= 513 millions USD), are known, then it's (**D**= Dividend Paid planned) is:

$$D = d[1-t][\$-V-F-I]$$
$$= 0.3000[1-0.3000][51,294-9,746 -29,750-513]$$
$$= 2,370 \text{ millions USD}$$

Law-1605:
If (**F**= Fixed Cost= 29,750 millions USD), (**V**= Variable Cost= 9,746 millions USD), (**d**= D/A=Dividend Portion or Payout= 30.00%), (**D**= Dividend Paid= 2,370 millions USD), (**t**= T/B= Tax Rate= 30.00%), and (**I**= Interest Expense= 513 millions USD), are known, then it's (**$**= Sales or Revenues planned) is:

$$\$ = V+F+I+D/\{d[1-t]\}$$
$$= 9,746+29,750+513+2,370/\{0.3000 [1-0.3000]\}$$
$$= 51,294 \text{ millions USD}$$

Law-1606:

If (**F**= Fixed Cost= 29,750 millions USD), (**S**= Sales or Revenues= 51,294 millions USD), (**d**= D/A= Dividend Portion or Payout= 30.00%), (**D**= Dividend Paid= 2,370 millions USD), (**t**= T/B= Tax Rate= 30.00%), and (**I**= Interest Expense= 513 millions USD), are known, then it's (**V**= Variable Cost planned) is:

$$V = S-F-I-D/\{d[1-t]\}$$
$$= 51,294-29,750-513-2,370/\{0.3000[1-0.3000]\}$$
$$= 9,746 \text{ millions USD}$$

Law-1607:

If (**V**= Variable Cost= 9,746 millions USD), (**S**= Sales or Revenues= 51,294 millions USD), (**d**= D/A=Dividend Portion or Payout= 30.00%), (**D**= Dividend Paid= 2,370 millions USD), (**t**= T/B= Tax Rate= 30.00%), and (**I**= Interest Expense= 513 millions USD), are known, then it's (**F**= Fixed Cost planned) is:

$$F = S-V-I-D/\{d[1-t]\}$$
$$= 51,294-9,746-513-2,370/\{0.3000[1-0.3000]\}$$
$$= 29,750 \text{ millions USD}$$

Law-1608:
If (**V**= Variable Cost= 9,746 millions USD), (**$**= Sales or Revenues= 51,294 millions USD), (**d**= D/A= Dividend Portion or Payout= 30.00%), (**D**= Dividend Paid= 2,370 millions USD), (**t**= T/B= Tax Rate= 30.00%), and (**F**= Fixed Cost= 29,750 millions USD), are known, then it's (**I**= Interest Expense planned) is:

$$I = \$-V-F-D/\{d[1-t]\}$$
$$= 51{,}294-9{,}746-29{,}750-2{,}370/\{0.3000\;[1-0.3000]\}$$
$$= \underline{513}\text{ millions USD}$$

Law-1609:
If (**V**= Variable Cost= 9,746 millions USD), (**$**= Sales or Revenues= 51,294 millions USD), (**d**= D/A= Dividend Portion or Payout= 30.00%), (**D**= Dividend Paid= 2,370 millions USD), (**I**= Interest Expense= 513 millions USD), and (**F**= Fixed Cost= 29,750 millions USD), are known, then it's (**t**= Tax Rate planned) is:

$$t = 1 - D/\{d[\$-V-F-I]\}$$
$$= 1 - 2{,}370/\{0.3000\;[51{,}294-9{,}746-29{,}750-513]\}$$
$$= \underline{30.00\%}$$

Law-1610:

If (**V**= Variable Cost= 9,746 millions USD), (**$**= Sales or Revenues= 51,294 millions USD), (**t**= T/B= Tax Rate= 30.00%), (**D**= Dividend Paid= 2,370 millions USD), (**I**= Interest Expense= 513 millions USD), and (**F**= Fixed Cost= 29,750 millions USD), are known, then it's (**d**= Dividend Portion or Payout planned) is:

$$d = D/\{[1-t][\$-V-F-I]\}$$
$$= 2,370/\{[1-0.3000][51,294-9,746-29,750-513]\}$$
$$= 30.00\%$$

Law-1611:

If (**V**= Variable Cost= 9,746 millions USD), (**$**= Sales or Revenues= 51,294 millions USD), (**t**= T/B= Tax Rate= 30.00%), (**d**= D/A= Dividend Portion or Payout= 30.00%), (**i**= I/S= Interest Portion= 1.00%), and (**F**= Fixed Cost= 29,750 millions USD), are known, then it's (**D**= Dividend Paid planned) is:

$$D = d[1-t]\{\$-V-F-\$i\} = d[1-t]\{\$[1-i]-V-F\}$$
$$= 0.3000[1-0.3000]\{51,294[1-0.0100]-9,746-29,750\}$$
$$= 2,370 \text{ millions USD}$$

Law-1612:
If (V= Variable Cost= 9,746 millions USD), (D= Dividend Paid= 2,370 millions USD), (t= T/B= Tax Rate= 30.00%), (d= D/A=Dividend Portion or Payout= 30.00%), (i= I/S= Interest Portion= 1.00%), and (F= Fixed Cost= 29,750 millions USD), are known, then it's (S= Sales or Revenues planned) is:

$S = (V+F+D/\{d[1-t]\})/[1-i]$
$= (9,746+29,750+2,370/\{0.3000$
$[1-0.3000]\})/[1-0.0100]$
$= \underline{51,294}$ millions USD

Law-1613:
If (V= Variable Cost= 9,746 millions USD), (D= Dividend Paid= 2,370 millions USD), (t= T/B= Tax Rate= 30.00%), (d= D/A= Dividend Portion or Payout= 30.00%), (S= Sales or Revenues= 51,294 millions USD), and (F= Fixed Cost= 29,750 millions USD), are known, then it's (i= Interest Portion planned) is:

$i = 1-(V+F+D/\{d[1-t]\})/S$
$= 1- (9,746+29,750+2,370/\{0.3000$
$[1-0.3000]\})/51,294$
$= \underline{1.00\%}$

Law-1614:

If (**i**= I/S= Interest Portion= 1.00%), (**D**= Dividend Paid= 2,370 millions USD), (**t**= T/B= Tax Rate= 30.00%), (**d**= D/A= Dividend Portion or Payout= 30.00%), (**$**= Sales or Revenues= 51,294 millions USD), and (**F**= Fixed Cost= 29,750 millions USD), are known, then it's (**V**= Variable Cost planned) is:

$$V = \$[1-i] - F - D/\{d[1-t]\}$$
$$= 51,294[1-0.0100] - 29,750 - 2,370 /\{0.3000[1-0.3000]\}$$
$$= 9,746 \text{ millions USD}$$

Law-1615:

If (**i**= I/S= Interest Portion= 1.00%), (**D**= Dividend Paid= 2,370 millions USD), (**t**= T/B= Tax Rate= 30.00%), (**d**= D/A= Dividend Portion or Payout= 30.00%), (**$**= Sales or Revenues= 51,294 millions USD), and (**V**= Variable Cost= 9,746 millions USD), are known, then it's (**F**= Fixed Cost planned) is:

$$F = \$[1-i] - V - D/\{d[1-t]\}$$
$$= 51,294[1-0.0100] - 9,746 - 2,370 /\{0.3000[1-0.3000]\}$$
$$= 29,750 \text{ millions USD}$$

Law-1616:
   If (i= I/S= Interest Portion= 1.00%), (D= Dividend Paid= 2,370 millions USD), (F= Fixed Cost= 29,750 millions USD), (d= D/A= Dividend Portion or Payout= 30.00%), ($\$$= Sales or Revenues= 51,294 millions USD), and (V= Variable Cost= 9,746 millions USD), are known, then it's (t= Tax Rate planned) is:
   t= 1-[D/d]/[$\$$-V-F-$\$i$]= 1-$D$/($d${$\$$[1-i]-V-F})
    = 1-2,370/(0.3000{51,294[1-0.0100]
     -9,746- 29,750})
    = 30.00%

Law-1617:
   If (i= I/S= Interest Portion= 1.00%), (D= Dividend Paid= 2,370 millions USD), (F= Fixed Cost= 29,750 millions USD), (t= T/B= Tax Rate= 30.00%), ($\$$= Sales or Revenues= 51,294 millions USD), and (V= Variable Cost= 9,746 millions USD), are known, then it's (d= Dividend Portion planned) is:
   d= D/{[1-t]/[$\$$-V-F-$\$i$]= D/([1-t]{$\$$[1-i]-V-F})
    = 2,370/([1-0.3000]{51,294[1-0.0100]
     -9,746-29,750 })
    = 30.00%

Law-1618:

If (**i**= I/S= Interest Portion= 1.00%), (**d**= D/A= Dividend Portion or Payout= 30.00%), (**F**= Fixed Cost= 29,750 millions USD), (**$'**= Sales of Past Year= 48,851 millions USD), (**t**= T/B= Tax Rate= 30.00%), (**s**= [S/S']-1= Sales Growth= 5.00%), (**$**= Sales or Revenues= 51,294 millions USD), and (**V**= Variable Cost= 9,746 millions USD), are known, then it's (**D**= Dividend Paid planned) is:

$$D= d[1-t]/\{\$-V-F-\$'i[1+s]\}$$
$$= 0.3000[1- 0.3000]\{51,294 -9,746$$
$$- 29,750-48,851*0.0100$$
$$[1+0.0500]\}$$
$$= 2,370 \text{ millions USD}$$

Law-1619:

If (**i**= I/S= Interest Portion= 1.00%), (**d**= D/A= Dividend Portion or Payout= 30.00%), (**F**= Fixed Cost= 29,750 millions USD), (**$'**= Sales of Past Year= 48,851 millions USD), (**t**= T/B= Tax Rate= 30.00%), (**s**= [S/S']-1= Sales Growth= 5.00%), (**D**= Dividend Paid= 2,370 millions USD), and (**V**= Variable Cost= 9,746 millions USD), are known, then it's (**$**= Sales or Revenues planned) is:

$$\$= V+F+\$'i[1+s]+D/\{d[1-t]\}$$
$$= 9,746+ 29,750+48,851*0.0100$$
$$[1+0.0500]\}+2,370/\{ 0.3000$$
$$[1- 0.3000]\}$$
$$= 51,294 \text{ millions USD}$$

Law-1620:

If (i= I/S= Interest Portion= 1.00%), (d= D/A= Dividend Portion or Payout= 30.00%), (F= Fixed Cost= 29,750 millions USD), ($\$'$= Sales of Past Year= 48,851 millions USD), (s= [S/S']-1= Sales Growth= 5.00%), (t= T/B= Tax Rate= 30.00%), (D= Dividend Paid= 2,370 millions USD), and ($\$$= Sales or Revenues= 51,294 millions USD), are known, then it's (V= Variable Cost planned) is:

$V = \$-F-\$'i[1+s]-D/\{d[1-t]\}$
= 51,294-29,750-48,851*0.0100
        [1+0.0500]-2,370/{0.3000
        [1- 0.3000]}
= 9,746 millions USD

Law-1621:

If (i= I/S= Interest Portion= 1.00%), (d= D/A= Dividend Portion or Payout= 30.00%), ($\$'$= Sales of Past Year= 48,851 millions USD), (t= T/B= Tax Rate= 30.00%), (V= Variable Cost= 9,746 millions USD), (s= [S/S']-1= Sales Growth= 5.00%), (D= Dividend Paid= 2,370 millions USD), and ($\$$= Sales or Revenues= 51,294 millions USD), are known, then it's (F= Fixed Cost planned) is:

$F = \$-V-\$'i[1+s]-D/\{d[1-t]\}$
= 51,294-9,746-48,851*0.0100
        [1+0.0500]-2,370/{0.3000
        [1- 0.3000]}
= 29,750 millions USD

Law-1622:
If (i= I/S= Interest Portion= 1.00%), (d= D/A= Dividend Portion or Payout= 30.00%), (F= Fixed Cost= 29,750 millions USD), (t= T/B= Tax Rate= 30.00%), (V= Variable Cost= 9,746 millions USD), (s= [S/S']-1= Sales Growth= 5.00%), (D= Dividend Paid= 2,370 millions USD), and ($\$$= Sales or Revenues= 51,294 millions USD), are known, then it's ($\$'$= Sales Past) must be:

$\$' = (\$-V-F-D/\{d[1-t]\})/\{i[1+s]\}$
$= (51{,}294-9{,}746-29{,}750-\{2{,}370/\{0.3000[1-0.3000]\}\})$
$/\{0.0100[1+0.0500]\}$
$= 48{,}851$ millions USD

Law-1623:
If ($\$'$= Sales of Past Year= 48,851 millions USD), (d= D/A= Dividend Portion or Payout= 30.00%), (F= Fixed Cost= 29,750 millions USD), (t= T/B= Tax Rate= 30.00%), (V= Variable Cost= 9,746 millions USD), (s= [S/S']-1= Sales Growth= 5.00%), (D= Dividend Paid= 2,370 millions USD), and ($\$$= Sales or Revenues= 51,294 millions USD), are known, then it's (i= Interest Portion planned) is:

$i = (\$-V-F-D/\{d[1-t]\})/\{\$'[1+s]\}$
$= (51{,}294-9{,}746-29{,}750-\{2{,}370/\{0.3000[1-0.3000]\}\})$
$/\{48{,}851[1+0.0500]\}$
$= 1.00\%$

Law-1624:

If ($'=$ Sales of Past Year= 48,851 millions USD), (**d**= D/A= Dividend Portion or Payout= 30.00%), (**F**= Fixed Cost= 29,750 millions USD), (**t**= T/B= Tax Rate= 30.00%), (**V**= Variable Cost= 9,746 millions USD), (**i**= I/S= Interest Portion= 1.00%), (**D**= Dividend Paid= 2,370 millions USD), and ($=$ Sales or Revenues= 51,294 millions USD), are known, then it's ($=$ Sales Growth planned) is:

$s = ($-V-F-D/\{d[1-t]\})/[$'i]-1$
$= (51,294-9,746-29,750-\{2,370/\{0.3000$
$[1- 0.3000]\})$
$/\{48,851*0.0100\}-1$
$= 5.00\%$

Law-1625:

If ($'=$ Sales of Past Year= 48,851 millions USD), (**d**= D/A= Dividend Portion or Payout= 30.00%), (**F**= Fixed Cost= 29,750 millions USD), (**s**= [S/S']-1= Sales Growth= 5.00%), (**V**= Variable Cost= 9,746 millions USD), (**i**= I/S= Interest Portion= 1.00%), (**D**= Dividend Paid= 2,370 millions USD), and ($=$ Sales or Revenues= 51,294 millions USD), are known, then it's (**t**= Tax Rate planned) is:

$t = 1-D/(d\{$-V-F-$'i[1+s]\})$
$= 1-2,370/(0.3000[1- 0.3000]\{51,294$
$-9,746-29,750-48,851*0.0100$
$[1+0.0500]\})$
$= 30.00\%$

Law-1626:
   If ($= Sales of Past Year= 48,851 millions USD), (t= T/B= Tax Rate= 30.00%), (F= Fixed Cost= 29,750 millions USD), (s= [S/S']-1= Sales Growth= 5.00%), (V= Variable Cost= 9,746 millions USD), (i= I/S= Interest Portion= 1.00%), (D= Dividend Paid= 2,370 millions USD), and ($= Sales or Revenues= 51,294 millions USD), are known, then it's (d= Dividend Portion planned) is:
   d= D/([1-t]{$-V-F-$'i[1+s]})
      = 2,370/([1- 0.3000]{ 51,294-9,746
         -29,750-48,851 *0.0100
         [1+0.0500]})
      = 30.00%

Law-1627:
   If (t= T/B= Tax Rate= 30.00%), (f= F/S=Fixed Portion= 58.00%), (V= Variable Cost= 9,746 millions USD), (I= Interest Expense= 513 millions USD), (d= D/A=Dividend Portion or Payout= 30.00%), and ($= Sales or Revenues= 51,294 millions USD), are known, then it's (D= Dividend Paid planned) is:
   D= d[1-t][$-V-$f-I]= d[1-t]{$[1-f]-V-I}
      = 0.3000[1- 0.3000]{51,294[1-0.5800]
         -9,746-513}
      = 2,370 millions USD

Law-1628:
If (**t**= T/B= Tax Rate= 30.00%), (**f**= F/S=Fixed Portion= 58.00%), (**V**= Variable Cost= 9,746 millions USD), (**I**= Interest Expense= 513 millions USD), (**d**= D/A=Dividend Portion or Payout= 30.00%), and (**D**= Dividend Paid= 2,370 millions USD), are known, then it's (**$**= Sales or Revenues planned) is:
$= (**V**+**I**+**D**/{**d**[1-**t**]})/[1-**f**]
 = (9,746+513+ 2,370/{0.300
  [1- 0.3000]})/{[1-0.5800]
 = 51,294 millions USD

Law-1629:
If (**t**= T/B= Tax Rate= 30.00%), (**$**= Sales or Revenues= 51,294 millions USD), (**V**= Variable Cost= 9,746 millions USD), (**I**= Interest Expense= 513 millions USD), (**d**= D/A=Dividend Portion or Payout= 30.00%), and (**D**= Dividend Paid= 2,370 millions USD), are known, then it's (**f**= Fixed Portion planned) is:
f= 1-(**V**+**I**+**D**/{**d**[1-**t**]})/**$**
 = 1-(9,746+513+ 2,370/{0.300
  [1- 0.3000]})/51,294
 = 58.00%

Law-1630:

If (t= T/B= Tax Rate= 30.00%), (S= Sales or Revenues= 51,294 millions USD), (f= F/S=Fixed Portion= 58.00%), (I= Interest Expense= 513 millions USD), (d= D/A=Dividend Portion or Payout= 30.00%), and (D= Dividend Paid= 2,370 millions USD), are known, then it's (V= Variable Cost planned) is:

$$V = S[1-f] - I - D/\{d[1-t]\}$$
$$= 51,294[1-0.5800] - 513 - 2,370 / \{0.300[1 - 0.3000]\}$$
$$= \underline{9,746} \text{ millions USD}$$

Law-1631:

If (t= T/B= Tax Rate= 30.00%), (S= Sales or Revenues= 51,294 millions USD), (f= F/S=Fixed Portion= 58.00%), (V= Variable Cost= 9,746 millions USD), (d=D/A=Dividend Portion or Payout= 30.00%), and (D= Dividend Paid= 2,370 millions USD), are known, then it's (I= Interest Expense planned) is:

$$I = S[1-f] - V - D/\{d[1-t]\}$$
$$= 51,294[1-0.5800] - 9,746 - 2,370 / \{0.3000[1 - 0.3000]\}$$
$$= \underline{513} \text{ millions USD}$$

Law-1632:

If (**I**= Interest Expense= 513 millions USD), (**$**= Sales or Revenues= 51,294 millions USD), (**f**= F/S=Fixed Portion= 58.00%), (**V**= Variable Cost= 9,746 millions USD), (**d**=D/A=Dividend Portion or Payout= 30.00%), and (**D**= Dividend Paid= 2,370 millions USD), are known, then it's (**t**= Tax Rate planned) is:

t= 1-[**D**/**d**]/[**$**-**V**-**$f**-**I**]= 1-**D**/(**d**{**$**[1-**f**]-**V**-**I**})
= 1-2,370/(0.3000{ 51,294[1-0.5800]
-9,746- 513})
= 30.00%

Law-1633:

If (**I**= Interest Expense= 513 millions USD), (**$**= Sales or Revenues= 51,294 millions USD), (**f**= F/S=Fixed Portion= 58.00%), (**V**= Variable Cost= 9,746 millions USD), (**t**= T/B= Tax Rate= 30.00%), and (**D**= Dividend Paid= 2,370 millions USD), are known, then it's (**d**= Dividend Portion planned) is:

d= **D**/(**d**[1-**t**]/[**$**-**V**-**$f**-**I**]= **D**/(**d**[1-**t**]{**$**[1-**f**]-**V**-**I**})
= 2,370/([1-0.3000]{51,294[1-0.5800]
-9,746-513})
= 30.00%

Law-1634:
If (**i**= I/S= Interest Portion= 1.00%), (**$**= Sales or Revenues= 51,294 millions USD), (**f**= F/S=Fixed Portion= 58.00%), (**V**= Variable Cost= 9,746 millions USD), (**t**= T/B= Tax Rate= 30.00%), and (**d**= D/A= Dividend Portion or Payout= 30.00%), are known, then it's (**D**= Dividend Paid planned) is:

$$D = d[1-t][\$-V-\$f-\$i] = d[1-t]\{\$[1-f-i]-V\}$$
$$= 0.3000[1-0.3000]\{51,294[1-0.5800-0.0100]-9,746\}$$
$$= 2,370 \text{ millions USD}$$

Law-1635:
If (**i**= I/S= Interest Portion= 1.00%), (**D**= Dividend Paid= 2,370 millions USD), (**f**= F/S=Fixed Portion= 58.00%), (**V**= Variable Cost= 9,746 millions USD), (**t**= T/B= Tax Rate= 30.00%), and (**d**= D/A= Dividend Portion or Payout= 30.00%), are known, then it's (**$**= Sales or Revenues planned) is:

$$\$ = V + D\{d[1-t]\}/[1-f-i]$$
$$= 9,746 + 2,370/\{0.3000[1-0.3000]\}/[1-0.5800-0.0100]$$
$$= 51,294 \text{ millions USD}$$

Finance Construction-2, *Tim Asikin, Steve Asikin, Indra Senihardja*

Law-1636:
If (i= I/S= Interest Portion= 1.00%), (**D**= Dividend Paid= 2,370 millions USD), ($= Sales or Revenues= 51,294 millions USD), (**V**= Variable Cost= 9,746 millions USD), (**t**= T/B= Tax Rate= 30.00%), and (**d**= D/A= Dividend Portion or Payout= 30.00%), are known, then it's (**f**= Fixed Portion planned) is:

$$f = 1-i-(V+D\{d[1-t]\})/\$$$
$$= 1-0.0100(9,746+2,370/\{0.3000[1-0.3000]\})/51,294$$
$$= 58.00\%$$

Law-1637:
If (**f**= F/S=Fixed Portion= 58.00%), (**D**= Dividend Paid= 2,370 millions USD), ($= Sales or Revenues= 51,294 millions USD), (**V**= Variable Cost= 9,746 millions USD), (**t**= T/B= Tax Rate= 30.00%), and (**d**= D/A=Dividend Portion or Payout= 30.00%), are known, then it's (**i**= Interest Portion planned) is:

$$i = 1-f-(V+D\{d[1-t]\})/\$$$
$$= 1-0.5800(9,746+2,370/\{0.3000[1-0.3000]\})/51,294$$
$$= 1.00\%$$

Law-1638:
If (f= F/S=Fixed Portion= 58.00%), (D= Dividend Paid= 2,370 millions USD), (S= Sales or Revenues= 51,294 millions USD), (i= I/S= Interest Portion= 1.00%), (t= T/B= Tax Rate= 30.00%), and (d= D/A=Dividend Portion or Payout= 30.00%), are known, then it's (V= Variable Cost planned) is:
$$V = S[1-f-i]-D/\{d[1-t]\}$$
$$= 51{,}294[1-0.5800-0.0100]-2{,}370$$
$$/\{0.3000[1-0.3000]\}$$
$$= 9{,}746 \text{ millions USD}$$

Law-1639:
If (f= F/S=Fixed Portion= 58.00%), (D= Dividend Paid= 2,370 millions USD), (S= Sales or Revenues= 51,294 millions USD), (i= I/S= Interest Portion= 1.00%), (V= Variable Cost= 9,746 millions USD), and (d= D/A= Dividend Portion or Payout= 30.00%), are known, then it's (t= Tax Rate planned) is:
$$t = 1-[D/d]/[S-V-Sf-Si] = 1-D/(d\{S[1-f-i]-V\})$$
$$= 1-2{,}370/(0.3000\{51{,}294[1-0.5800$$
$$-0.0100]-9{,}746\})$$
$$= 30.00\%$$

Law-1640:

If (**f**= F/S=Fixed Portion= 58.00%), (**D**= Dividend Paid= 2,370 millions USD), ($= Sales or Revenues= 51,294 millions USD), (**i**= I/S= Interest Portion= 1.00%), (**V**= Variable Cost= 9,746 millions USD), and (**t**= T/B= Tax Rate= 30.00%), are known, then it's (**d**= Dividend Portion planned) is:

$$d = D/[1-t]/[\$-V-\$f-\$i] = D/([1-t]\{\$[1-f-i]-V\})$$
$$= 2,370/([1-0.3000]\{51,294[1-0.5800$$
$$-0.0100]-9,746\})$$
$$= 30.00\%$$

Law-1641:

If (**f**= F/S=Fixed Portion= 58.00%), (**t**= T/B= Tax Rate= 30.00%), ($'= Sales of Past Year= 48,851 millions USD), (**s**= [S/S']-1= Sales Growth= 5.00%), ($= Sales or Revenues= 51,294 millions USD), (**i**= I/S= Interest Portion= 1.00%), (**V**= Variable Cost= 9,746 millions USD), and (**d**= D/A= Dividend Portion or Payout= 30.00%), are known, then it's (**D**= Dividend Paid planned) is:

$$D = d[1-t]/\{\$-V-\$f-\$i[1+s]\}$$
$$= d[1-t]\{\$[1-f]-V-S'i[1+s]\}$$
$$= 0.3000[1-0.3000]\{51,294[1-0.5800]$$
$$-9,746-48,851*0.0100$$
$$[1+0.0500]\}$$
$$= 2,370 \text{ millions USD}$$

Law-1642:
If (f= F/S=Fixed Portion= 58.00%), (t= T/B= Tax Rate= 30.00%), (S'= Sales of Past Year= 48,851 millions USD), (s= [S/S']-1= Sales Growth= 5.00%), (D= Dividend Paid= 2,370 millions USD), (i= I/S= Interest Portion= 1.00%), (V= Variable Cost= 9,746 millions USD), and (d= D/A=Dividend Portion or Payout= 30.00%), are known, then it's (S= Sales or Revenues planned) is:

$$S = (V + S'i[1+s] + D\{d[1-t]\})/[1-f]$$
$$= 9{,}746 + 48{,}851*0.0100[1+0.0500]$$
$$+ 2{,}370/\{0.3000[1-0.3000]\}$$
$$/[1-0.5800]$$
$$= \underline{51{,}294} \text{ millions USD}$$

Law-1643:
If (f= F/S=Fixed Portion= 58.00%), (t= T/B= Tax Rate= 30.00%), (S'= Sales of Past Year= 48,851 millions USD), (s= [S/S']-1= Sales Growth= 5.00%), (D= Dividend Paid= 2,370 millions USD), (i= I/S= Interest Portion= 1.00%), (S= Sales or Revenues= 51,294 millions USD), and (d= D/A=Dividend Portion or Payout= 30.00%), are known, then it's (V= Variable Cost planned) is:

$$V = S[1-f] - S'i[1+s] - D\{d[1-t]\}$$
$$= 51{,}294[1-0.5800] - 48{,}851*0.0100$$
$$[1+0.0500] - 2{,}370/\{0.3000$$
$$[1-0.3000]\}$$
$$= \underline{9{,}746} \text{ millions USD}$$

Law-1644:

If (**f**= F/S=Fixed Portion= 58.00%), (**t**= T/B= Tax Rate= 30.00%), (**V**= Variable Cost= 9,746 millions USD), (**s**= [S/S']-1= Sales Growth= 5.00%), (**D**= Dividend Paid= 2,370 millions USD), (**i**= I/S= Interest Portion= 1.00%), (**$**= Sales or Revenues= 51,294 millions USD), and (**d**= D/A= Dividend Portion or Payout= 30.00%), are known, then it's (**$'**= Sales Past) must be:

$'= ($[1-**f**]-**V**-**D**{**d**[1-**t**]}/{**i**[1+**s**]}
= (51,294[1-0.5800]-9,746-2,370
   /{0.3000[1-0.3000]})
   /{0.0100[1+0.0500]}
= 48,851 millions USD

Law-1645:

If (**$'**= Sales of Past Year= 48,851 millions USD), (**t**= T/B= Tax Rate= 30.00%), (**V**= Variable Cost= 9,746 millions USD), (**s**= [S/S']-1= Sales Growth= 5.00%), (**D**= Dividend Paid= 2,370 millions USD), (**i**= I/S= Interest Portion= 1.00%), (**$**= Sales or Revenues= 51,294 millions USD), and (**d**= D/A= Dividend Portion or Payout= 30.00%), are known, then it's (**f**= Fixed Portion planned) must be:

f= 1-(**V**+**$'i**[1+**s**]+**D**{**d**[1-**t**]})/**$**
= 1-(9,746+48,851*0.0100[1+0.0500]
   +2,370/{0.3000[1-0.3000]})
   /51,294
= 58.00%

Law-1646:

If ($= Sales of Past Year= 48,851 millions USD), (**t**= T/B= Tax Rate= 30.00%), (**V**= Variable Cost= 9,746 millions USD), (**f**= F/S=Fixed Portion= 58.00%), (**D**= Dividend Paid= 2,370 millions USD), (**i**= I/S= Interest Portion= 1.00%), ($= Sales or Revenues= 51,294 millions USD), and (**d**= D/A=Dividend Portion or Payout= 30.00%), are known, then it's (**s**= Sales Growth planned) is:

$s = (\$[1\text{-}f]\text{-}V\text{-}D\{d[1\text{-}t]\})/[\$'i]\text{-}1$
$= (51,294[1\text{-}0.5800]\text{-}9,746\text{-}2,370$
$/\{0.3000[1\text{-}0.3000]\})$
$/[48,851*0.0100]\text{-}1$
$= 5.00\%$

Law-1647:

If ($'= Sales of Past Year= 48,851 millions USD), (**t**= T/B= Tax Rate= 30.00%), (**V**= Variable Cost= 9,746 millions USD), (**f**= F/S=Fixed Portion= 58.00%), (**D**= Dividend Paid= 2,370 millions USD), (**s**= [S/S']-1= Sales Growth= 5.00%), ($= Sales or Revenues= 51,294 millions USD), and (**d**= D/A=Dividend Portion or Payout= 30.00%), are known, then it's (**i**= Interest Portion planned) is:

$i = (\$[1\text{-}f]\text{-}V\text{-}D\{d[1\text{-}t]\})/\{\$'[1+s]\}$
$= (51,294[1\text{-}0.5800]\text{-} 9,746\text{-}2,370$
$/\{0.3000[1\text{-}0.3000]\})$
$/[48,851[1+0.0500]\}$
$= 1.00\%$

Law-1648:
If ($'$= Sales of Past Year= 48,851 millions USD), (i= I/S= Interest Portion= 1.00%), (V= Variable Cost= 9,746 millions USD), (f= F/S=Fixed Portion= 58.00%), (D= Dividend Paid= 2,370 millions USD), (s= [S/S']-1= Sales Growth= 5.00%), (S= Sales or Revenues= 51,294 millions USD), and (d= D/A= Dividend Portion or Payout= 30.00%), are known, then it's (t= Tax Rate planned) is:

t = 1-[D/d]/{S-V-Sf-$S'i$[1+s]}
  = 1-D/(d{S[1-f]-V-S'[1+s]})
  = 1-2,370/(0.3000{51,294[1-0.5800]
    -9,746-48,851[1+0.0500]})
  = 30.00%

Law-1649:
If ($'$= Sales of Past Year= 48,851 millions USD), (i= I/S= Interest Portion= 1.00%), (V= Variable Cost= 9,746 millions USD), (f= F/S=Fixed Portion= 58.00%), (D= Dividend Paid= 2,370 millions USD), (s= [S/S']-1= Sales Growth= 5.00%), (S= Sales or Revenues= 51,294 millions USD), and (t= T/B= Tax Rate= 30.00%), are known, then it's (d= Dividend Portion or Payout planned) is:

d = D([1-t]{S-V-Sf-$S'i$[1+s]}
  = 1-D/([1-t]{S[1-f]-V-$S'i$[1+s]})
  = 2,370/([1-0.3000]{51,294[1-0.5800]
    -9,746-48,851*0.0100
    [1+0.0500]})
  = 30.00%

Law-1650:
If ($'= Sales of Past Year= 48,851 millions USD), (I= Interest Expense= 513 millions USD), (V= Variable Cost= 9,746 millions USD), (f= F/S=Fixed Portion= 58.00%), (d= D/A= Dividend Portion or Payout= 30.00%), (s= [S/S']-1= Sales Growth= 5.00%), ($= Sales or Revenues= 51,294 millions USD), and (t= T/B= Tax Rate= 30.00%), are known, then it's (D= Dividend Paid planned) is:

$\quad$ D= d[1-t]{$-V-I-$'f[1+s]}
$\quad\quad$ = 0.3000[1- 0.3000]{51,294 -9,746 -513
$\quad\quad\quad$ -48,851*0.5800[1+0.0500]})
$\quad\quad$ = 2,370 millions USD

Law-1651:
If ($'= Sales of Past Year= 48,851 millions USD), (I= Interest Expense= 513 millions USD), (V= Variable Cost= 9,746 millions USD), (f= F/S=Fixed Portion= 58.00%), (d= D/A= Dividend Portion or Payout= 30.00%), (s= [S/S']-1= Sales Growth= 5.00%), (D= Dividend Paid= 2,370 millions USD), and (t= T/B= Tax Rate= 30.00%), are known, then it's ($= Sales or Revenues planned) is:

$\quad$ $= $'f[1+s]+V+D/d[1-t]}
$\quad\quad$ = 48,851*0.5800[1+0.0500]+9,746+513
$\quad\quad\quad$ +2,370/{0.3000[1- 0.3000]}
$\quad\quad$ = 51,294 millions USD

Law-1652:
If ($= Sales of Past Year= 48,851 millions USD), (I= Interest Expense= 513 millions USD), ($= Sales or Revenues= 51,294 millions USD), (f= F/S=Fixed Portion= 58.00%), (d= D/A= Dividend Portion or Payout= 30.00%), (s= [S/S']-1= Sales Growth= 5.00%), (D= Dividend Paid= 2,370 millions USD), and (t= T/B= Tax Rate= 30.00%), are known, then it's (V=Variable Cost planned) is:

$$V = \$ - I - \$'f[1+s] - D/\{d[1-t]\}$$
$$= 51,294 - 513 - 48,851*0.5800[1+0.0500]$$
$$-2,370/\{0.3000[1-0.3000]\}$$
$$= 9,746 \text{ millions USD}$$

Law-1653:
If (V= Variable Cost= 9,746 millions USD), (I= Interest Expense= 513 millions USD), ($= Sales or Revenues= 51,294 millions USD), (f= F/S=Fixed Portion= 58.00%), (d= D/A= Dividend Portion or Payout= 30.00%), (s= [S/S']-1= Sales Growth= 5.00%), (D= Dividend Paid= 2,370 millions USD), and (t= T/B= Tax Rate= 30.00%), are known, then it's ($'= Sales Past) must be:

$$\$' = \$ - V - I - D/\{d[1-t]\}/\{f[1+s]\}$$
$$= 51,294 - 9,746 - 513 - 2,370/\{0.3000$$
$$[1-0.3000]\}$$
$$/\{0.5800[1+0.0500]\}$$
$$= 48,851 \text{ millions USD}$$

Law-1654:

If (V= Variable Cost= 9,746 millions USD), (I= Interest Expense= 513 millions USD), (S= Sales or Revenues= 51,294 millions USD), (S'= Sales of Past Year= 48,851 millions USD), (d= D/A= Dividend Portion or Payout= 30.00%), (s= [S/S']-1= Sales Growth= 5.00%), (D= Dividend Paid= 2,370 millions USD), and (t= T/B= Tax Rate= 30.00%), are known, then it's (f= Fixed Portion planned) is:

$f = S-V-I- D/\{d[1-t]\}/\{S'[1+s]\}$
  = 51,294-9,746-513-2,370/{0.3000
     [1- 0.3000]}
     /{ 48,851[1+0.0500]}
  = 58.00%

Law-1655:

If (V= Variable Cost= 9,746 millions USD), (I= Interest Expense= 513 millions USD), (S= Sales or Revenues= 51,294 millions USD), (S'= Sales of Past Year= 48,851 millions USD), (d= D/A= Dividend Portion or Payout= 30.00%), (f= F/S= Fixed Portion= 58.00%), (D= Dividend Paid= 2,370 millions USD), and (t= T/B= Tax Rate= 30.00%), are known, then it's (s= Sales Growth planned) is:

$s = S-V-I- D/\{d[1-t]\}/[S'f]-1$
  = 51,294-9,746-513-2,370/{0.3000
     [1- 0.3000]}/[48,851*0.5800]-1
  = 5.00%

Law-1656:

If (V= Variable Cost= 9,746 millions USD), (s= [S/S']-1= Sales Growth= 5.00%), ($\$$= Sales or Revenues= 51,294 millions USD), ($\$'$= Sales of Past Year= 48,851 millions USD), (d= D/A= Dividend Portion or Payout= 30.00%), (f= F/S= Fixed Portion= 58.00%), (D= Dividend Paid= 2,370 millions USD), and (t= T/B= Tax Rate= 30.00%), are known, then it's (I= Interest Expense planned) is:

$I = \$-V-\$'f[1+s]-D/\{d[1-t]\}$
$= 51,294-9,746-48,851*0.5800]$
$[1+0.0500]-2,370$
$/\{0.3000[1-0.3000]\}$
$= 513$ millions USD

Law-1657:

If (V= Variable Cost= 9,746 millions USD), (s= [S/S']-1= Sales Growth= 5.00%), ($\$$= Sales or Revenues= 51,294 millions USD), ($\$'$= Sales of Past Year= 48,851 millions USD), (d= D/A= Dividend Portion or Payout= 30.00%), (f= F/S= Fixed Portion= 58.00%), (D= Dividend Paid= 2,370 millions USD), and (I= Interest Expense= 513 millions USD), are known, then it's (t= Tax Rate planned) is:

$t = 1-D/(d\{\$-V-I-\$'f[1+s]\})$
$= 1-2,370/(0.3000\{51,294-9,746-513$
$-48,851*0.5800][1+0.0500]\}$
$= 30.00\%$

Law-1658:

If (**V**= Variable Cost= 9,746 millions USD), (**s**= [S/S']-1= Sales Growth= 5.00%), (**$**= Sales or Revenues= 51,294 millions USD), (**$'**= Sales of Past Year= 48,851 millions USD), (**t**= T/B= Tax Rate= 30.00%), (**f**= F/S= Fixed Portion= 58.00%), (**D**= Dividend Paid= 2,370 millions USD), and (**I**= Interest Expense= 513 millions USD), are known, then it's (**d**= Dividend Portion or Payout planned) is:

$$d = D/([1-t]\{\$-V-I-\$'f[1+s]\})$$
$$= 2,370/([1-0.3000]\{51,294-9,746-513$$
$$-48,851*0.5800][1+0.0500]\}$$
$$= 30.00\%$$

Law-1659:

If (**V**= Variable Cost= 9,746 millions USD), (**s**= [S/S']-1= Sales Growth= 5.00%), (**$**= Sales or Revenues= 51,294 millions USD), (**$'**= Sales of Past Year= 48,851 millions USD), (**t**= T/B= Tax Rate= 30.00%), (**f**= F/S= Fixed Portion= 58.00%), (**d**= D/A= Dividend Portion or Payout= 30.00%), and (**I**= Interest Expense= 513 millions USD), are known, then it's (**D**= Dividend Paid planned) is:

$$D = d[1-t]\{\$-V-I-\$'f[1+s]\}$$
$$= 0.3000[1-0.3000]\{51,294-9,746-513$$
$$-48,851*0.5800][1+0.0500]\}$$
$$= 2,370 \text{ millions USD}$$

Law-1660:

If (V= Variable Cost= 9,746 millions USD), (s= [S/S']-1= Sales Growth= 5.00%), (D= Dividend Paid= 2,370 millions USD), ($\$'$= Sales of Past Year= 48,851 millions USD), (t= T/B= Tax Rate= 30.00%), (f= F/S= Fixed Portion= 58.00%), (d= D/A= Dividend Portion or Payout= 30.00%), and (i= I/S= Interest Portion= 1.00%), are known, then it's ($\$$= Sales or Revenues planned) is:

$$\$ = \$'f[1+s]+V+D/\{d[1-t]\}/[1-i]$$
$$= 48,851*0.5800][1+0.0500]+9,746$$
$$+ 2,370/\{ 0.3000[1-0.3000]\}$$
$$/[1-0.0100]$$
$$= 51,294 \text{ millions US}$$

Law-1661:

If (V= Variable Cost= 9,746 millions USD), (s= [S/S']-1= Sales Growth= 5.00%), (D= Dividend Paid= 2,370 millions USD), ($\$'$= Sales of Past Year= 48,851 millions USD), (t= T/B= Tax Rate= 30.00%), (f= F/S= Fixed Portion= 58.00%), (d= D/A= Dividend Portion or Payout= 30.00%), and ($\$$= Sales or Revenues= 51,294 millions USD), are known, then it's (i= Interest Portion planned) is:

$$i=1-(\$'f[1+s]+V+D/\{d[1-t]\}/\$$$
$$= 48,851*0.5800][1+0.0500]+9,746$$
$$+ 2,370/\{0.3000[1-0.3000]\}$$
$$/51,294$$
$$= 1.00\%$$

Law-1662:

If (**i**= I/S= Interest Portion= 1.00%), (**s**= [S/S']-1= Sales Growth= 5.00%), (**D**= Dividend Paid= 2,370 millions USD), (**$'**= Sales of Past Year= 48,851 millions USD), (**t**= T/B= Tax Rate= 30.00%), (**f**= F/S= Fixed Portion= 58.00%), (**d**= D/A= Dividend Portion or Payout= 30.00%), and (**$**= Sales or Revenues= 51,294 millions USD), are known, then it's (**V**= Variable Cost planned) is:

$V = \$[1-i] - \$'f[1+s] - D\{d[1-t]\}$
   $= 51,294[1-0.0100] - 48,851*0.5800]$
   $[1+0.0500] - 2,370/\{0.3000$
   $[1-0.3000]\}$
   $= \underline{9,746}$ milions USD

Law-1663:

If (**i**= I/S= Interest Portion= 1.00%), (**s**= [S/S']-1= Sales Growth= 5.00%), (**D**= Dividend Paid= 2,370 millions USD), (**V**= Variable Cost= 9,746 millions USD), (**t**= T/B= Tax Rate= 30.00%), (**f**= F/S= Fixed Portion= 58.00%), (**d**= D/A= Dividend Portion or Payout= 30.00%), and (**$**= Sales or Revenues= 51,294 millions USD), are known, then it's (**$'**= Sales Past) must be:

$\$' = (\$[1-i] - V - D\{d[1-t]\}) / \{f[1+s]\}$
   $= 51,294[1-0.0100] - 9,746 - 2,370$
   $/\{0.3000[1-0.3000]\}$
   $/\{0.5800[1+0.0500]\}$
   $= \underline{48,851}$ milions USD

Law-1664:
If (i= I/S= Interest Portion= 1.00%), (s= [S/S']-1= Sales Growth= 5.00%), (D= Dividend Paid= 2,370 millions USD), (V= Variable Cost= 9,746 millions USD), (t= T/B= Tax Rate= 30.00%), (S'= Sales of Past Year= 48,851 millions USD), (d= D/A= Dividend Portion or Payout= 30.00%), and (S= Sales or Revenues= 51,294 millions USD), are known, then it's (f= Fixed Portion planned) is:

$f = (S[1-i] - V - D\{d[1-t]\})/\{S'[1+s]\}$
　　= 51,294[1-0.0100]- 9,746- 2,370
　　　/{0.3000[1-0.3000]}
　　　/{48,851[1+0.0500]}
　　= 58.00%

Law-1665:
If (i= I/S= Interest Portion= 1.00%), (f= F/S= Fixed Portion= 58.00%), (D= Dividend Paid= 2,370 millions USD), (V= Variable Cost= 9,746 millions USD), (t= T/B= Tax Rate= 30.00%), (S'= Sales of Past Year= 48,851 millions USD), (d= D/A= Dividend Portion or Payout= 30.00%), and (S= Sales or Revenues= 51,294 millions USD), are known, then it's (s= Sales Growth planned) is:

$s = (S[1-i] - V - D\{d[1-t]\})/[S'f] - 1$
　　= 51,294[1-0.0100]-9,746- 2,370
　　　/{0.3000[1-0.3000]
　　　}/[48,851*0.5800]-1
　　= 5.00%

Law-1666:

If (i = I/S = Interest Portion = 1.00%), (f = F/S = Fixed Portion = 58.00%), (D = Dividend Paid = 2,370 millions USD), (V = Variable Cost = 9,746 millions USD), (s = [S/S']-1 = Sales Growth = 5.00%), (S' = Sales of Past Year = 48,851 millions USD), (d = D/A = Dividend Portion or Payout = 30.00%), and (S = Sales or Revenues = 51,294 millions USD), are known, then it's (t = Tax Rate planned) is:

t = 1-[D/d]/{S-V-$S'f$[1+s]-Si}
= 1-D/(d{S[1-i]-V-$S'f$[1+s]})
= 1-2,370/(0.3000{51,294[1-0.0100]
-9,746- 48,851*0.5800]
[1+0.0500]})
= 30.00%

Law-1667:

If (i = I/S = Interest Portion = 1.00%), (f = F/S = Fixed Portion = 58.00%), (D = Dividend Paid = 2,370 millions USD), (V = Variable Cost = 9,746 millions USD), (s = [S/S']-1 = Sales Growth = 5.00%), (S' = Sales of Past Year = 48,851 millions USD), (t = T/B = Tax Rate = 30.00%), and (S = Sales or Revenues = 51,294 millions USD), are known, then it's (d = Dividend Portion or Payout planned) is:

d = D/([1-t]/{S-V-$S'f$[1+s]-Si}
= D/([1-t]{S[1-i]-V-$S'f$[1+s]})
= 2,370/([1-0.3000]{51,294[1-0.0100]
-9,746- 48,851*0.5800]
[1+0.0500]})
= 30.00%

Law-1668:
If (**i**= I/S= Interest Portion= 1.00%), (**f**= F/S= Fixed Portion= 58.00%), (**d**= D/A= Dividend Portion or Payout= 30.00%), (**V**= Variable Cost= 9,746 millions USD), (**s**= [S/S']-1= Sales Growth= 5.00%), (**$'**= Sales of Past Year= 48,851 millions USD), (**t**= T/B= Tax Rate= 30.00%), and (**$**= Sales or Revenues= 51,294 millions USD), are known, then it's (**D**= Dividend Paid planned) is:

$$D = d[1-t]\{\$-V-\$'f[1+s]-\$'i[1+s]\}$$
$$= d[1-t]\{\$-V-\$'[1+s][f+i]\}$$
$$= 0.3000[1-0.3000]\{51,294 - 9,746$$
$$-48,851[1+0.0500]$$
$$[0.5800+0.0100]\}$$
$$= 2,370 \text{ millions USD}$$

Law-1669:
If (**i**= I/S= Interest Portion= 1.00%), (**f**= F/S= Fixed Portion= 58.00%), (**d**= D/A= Dividend Portion or Payout= 30.00%), (**V**= Variable Cost= 9,746 millions USD), (**s**= [S/S']-1= Sales Growth= 5.00%), (**$'**= Sales of Past Year= 48,851 millions USD), (**t**= T/B= Tax Rate= 30.00%), and (**D**= Dividend Paid= 2,370 millions USD), are known, then it's (**$**= Sales or Revenues planned) is:

$$\$ = V+\$'[1+s][f+i]+D/\{d[1-t]\}$$
$$= 9,746 + 48,851[1+0.0500] [0.5800$$
$$+0.0100\}+2,370$$
$$/\{0.3000[1-0.3000]\}$$
$$= 51,294 \text{ millions USD}$$

Law-1671:

If (i= I/S= Interest Portion= 1.00%), (f= F/S= Fixed Portion= 58.00%), (d= D/A= Dividend Portion or Payout= 30.00%), ($\$$= Sales or Revenues= 51,294 millions USD), (s= [S/S']-1= Sales Growth= 5.00%), (V= Variable Cost= 9,746 millions USD), (t= T/B= Tax Rate= 30.00%), and (D= Dividend Paid= 2,370 millions USD), are known, then it's ($\$'$= Sales Past) must be:

$\$' = (\$-V-D/\{d[1-t]\})/\{[1+s][f+i]\}$
    = (51,294- 9,746 -2,370/{0.3000 [1-0.3000]})/{ [1+0.0500] [0.5800+0.0100]}
    = 48,851 millions USD

Law-1672:

If (i= I/S= Interest Portion= 1.00%), (f= F/S= Fixed Portion= 58.00%), (d= D/A= Dividend Portion or Payout= 30.00%), ($\$$= Sales or Revenues= 51,294 millions USD), ($\$'$= Sales of Past Year= 48,851 millions USD), (V= Variable Cost= 9,746 millions USD), (t= T/B= Tax Rate= 30.00%), and (D= Dividend Paid= 2,370 millions USD), are known, then it's (s= Sales Growth planned) is:

$s = (\$-V-D/\{d[1-t]\})/\{\$'[f+i]\}-1$
   = (51,294-9,746 -2,370/{0.3000 [1-0.3000]})/{48,851 [0.5800+0.0100]}-1
   = 5.00%

Law-1673:

If (i= I/S= Interest Portion= 1.00%), (s= [S/S']-1= Sales Growth= 5.00%), (d= D/A= Dividend Portion or Payout= 30.00%), (S= Sales or Revenues= 51,294 millions USD), (S'= Sales of Past Year= 48,851 millions USD), (V= Variable Cost= 9,746 millions USD), (t= T/B= Tax Rate= 30.00%), and (D= Dividend Paid= 2,370 millions USD), are known, then it's (f= Fixed Cost planned) is:

$f = (S-V-D/\{d[1-t]\})/\{S'[1+s]\} - i$
$= (51,294 - 9,746 - 2,370/\{0.3000[1-0.3000]\})/\{48,851[1+0.0500]\} - 0.0100$
$= 58.00\%$

Law-1674:

If (f= F/S=Fixed Portion= 58.00%), (s= [S/S']-1= Sales Growth= 5.00%), (d= D/A= Dividend Portion or Payout= 30.00%), (S= Sales or Revenues= 51,294 millions USD), (S'= Sales of Past Year= 48,851 millions USD), (V= Variable Cost= 9,746 millions USD), (t= T/B= Tax Rate= 30.00%), and (D= Dividend Paid= 2,370 millions USD), are known, then it's (i= Interest Portion planned) is:

$i = (S-V-D/\{d[1-t]\})/\{S'[1+s]\} - f$
$= (51,294 - 9,746 - 2,370/\{0.3000[1-0.3000]\})/\{48,851[1+0.0500]\} - 0.5800$
$= 1.00\%$

Law-1675:
If (f= F/S=Fixed Portion= 58.00%), (s= [S/S']-1= Sales Growth= 5.00%), (d= D/A= Dividend Portion or Payout= 30.00%), (S= Sales or Revenues= 51,294 millions USD), (S'= Sales of Past Year= 48,851 millions USD), (V= Variable Cost= 9,746 millions USD), (i= I/S= Interest Portion= 1.00%), and (D= Dividend Paid= 2,370 millions USD), are known, then it's (t= Tax Rate planned) is:

t= 1-[D/d]{S-V-$S'f$[1+s]-$S'i$[1+s]}
  = 1-D/(d{S-V-S'[1+s][f+i]})
  = 1-2,370/(0.3000{51,294- 9,746
       -{48,851[1+0.0500]
       [0.5800+0.0100]})
  = 30.00%

Law-1676:
If (f= F/S= Fixed Portion= 58.00%), (s= [S/S']-1= Sales Growth= 5.00%), (t= T/B= Tax Rate= 30.00%), (S= Sales or Revenues= 51,294 millions USD), (S'= Sales of Past Year= 48,851 millions USD), (V= Variable Cost= 9,746 millions USD), (i= I/S= Interest Portion= 1.00%), and (D= Dividend Paid= 2,370 millions USD), are known, then it's (d= Dividend Portion or Payout planned) is:

d= D/([1-t]{S-V-$S'f$[1+s]-$S'i$[1+s]}
  = D/([1-t]{S-V-S'[1+s][f+i]})
  = 2,370/([1-0.300]{51,294- 9,746
       -{48,851[1+0.0500]
       [0.5800+0.0100]})
  = 30.00%

Law-1677:

If (**F**= Fixed Cost= 29,750 millions USD), (**t**= T/B= Tax Rate= 30.00%), (**$**= Sales or Revenues= 51,294 millions USD), (**v**= V/S= Variable Portion= 19.00%), (**I**= Interest Expense= 513 millions USD), and (**d**= D/A=Dividend Portion or Payout= 30.00%), are known, then it's (**D**= Dividend Paid planned) is:

$$D = d[1-t][\$-\$v-F-I] = d[1-t]\{\$[1-v]-F-I\}$$
$$= 0.3000[1-0.3000]\{51,294[1-0.1900] -29,750-513\}$$
$$= \underline{2,370} \text{ millions USD}$$

Law-1678:

If (**F**= Fixed Cost= 29,750 millions USD), (**t**= T/B= Tax Rate= 30.00%), (**D**= Dividend Paid= 2,370 millions USD), (**v**= V/S= Variable Portion= 19.00%), (**I**= Interest Expense= 513 millions USD), and (**d**= D/A=Dividend Portion or Payout= 30.00%), are known, then it's (**$**= Sales or Revenues planned) is:

$$\$ = (F+I+D/\{d[1-t]\}) [1-v]$$
$$= (29,750+513+2,370/\{0.3000 [1-0.3000]\})/[1-0.1900]$$
$$= \underline{51,294} \text{ millions USD}$$

Law-1679:

If (**F**= Fixed Cost= 29,750 millions USD), (**t**= T/B= Tax Rate= 30.00%), (**D**= Dividend Paid= 2,370 millions USD), (**$**= Sales or Revenues= 51,294 millions USD), (**I**= Interest Expense= 513 millions USD), and (**d**= D/A= Dividend Portion or Payout= 30.00%), are known, then it's (**v**= Variable Portion planned) is:

$$v = 1-(F+I+D/\{d[1-t]\})/\$$$
$$= 1-(29,750+513+2,370/\{0.3000[1-0.3000]\})/51,294$$
$$= 19.00\%$$

Law-1680:

If (**v**= V/S= Variable Portion= 19.00%), (**t**= T/B= Tax Rate= 30.00%), (**D**= Dividend Paid= 2,370 millions USD), (**$**= Sales or Revenues= 51,294 millions USD), (**I**= Interest Expense= 513 millions USD), and (**d**= D/A= Dividend Portion or Payout= 30.00%), are known, then it's (**F**= Fixed Cost planned) is:

$$F = \$[1-v]-I-D/\{d[1-t]\}$$
$$= 51,294[1-0.1900]-513-2,370/\{0.3000[1-0.3000]\}$$
$$= 29,750 \text{ millions USD}$$

Law-1681:

If (v= V/S= Variable Portion= 19.00%), (t= T/B= Tax Rate= 30.00%), (D= Dividend Paid= 2,370 millions USD), (S= Sales or Revenues= 51,294 millions USD), (F= Fixed Cost= 29,750 millions USD), and (d= D/A= Dividend Portion or Payout= 30.00%), are known, then it's (I= Interest Expense planned) is:

$I = S[1-v] - F - D/\{d[1-t]\}$
$= 51,294[1-0.1900] - 29,750 - 2,370$
$/\{0.3000[1-0.3000]\}$
$= \underline{513}$ millions USD

Law-1682:

If (v= V/S= Variable Portion= 19.00%), (I= Interest Expense= 513 millions USD), (D= Dividend Paid= 2,370 millions USD), (S= Sales or Revenues= 51,294 millions USD), (F= Fixed Cost= 29,750 millions USD), and (d= D/A= Dividend Portion or Payout= 30.00%), are known, then it's (t= Tax Rate planned) is:

$t = 1 - [D/d]/[S - Sv - F - I] = 1 - D/d\{S[1-v] - F - I\}$
$= 1 - 2,370/\{ 0.3000\{51,294[1-0.1900]$
$-29,750-513\}$
$= \underline{30.00}\%$

Law-1683:

If (v= V/S= Variable Portion= 19.00%), (I= Interest Expense= 513 millions USD), (D= Dividend Paid= 2,370 millions USD), (S= Sales or Revenues= 51,294 millions USD), (F= Fixed Cost= 29,750 millions USD), and (t= T/B= Tax Rate= 30.00%), are known, then it's (d= Dividend Portion or Payout planned) is:

$$d= D/([1-t][S-Sv-F-I]= D/[1-t]\{S[1-v]-F-I\}$$
$$= 2,370/\{0.3000\{51,294[1-0.1900] -29,750-513\}$$
$$= 30.00\%$$

Law-1684:

If (v= V/S= Variable Portion= 19.00%), (d= D/A= Dividend Portion or Payout= 30.00%), (i= I/S= Interest Portion= 1.00%), (S= Sales or Revenues= 51,294 millions USD), (F= Fixed Cost= 29,750 millions USD), and (t= T/B= Tax Rate= 30.00%), are known, then it's (D= Dividend Paid planned) is:

$$D= d[1-t]\{S-Sv-F-Si\}= d[1-t]\{S[1-v-i]-F\}$$
$$= 0.3000[1- 0.3000]\{51,294[1-0.1900 -0.0100]-29,750\}$$
$$= 2,370 \text{ millions USD}$$

Law-1685:
If (**D**= Dividend Paid= 2,370 millions USD), (**v**= V/S= Variable Portion= 19.00%), (**d**= D/A= Dividend Portion or Payout= 30.00%), (**i**= I/S= Interest Portion= 1.00%), (**F**= Fixed Cost= 29,750 millions USD), and (**t**= T/B= Tax Rate= 30.00%), are known, then it's (**$**= Sales or Revenues planned) is:
$$\$ = (\mathbf{F}+\mathbf{D}/\{\mathbf{d}[1-\mathbf{t}]\})/[1-\mathbf{v}-\mathbf{i}]$$
$$= (29{,}750+2{,}370/\{0.3000[1-0.3000]\}/[1-0.1900-0.0100]$$
$$= 51{,}294 \text{ millions USD}$$

Law-1686:
If (**D**= Dividend Paid= 2,370 millions USD), (**$**= Sales or Revenues= 51,294 millions USD), (**d**= D/A= Dividend Portion or Payout= 30.00%), (**i**= I/S= Interest Portion= 1.00%), (**F**= Fixed Cost= 29,750 millions USD), and (**t**= T/B= Tax Rate= 30.00%), are known, then it's (**v**= Variable Portion planned) is:
$$\mathbf{v} = 1-\mathbf{i}-(\mathbf{F}+\mathbf{D}/\{\mathbf{d}[1-\mathbf{t}]\})/\$$$
$$= 1-0.0100-(29{,}750+2{,}370/\{0.3000[1-0.3000]\}/51{,}294$$
$$= 19.00\%$$

Law-1687:

If ($= Sales or Revenues= 51,294 millions USD), (v= V/S= Variable Portion= 19.00%), (D= Dividend Paid= 2,370 millions USD), (d= D/A= Dividend Portion or Payout= 30.00%), (F= Fixed Cost= 29,750 millions USD), and (t= T/B= Tax Rate= 30.00%), are known, then it's (i= Interest Portion planned) is:

$$i = 1-v-(F+D/\{d[1-t]\})/\$$$
$$= 1-0.1900- (29,750+2,370/\{0.3000$$
$$[1- 0.3000]\}/51,294$$
$$= \underline{1.00\%}$$

Law-1688:

If ($= Sales or Revenues= 51,294 millions USD), (v= V/S= Variable Portion= 19.00%), (D= Dividend Paid= 2,370 millions USD), (d= D/A= Dividend Portion or Payout= 30.00%), (i= I/S= Interest Portion= 1.00%), and (t= T/B= Tax Rate= 30.00%), are known, then it's (F= Fixed Cost planned) is:

$$F = \$[1-v-i]-D/\{d[1-t]\}$$
$$= 51,294[1-0.1900-0.0100]-2,370$$
$$/\{0.3000[1- 0.3000]\}$$
$$= \underline{29,750} \text{ millions USD}$$

Law-1689:

If ($\$$= Sales or Revenues= 51,294 millions USD), (v= V/S= Variable Portion= 19.00%), (D= Dividend Paid= 2,370 millions USD), (d= D/A= Dividend Portion or Payout= 30.00%), (i= I/S= Interest Portion= 1.00%), and (F= Fixed Cost= 29,750 millions USD), are known, then it's (t= Tax Rate planned) is:

t = 1-[D/d]/[$\$$-$\$v$-$F$-$\$i$]= 1-$D$/($d${$\$$[1-v-i]-F})
= 1-2,370/{0.3000{51,294[1-0.1900
-0.0100]-29,750})\
= 30.00%

Law-1690:

If ($\$$= Sales or Revenues= 51,294 millions USD), (v= V/S= Variable Portion= 19.00%), (D= Dividend Paid= 2,370 millions USD), (t= T/B= Tax Rate= 30.00%), (i= I/S= Interest Portion= 1.00%), and (F= Fixed Cost= 29,750 millions USD), are known, then it's (d= Dividend Portion or Payout planned) is:

d = D/[1-t]{$\$$-$\$v$-$F$-$\$i$}= D/([1-t]{$\$$[1-v-i]-F})
= 2,370/([1-0.3000]{51,294[1-0.1900
-0.0100]- 29,750})
= 30.00%

Law-1691:

If ($= Sales or Revenues= 51,294 millions USD), (v= V/S= Variable Portion= 19.00%), (t= T/B= Tax Rate= 30.00%), ($'= Sales of Past Year= 48,851 millions USD), (s= [S/S']-1= Sales Growth= 5.00%), (d= D/A= Dividend Portion or Payout= 30.00%), (i= I/S= Interest Portion= 1.00%), and (F= Fixed Cost= 29,750 millions USD), are known, then it's (D= Dividend Paid planned) is:

$$D = d[1-t]\{\$-\$v-F-\$'i[1+s]$$
$$= d[1-t]\{\$[1-v]-F-\$'i[1+s]$$
$$= 0.3000[1-0.3000]\{51,294[1-0.1900]$$
$$-29,750 -48,851*0.0100$$
$$[1+0.0500]\}$$
$$= 2,370 \text{ millions USD}$$

Law-1692:

If (D= Dividend Paid= 2,370 millions USD), (v= V/S= Variable Portion= 19.00%), (t= T/B= Tax Rate= 30.00%), ($'= Sales of Past Year= 48,851 millions USD), (s= [S/S']-1= Sales Growth= 5.00%), (d= D/A= Dividend Portion or Payout= 30.00%), (i= I/S= Interest Portion= 1.00%), and (F= Fixed Cost= 29,750 millions USD), are known, then it's ($= Sales or Revenues planned) is:

$$\$ = F+\$'i[1+s]+D/\{d[1-t]\}/[1-v]$$
$$= (29,750 +48,851*0.0100[1+0.0500]$$
$$+2,370/\{0.3000[1-0.3000]\}$$
$$/[1-0.1900]$$
$$= 51,294 \text{ millions USD}$$

Law-1693:
If (**D**= Dividend Paid= 2,370 millions USD), (**$**= Sales or Revenues= 51,294 millions USD), (**t**= T/B= Tax Rate= 30.00%), (**$'**= Sales of Past Year= 48,851 millions USD), (**s**= [S/S']-1= Sales Growth= 5.00%), (**d**= D/A= Dividend Portion or Payout= 30.00%), (**i**= I/S= Interest Portion= 1.00%), and (**F**= Fixed Cost= 29,750 millions USD), are known, then it's (**v**= Sales or Revenues planned) is:

$$v = 1-(F+\$'i[1+s]+D/\{d[1-t]\})/\$$$
$$= 1-(29,750 +48,851*0.0100[1+0.0500]$$
$$+2,370/\{0.3000[1-0.3000]\}$$
$$/ 51,294$$
$$= 19.00\%$$

Law-1694:
If (**D**= Dividend Paid= 2,370 millions USD), (**$**= Sales or Revenues= 51,294 millions USD), (**t**= T/B= Tax Rate= 30.00%), (**$'**= Sales of Past Year= 48,851 millions USD), (**s**= [S/S']-1= Sales Growth= 5.00%), (**d**= D/A= Dividend Portion or Payout= 30.00%), (**i**= I/S= Interest Portion= 1.00%), and (**v**= V/S= Variable Portion= 19.00%), are known, then it's (**F**= Fixed Cost planned) is:

$$F = \$[1-v]-\$'i[1+s]-D/\{d[1-t]\}$$
$$= 51,294[1-0.1900]-48,851*0.0100$$
$$[1+0.0500]-2,370/\{0.3000$$
$$[1-0.3000]\}$$
$$= 29,750 \text{ millions USD}$$

Law-1695:

If (**D**= Dividend Paid= 2,370 millions USD), (**$**= Sales or Revenues= 51,294 millions USD), (**t**= T/B= Tax Rate= 30.00%), (**F**= Fixed Cost= 29,750 millions USD), (**s**= [S/S']-1= Sales Growth= 5.00%), (**d**= D/A= Dividend Portion or Payout= 30.00%), (**i**= I/S= Interest Portion= 1.00%), and (**v**= V/S= Variable Portion= 19.00%), are known, then it's (**$'**= Sales Past) must be:

$$\$' = \$[1-v]\text{-}F\text{-}D/\{d[1-t]\}/\{i[1+s]\}$$
$$= 51{,}294[1-0.1900]-29{,}750-2{,}370$$
$$/\{0.3000[1-0.3000]\}$$
$$/\{0.0100[1+0.0500]\}$$
$$= 48{,}851 \text{ millions USD}$$

Law-1696:

If (**D**= Dividend Paid= 2,370 millions USD), (**$**= Sales or Revenues= 51,294 millions USD), (**t**= T/B= Tax Rate= 30.00%), (**F**= Fixed Cost= 29,750 millions USD), (**s**= [S/S']-1= Sales Growth= 5.00%), (**d**= D/A= Dividend Portion or Payout= 30.00%), (**$'**= Sales of Past Year= 48,851 millions USD), and (**v**= V/S= Variable Portion= 19.00%), are known, then it's (**i**= Interest Portion planned) is:

$$i = (\$[1-v]\text{-}F\text{-}D/\{d[1-t]\})/\{\$'[1+s]\}$$
$$= 51{,}294[1-0.1900]-29{,}750-2{,}370$$
$$/\{0.3000[1-0.3000]\}$$
$$/\{48{,}851[1+0.0500]\}$$
$$= 1.00\%$$

Law-1697:

If (**D**= Dividend Paid= 2,370 millions USD), (**$**= Sales or Revenues= 51,294 millions USD), (**t**= T/B= Tax Rate= 30.00%), (**F**= Fixed Cost= 29,750 millions USD), (**i**= I/S= Interest Portion= 1.00%), (**d**= D/A= Dividend Portion or Payout= 30.00%), (**$'**= Sales of Past Year= 48,851 millions USD), and (**v**= V/S= Variable Portion= 19.00%), are known, then it's (**s**= Sales Growth planned) is:

$$s = (\$[1-v]-F-D/\{d[1-t]\}/[\$'i] -1$$
$$= 51,294[1-0.1900]-29,750-2,370$$
$$/\{0.3000[1-0.3000]\}$$
$$/[48,851*0.0100]-1$$
$$= 5.00\%$$

Law-1698:

If (**D**= Dividend Paid= 2,370 millions USD), (**$**= Sales or Revenues= 51,294 millions USD), (**s**= [S/S']-1= Sales Growth= 5.00%), (**F**= Fixed Cost= 29,750 millions USD), (**i**= I/S= Interest Portion= 1.00%), (**d**= D/A= Dividend Portion or Payout= 30.00%), (**$'**= Sales of Past Year= 48,851 millions USD), and (**v**= V/S= Variable Portion= 19.00%), are known, then it's (**t**= Tax Rate planned) is:

$$t = 1-[D/d]\{\$-\$v-F-\$'i[1+s]\}$$
$$= 1-D/(d\{\$[1-v]-F-\$'i[1+s]\})$$
$$= 1-2,370/(0.3000\{51,294[1-0.1900]$$
$$-29,750- 48,851*0.0100$$
$$[1+0.0500]\})$$
$$= 30.00\%$$

Law-1699:
If (**D**= Dividend Paid= 2,370 millions USD), (**$**= Sales or Revenues= 51,294 millions USD), (**s**= [S/S']-1= Sales Growth= 5.00%), (**F**= Fixed Cost= 29,750 millions USD), (**i**= I/S= Interest Portion= 1.00%), (**t**= T/B= Tax Rate= 30.00%), (**$'**= Sales of Past Year= 48,851 millions USD), and (**v**= V/S= Variable Portion= 19.00%), are known, then it's (**d**= Dividend Portion planned) is:

$$d = D/([1-t]\{\$-\$v-F-\$'i[1+s]\})$$
$$= D/([1-t]\{\$[1-v]-F-\$'i[1+s]\})$$
$$= 2{,}370/([1-0.3000]\{51{,}294\,[1-0.1900] -29{,}750 - 48{,}851 * 0.0100 \,[1+0.0500]\})$$
$$= 30.00\%$$

Law-1700:
If (**d**= D/A= Dividend Portion or Payout= 30.00%), (**$**= Sales or Revenues= 51,294 millions USD), (**f**= F/S= Fixed Portion= 58.00%), (**I**= Interest Expense= 513 millions USD), (**t**= T/B= Tax Rate= 30.00%), and (**v**= V/S= Variable Portion= 19.00%), are known, then it's (**D**= Dividend Paid planned) is:

$$D = d[1-t][\$-\$v-\$f-I] = d[1-t]\{\$[1-v-f]-I\}$$
$$= 0.3000[1-0.3000]\{51{,}294\,[1-0.1900 -0.5800] - 513\}$$
$$= 2{,}370 \text{ millions USD}$$

Law-1701:

If (d= D/A=Dividend Portion or Payout= 30.00%), (D= Dividend Paid= 2,370 millions USD), (f= F/S= Fixed Portion= 58.00%), (I= Interest Expense= 513 millions USD), (t= T/B= Tax Rate= 30.00%), and (v= V/S= Variable Portion= 19.00%), are known, then it's ($\$$= Sales or Revenues planned) is:

$$\$ = I+D(d[1-t])\}/[1-v-f]$$
$$= 513+2,370(0.3000[1-0.3000])\}$$
$$/[1-0.1900-0.5800]$$
$$= 51,294 \text{ millions USD}$$

Law-1702:

If (d= D/A=Dividend Portion or Payout= 30.00%), (D= Dividend Paid= 2,370 millions USD), (f= F/S= Fixed Portion= 58.00%), (I= Interest Expense= 513 millions USD), (t= T/B= Tax Rate= 30.00%), and ($\$$= Sales or Revenues= 51,294 millions USD), are known, then it's (v= Variable Portion planned) is:

$$v = 1-f-(I+D(d[1-t])\})/\$$$
$$= 1-0.5800- (513+2,370(0.3000$$
$$[1-0.3000])\})/51,294$$
$$= 19.00\%$$

Law-1703:
   If (**d**= D/A=Dividend Portion or Payout= 30.00%), (**D**= Dividend Paid= 2,370 millions USD), (**v**= V/S= Variable Portion= 19.00%), (**I**= Interest Expense= 513 millions USD), (**t**= T/B= Tax Rate= 30.00%), and (**$**= Sales or Revenues= 51,294 millions USD), are known, then it's (**f**= Fixed Portion planned) is:
   f= 1-**v**-(**I**+**D**(**d**[1-**t**]})/**$**
   = 1-0.1900-(513+2,370(0.3000
   [1-0.3000]0.3000]})/51,294
   = 58.00%

Law-1704:
   If (**d**= D/A=Dividend Portion or Payout= 30.00%), (**D**= Dividend Paid= 2,370 millions USD), (**v**= V/S= Variable Portion= 19.00%), (**f**= F/S=Fixed Portion= 58.00%), (**t**= T/B= Tax Rate= 30.00%), and (**$**= Sales or Revenues= 51,294 millions USD), are known, then it's (**I**= Interest Expense planned) is:
   I= **$**[1-**v**-**f**]-**D**/{**d**[1-**t**]}
   = 51,294[1-0.1900-0.5800]-2,370
   /{0.3000[1-0.3000]}
   = 513 millions USD

Law-1705:
If (**d**= D/A=Dividend Portion or Payout= 30.00%), (**D**= Dividend Paid= 2,370 millions USD), (**v**= V/S= Variable Portion= 19.00%), (**f**= F/S= Fixed Portion= 58.00%), (**I**= Interest Expense= 513 millions USD), and (**$**= Sales or Revenues= 51,294 millions USD), are known, then it's (**t**= Tax Rate planned) is:

$$t = 1-[D/d]/[\$-\$v-\$f-I] = 1-D/(d\{\$[1-v-f]-I\})$$
$$= 1-2,370/(0.3000\{51,294[1-0.1900 -0.5800]-513\})$$
$$= 30.00\%$$

Law-1706:
If (**t**= T/B= Tax Rate= 30.00%), (**d**= D/A=Dividend Portion or Payout= 30.00%), (**v**= V/S= Variable Portion= 19.00%), (**f**= F/S= Fixed Portion= 58.00%), (**i**= I/S= Interest Portion= 1.00%), and (**$**= Sales or Revenues= 51,294 millions USD), are known, then it's (**D**= Dividend Pid planned) is:

$$D = d[1-t][\$-\$v-\$f-\$i]$$
$$= D[1-t]\{\$[1-v-f-i]\})$$
$$= 0.3000 \ [1-0.3000]\{51,294[\ 1-0.1900 -0.5800-0.0100]\})$$
$$= 2,370 \text{ millions USD}$$

Law-1707:
If ($t=$ T/B= Tax Rate= 30.00%), ($d=$ D/A=Dividend Portion or Payout= 30.00%), ($v=$ V/S= Variable Portion= 19.00%), ($f=$ F/S= Fixed Portion= 58.00%), ($i=$ I/S= Interest Portion= 1.00%), and ($S=$ Sales or Revenues= 51,294 millions USD), are known, then it's ($D=$ Dividend Paid planned) is:

$D = d[1-t][S-Sv-Sf-Si] = D[1-t]\{S[1-v-f-i]\}$
 = 2,370 [1-0.3000]{51,294[1-0.1900
    -0.5800-0.0100]})
 = 2,370 millions USD

Law-1708:
If ($t=$ T/B= Tax Rate= 30.00%), ($d=$ D/A=Dividend Portion or Payout= 30.00%), ($v=$ V/S= Variable Portion= 19.00%), ($f=$ F/S= Fixed Portion= 58.00%), ($i=$ I/S= Interest Portion= 1.00%), and ($D=$ Dividend Paid= 2,370 millions USD), are known, then it's ($S=$ Sales or Revenues planned) is:

$S = D/\{d[1-t][1-v-f-i]\}$
 = 2,370/{0.3000 [1-0.3000] [1-0.1900
    -0.5800-0.0100]
 = 51,294 millions USD

Finance Construction-2, *Tim Asikin, Steve Asikin, Indra Senihardja*

Law-1709:
If (t= T/B= Tax Rate= 30.00%), (d= D/A=Dividend Portion or Payout= 30.00%), ($\$$= Sales or Revenues= 51,294 millions USD), (f= F/S= Fixed Portion= 58.00%), (i= I/S= Interest Portion= 1.00%), and (D= Dividend Paid= 2,370 millions USD), are known, then it's (v= Variable Portion planned) is:

v= 1-f-i-D/{$d\$$[1-t]}
 = 1-0.5800-0.0100-2,370/{0.3000
      *51,294[1-0.3000]}
 = 19.00%

Law-1710:
If (t= T/B= Tax Rate= 30.00%), (d= D/A=Dividend Portion or Payout= 30.00%), ($\$$= Sales or Revenues= 51,294 millions USD), (v= V/S= Variable Portion= 19.00%), (i= I/S= Interest Portion= 1.00%), and (D= Dividend Paid= 2,370 millions USD), are known, then it's (f= Fixed Portion planned) is:

f=1-v-i-D/{$d\$$[1-t]}
 = 1-0.1900-0.0100-2,370/{0.3000
      *51,294[1-0.3000]}
 = 58.00%

Law-1711:

If (t= T/B= Tax Rate= 30.00%), (d= D/A=Dividend Portion or Payout= 30.00%), ($\$$= Sales or Revenues= 51,294 millions USD), (v= V/S= Variable Portion= 19.00%), (f= F/S= Fixed Portion= 58.00%), and (D= Dividend Paid= 2,370 millions USD), are known, then it's (i= Interest Portion planned) is:

$i = 1 - f - v - D/\{d\$[1-t]\}$
$\quad = 1 - 0.5800 - 0.1900 - 2,370/\{0.3000$
$\quad\quad\quad *51,294[1-0.3000]\}$
$\quad = 1.00\%$

Law-1712:

If (i= I/S= Interest Portion= 1.00%), (d= D/A= Dividend Portion or Payout= 30.00%), ($\$$= Sales or Revenues= 51,294 millions USD), (v= V/S= Variable Portion= 19.00%), (f= F/S= Fixed Portion= 58.00%), and (D= Dividend Paid= 2,370 millions USD), are known, then it's (t= Tax Rate planned) is:

$t = 1 - [D/d]/[\$ - \$v - \$f - \$i] = 1 - D/\{d\$[1-v-f-i]\}$
$\quad = 1 - 2,370/\{0.3000*51,294[1-0.1900$
$\quad\quad\quad -0.5800-0.0100]\}$
$\quad = 30.00\%$

Law-1713:

If (i= I/S= Interest Portion= 1.00%), (t= T/B= Tax Rate= 30.00%), (S= Sales or Revenues= 51,294 millions USD), (v= V/S= Variable Portion= 19.00%), (f= F/S= Fixed Portion= 58.00%), and (D= Dividend Paid= 2,370 millions USD), are known, then it's (d= Dividend Portion planned) is:

d= $D/[1-t][S-Sv-Sf-Si]$= $D/\{S[1-t][1-v-f-i]\}$
 = 2,370/{51,294[1-0.3000][1-0.1900
   -0.5800-0.0100]}
 = 30.00%

Law-1714:

If (i= I/S= Interest Portion= 1.00%), (t= T/B= Tax Rate= 30.00%), (S= Sales or Revenues= 51,294 millions USD), (v= V/S= Variable Portion= 19.00%), (f= F/S= Fixed Portion= 58.00%), and (d= D/A= Dividend Portion or Payout= 30.00%), are known, then it's (D= Dividend Paid planned) is:

D= $d[1-t][S-Sv-Sf-Si]$= $d[1-t]\{S[1-v-f]-S'i[1+s]\}$
 = 0.3000[1-0.3000]{51,294[1-0.1900
   -0.5800]-48,851*0.0100
   [1+0.0500]}
 = 2,370 millions USD

Law-1715:

If (**i**= I/S= Interest Portion= <u>1.00</u>%), (**t**= T/B= Tax Rate= <u>30.00</u>%), (**D**= Dividend Paid= <u>2,370</u> millions USD), (**v**= V/S= Variable Portion= <u>19.00</u>%), (**f**= F/S= Fixed Portion= <u>58.00</u>%), and (**d**= D/A= Dividend Portion or Payout= <u>30.00</u>%), are known, then it's (**$**= Sales or Revenues planned) is:

$$\$ = (\$'i[1+s]+D/\{d[1-t]\})/[1-v-f]$$
$$= (48,851*0.0100[1+0.0500]+2,370$$
$$/\{0.3000[1-0.3000]\})$$
$$/[1-0.1900-0.5800]$$
$$= \underline{51,294} \text{ millions USD}$$

Law-1716:

If (**i**= I/S= Interest Portion= <u>1.00</u>%), (**t**= T/B= Tax Rate= <u>30.00</u>%), (**D**= Dividend Paid= <u>2,370</u> millions USD), (**$**= Sales or Revenues= <u>51,294</u> millions USD), (**f**= F/S= Fixed Portion= <u>58.00</u>%), and (**d**= D/A= Dividend Portion or Payout= <u>30.00</u>%), are known, then it's (**v**=Variable Portion planned) is:

$$v = 1-f-(\$'i[1+s]+D/\{d[1-t]\})/\$$$
$$= 1-0.5800-(48,851*0.0100[1+0.0500]$$
$$+2,370/\{0.3000[1-0.3000]\})$$
$$/51,294$$
$$= \underline{19.00}\%$$

Law-1717:

If (**i**= I/S= Interest Portion= 1.00%), (**t**= T/B= Tax Rate= 30.00%), (**D**= Dividend Paid= 2,370 millions USD), (**$**= Sales or Revenues= 51,294 millions USD), (**v**= V/S= Variable Portion= 19.00%), and (**d**= D/A= Dividend Portion or Payout= 30.00%), are known, then it's (**f**= Fixed Portion planned) is:

$$f = 1-v-(\$'i[1+s]+D/\{d[1-t]\})/\$$$
$$= 1-0.1900-(48,851*0.0100[1+0.0500]$$
$$+2,370/\{0.3000[1-0.3000]\})$$
$$/51,294$$
$$= 58.00\%$$

Law-1718:

If (**i**= I/S= Interest Portion= 1.00%), (**t**= T/B= Tax Rate= 30.00%), (**D**= Dividend Paid= 2,370 millions USD), (**$**= Sales or Revenues= 51,294 millions USD), (**f**= F/S= Fixed Portion= 58.00%), (**s**= [S/S']-1= Sales Growth= 5.00%), (**v**= V/S= Variable Portion= 19.00%), and (**d**= D/A= Dividend Portion or Payout= 30.00%), are known, then it's (**$'**=Sales Past) must be:

$$\$' = (\$[1-v-f]-D/\{d[1-t]\})/\{i[1+s]\}$$
$$= (51,294[1-0.1900-0.5800]-2,370$$
$$/\{0.3000[1-0.3000]\})$$
$$/\{0.0100[1+0.0500]\}$$
$$= 48,851 \text{ millions USD}$$

Law-1719:

If ($'=$ Sales of Past Year= 48,851 millions USD), (**t**= T/B= Tax Rate= 30.00%), (**D**= Dividend Paid= 2,370 millions USD), ($=$ Sales or Revenues= 51,294 millions USD), (**f**= F/S= Fixed Portion= 58.00%), (**s**= [S/S']-1= Sales Growth= 5.00%), (**v**= V/S= Variable Portion= 19.00%), and (**d**= D/A= Dividend Portion or Payout= 30.00%), are known, then it's (**i**=Interest Portion planned) is:

$$i = (\$[1-v-f]-D/\{d[1-t]\})/\{\$'[1+s]\}$$
$$= (51,294[1-0.1900-0.5800]-2,370$$
$$/\{0.3000[1-0.3000]\})$$
$$/\{48,851[1+0.0500]\}$$
$$= 1.00\%$$

Law-1720:

If ($'=$ Sales of Past Year= 48,851 millions USD), (**t**= T/B= Tax Rate= 30.00%), (**D**= Dividend Paid= 2,370 millions USD), ($=$ Sales or Revenues= 51,294 millions USD), (**f**= F/S= Fixed Portion= 58.00%), (**i**= I/S= Interest Portion= 1.00%), (**v**= V/S= Variable Portion= 19.00%), and (**d**= D/A= Dividend Portion or Payout= 30.00%), are known, then it's (**s**= Sales Growth Planned) is:

$$s = (\$[1-v-f]-D/\{d[1-t]\})/[\$'i]-1$$
$$= (51,294[1-0.1900-0.5800]-2,370$$
$$/\{0.3000[1-0.3000]\})$$
$$/[48,851*0.0100]-1$$
$$= 5.00\%$$

Law-1721:

If ($\$'$= Sales of Past Year= 48,851 millions USD), ($\$$= [S/S']-1= Sales Growth= 5.00%), (**D**= Dividend Paid= 2,370 millions USD), ($\$$= Sales or Revenues= 51,294 millions USD), (**f**= F/S= Fixed Portion= 58.00%), (**i**= I/S= Interest Portion= 1.00%), (**v**= V/S= Variable Portion= 19.00%), and (**d**= D/A= Dividend Portion or Payout= 30.00%), are known, then it's (**t**= Tax Rate Planned) is:

$$t = 1-[D/d]/\{\$-\$v-\$f-\$'i[1+s]\}$$
$$= 1-D/(d\{\$[1-v-f]- \$'i[1+s]\})$$
$$= 1-2,370/(0.3000\{51,294[1-0.1900 -0.5800]- 48,851*0.0100 [1+0.0500]\})$$
$$= 30.00\%$$

Law-1722:

If ($\$'$= Sales of Past Year= 48,851 millions USD), ($\$$= [S/S']-1= Sales Growth= 5.00%), (**D**= Dividend Paid= 2,370 millions USD), ($\$$= Sales or Revenues= 51,294 millions USD), (**f**= F/S= Fixed Portion= 58.00%), (**i**= I/S= Interest Portion= 1.00%), (**v**= V/S= Variable Portion= 19.00%), and (**t**= T/B= Tax Rate= 30.00%), are known, then it's (**d**= Dividend Portion Planned) is:

$$d = D/[1-t]\{\$-\$v-\$f-\$'i[1+s]\}$$
$$= D/([1-t]\{\$[1-v-f]-\$'i[1+s]\})$$
$$= 2,370/([1-0.3000]\{51,294[1-0.1900 -0.5800]- 48,851*0.0100 [1+0.0500]\})$$
$$= 30.00\%$$

Law-1723:

If (S'= Sales of Past Year= 48,851 millions USD), (s= [S/S']-1= Sales Growth= 5.00%), (d= D/A= Dividend Portion or Payout= 30.00%), (S= Sales or Revenues= 51,294 millions USD), (f= F/S= Fixed Portion= 58.00%), (I= Interest Expense= 513 millions USD), (v= V/S= Variable Portion= 19.00%), and (t= T/B= Tax Rate= 30.00%), are known, then it's (D= Dividend Paid Planned) is:

$$D = d[1-t]\{S-Sv-S'f[1+s]-I\}$$
$$= d[1-t]\{S[1-v]-S'f[1+s]-I\})$$
$$= 0.3000[1-0.3000]\{51,294[1-0.1900]$$
$$-48,851*0.5800[1+0.0500]-513\}$$
$$= 2,370 \text{ millions USD}$$

Law-1724:

If (S'= Sales of Past Year= 48,851 millions USD), (s= [S/S']-1= Sales Growth= 5.00%), (d= D/A= Dividend Portion or Payout= 30.00%), (D= Dividend Paid= 2,370 millions USD), (f= F/S= Fixed Portion= 58.00%), (I= Interest Expense= 513 millions USD), (v= V/S= Variable Portion= 19.00%), and (t= T/B= Tax Rate= 30.00%), are known, then it's (S=Sales or Revenues Planned) is:

$$S = I+S'f[1+s]+D/\{d[1-t]\}/[1-v]\}$$
$$= 513+48,851*0.5800[1+0.0500]+2,370$$
$$/\{0.3000[1-0.3000]\}/[1-0.1900]$$
$$= 51,294 \text{ millions USD}$$

Law-1725:
If ($\$'$= Sales of Past Year= 48,851 millions USD), ($\bf{s}$= [S/S']-1= Sales Growth= 5.00%), ($\bf{d}$= D/A= Dividend Portion or Payout= 30.00%), ($\bf{D}$= Dividend Paid= 2,370 millions USD), ($\bf{f}$= F/S= Fixed Portion= 58.00%), ($\bf{I}$= Interest Expense= 513 millions USD), ($\$$= Sales or Revenues= 51,294 millions USD), and ($\bf{t}$= T/B= Tax Rate= 30.00%), are known, then it's ($\bf{v}$=Variable Cost Planned) is:

$\bf{v}$= 1-($\bf{I}$+$\$'\bf{f}$[1+$\bf{s}$]+$\bf{D}$/{$\bf{d}$[1-$\bf{t}$]})/$\$$
= 1- (513+48,851*0.5800[1+0.0500]
+2,370/{0.3000[1-0.3000]})
/51,294
= 19.00%

Law-1726:
If ($\bf{v}$= V/S= Variable Portion= 19.00%), ($\bf{s}$= [S/S']-1= Sales Growth= 5.00%), ($\bf{d}$= D/A= Dividend Portion or Payout= 30.00%), ($\bf{D}$= Dividend Paid= 2,370 millions USD), ($\bf{f}$= F/S= Fixed Portion= 58.00%), ($\bf{I}$= Interest Expense= 513 millions USD), ($\$$= Sales or Revenues= 51,294 millions USD), and ($\bf{t}$= T/B= Tax Rate= 30.00%), are known, then it's ($\$'$= Sales Past) must be:

$\$'$= ($\$$[1-$\bf{v}$]-$\bf{I}$-$\bf{D}$/{$\bf{d}$[1-$\bf{t}$]})/{$\bf{f}$[1+$\bf{s}$]}
= 51,294[1-0.1900] -513-2,370/{ 0.3000
[1-0.3000]}/{0.5800[1+0.0500]}
= 48,851 millions USD

Law-1727:

If (v= V/S= Variable Portion= 19.00%), (s= [S/S']-1= Sales Growth= 5.00%), (d= D/A= Dividend Portion or Payout= 30.00%), (D= Dividend Paid= 2,370 millions USD), (S'= Sales of Past Year= 48,851 millions USD), (I= Interest Expense= 513 millions USD), (S= Sales or Revenues= 51,294 millions USD), and (t= T/B= Tax Rate= 30.00%), are known, then it's (f= Fixed Portion planned) is:

$f = (S[1-v]-I-D/\{d[1-t]\})/\{S'[1+s]\}$
  = 51,294[1-0.1900] -513-2,370/{0.3000
       [1-0.3000]}/{48,851[1+0.0500]}
  = 58.00%

Law-1728:

If (v= V/S= Variable Portion= 19.00%), (f= F/S=Fixed Portion= 58.00%), (d= D/A= Dividend Portion or Payout= 30.00%), (D= Dividend Paid= 2,370 millions USD), (S'= Sales of Past Year= 48,851 millions USD), (I= Interest Expense= 513 millions USD), (S= Sales or Revenues= 51,294 millions USD), and (t= T/B= Tax Rate= 30.00%), are known, then it's (s= Sales Growth planned) is:

$s = (S[1-v]-I-D/\{d[1-t]\})/[S'f]-1$
  = 51,294[1-0.1900] -513-2,370/{0.3000
       [1-0.3000]}/[48,851*0.5800] -1
  = 5.00%

Law-1729:

If (v= V/S= Variable Portion= 19.00%), (f= F/S=Fixed Portion= 58.00%), (d= D/A= Dividend Portion or Payout= 30.00%), (D= Dividend Paid= 2,370 millions USD), ($\$'$= Sales of Past Year= 48,851 millions USD), (s= [S/S']-1= Sales Growth= 5.00%), ($\$$= Sales or Revenues= 51,294 millions USD), and (t= T/B= Tax Rate= 30.00%), are known, then it's (I=Interest Expense planned) is:

I = $\$[1-v]-[\$'f[1+s]-D/\{d[1-t]\}$
  = 51,294[1-0.1900]-48,851*0.5800
    [1+0.0500] -2,370/{ 0.3000
    [1-0.3000]}
  = 513 millions USD

Law-1730:

If (v= V/S= Variable Portion= 19.00%), (f= F/S= Fixed Portion= 58.00%), (d= D/A= Dividend Portion or Payout= 30.00%), (D= Dividend Paid= 2,370 millions USD), ($\$'$= Sales of Past Year= 48,851 millions USD), (s= [S/S']-1= Sales Growth= 5.00%), ($\$$= Sales or Revenues= 51,294 millions USD), and (I= Interest Expense= 513 millions USD), are known, then it's (t= Interest Expense planned) is:

t= 1-[D/d]/{$\$$-$\$v$-$\$'f$[1+s]-$I$}
  = 1-D/(d{$\$$[1-v]-I-$\$'f$[1+s]]})
  = 1-2,370/(0.3000{ 51,294[1-0.1900]
    -513-48,851*0.5800
    [1+0.0500]})
  = 30.00%

Law-1731:

If (v= V/S= Variable Portion= 19.00%), (f= F/S=Fixed Portion= 58.00%), (t= T/B= Tax Rate= 30.00%), (D= Dividend Paid= 2,370 millions USD), (S'= Sales of Past Year= 48,851 millions USD), (s= [S/S']-1= Sales Growth= 5.00%), (S= Sales or Revenues= 51,294 millions USD), and (I= Interest Expense= 513 millions USD), are known, then it's (d= Dividend Portion planned) is:

$d = D/([1-t]\{S-Sv-S'f[1+s]-I\}$
$\quad = D/([1-t]\{S[1-v]-I-S'f[1+s]]\})$
$\quad = 2,370/([1- 0.3000]\{ 51,294[1-0.1900]$
$\qquad -513-48,851*0.5800$
$\qquad [1+0.0500]\})$
$\quad = 30.00\%$

Law-1732:

If (v= V/S= Variable Portion= 19.00%), (f= F/S=Fixed Portion= 58.00%), (t= T/B= Tax Rate= 30.00%), (d= D/A= Dividend Portion or Payout= 30.00%), (S'= Sales of Past Year= 48,851 millions USD), (s= [S/S']-1= Sales Growth= 5.00%), (S= Sales or Revenues= 51,294 millions USD), and (i= I/S= Interest Portion= 1.00%), are known, then it's (D= Dividend Paid planned) is:

$D = d[1-t]\{S-Sv-S'f[1+s]-Si\}$
$\quad = d[1-t]\{S[1-v-i]-S'f[1+s]]\}$
$\quad = 0.3000[1- 0.3000]\{ 51,294[1-0.1900$
$\qquad -0.0100]-48,851*0.5800$
$\qquad [1+0.0500]\}$
$\quad = 2,370$ millions USD

Law-1733:

If (v= V/S= Variable Portion= 19.00%), (f= F/S=Fixed Portion= 58.00%), (t= T/B= Tax Rate= 30.00%), (d= D/A= Dividend Portion or Payout= 30.00%), (S'= Sales of Past Year= 48,851 millions USD), (s= [S/S']-1= Sales Growth= 5.00%), (D= Dividend Paid= 2,370 millions USD), and (i= I/S= Interest Portion= 1.00%), are known, then it's (S= Sales or Revenues planned) is:

$S = (S'f[1+s]+D/\{d[1-t]\})/[1-v-i]$
  = (48,851*0.5800[1+0.0500]}+2,370
    /{0.3000[1- 0.3000]}
    [1-0.1900-0.0100]
  = 51,294 millions USD

Law-1734:

If (S= Sales or Revenues= 51,294 millions USD), (f= F/S= Fixed Portion= 58.00%), (t= T/B= Tax Rate= 30.00%), (d= D/A= Dividend Portion or Payout= 30.00%), (S'= Sales of Past Year= 48,851 millions USD), (s= [S/S']-1= Sales Growth= 5.00%), (D= Dividend Paid= 2,370 millions USD), and (i= I/S= Interest Portion= 1.00%), are known, then it's (v= Variable Portion planned) is:

$v = 1-i- (S'f[1+s]+D/\{d[1-t]\})/S$
  = 1-0.0100-(48,851*0.5800[1+0.0500]}
    +2,370/{0.3000[1- 0.3000]}
    /51,294
  = 19.00%

Law-1735:

If ($= Sales or Revenues= 51,294 millions USD), (f= F/S= Fixed Portion= 58.00%), (t= T/B= Tax Rate= 30.00%), (d= D/A= Dividend Portion or Payout= 30.00%), ($'= Sales of Past Year= 48,851 millions USD), (s= [S/S']-1= Sales Growth= 5.00%), (D= Dividend Paid= 2,370 millions USD), and (v= V/S= Variable Portion= 19.00%), are known, then it's (i= Interest Portion planned) is:

$$i = 1-v-(\$'f[1+s]+D/\{d[1-t]\})/\$$$
$$= 1-0.1900-(48,851*0.5800[1+0.0500]$$
$$+2,370/\{0.3000[1-0.3000]\}$$
$$/51,294$$
$$= 1.00\%$$

Law-1736:

If ($= Sales or Revenues= 51,294 millions USD), (f= F/S= Fixed Portion= 58.00%), (t= T/B= Tax Rate= 30.00%), (d= D/A= Dividend Portion or Payout= 30.00%), (i= I/S= Interest Portion= 1.00%), (s= [S/S']-1= Sales Growth= 5.00%), (D= Dividend Paid= 2,370 millions USD), and (v= V/S= Variable Portion= 19.00%), are known, then it's ($'= Sales Past) must be:

$$\$' = (\$[1-v-i]-D/\{d[1-t]\})/\{f[1+s]\}$$
$$= (51,294[1-0.1900-0.0100] -2,370$$
$$/\{0.3000[1-0.3000]\}$$
$$/\{0.5800[1+0.0500]\}$$
$$= 48,851 \text{ millions USD}$$

Law-1737:
If ($= Sales or Revenues= 51,294 millions USD), ($'= Sales of Past Year= 48,851 millions USD), (t= T/B= Tax Rate= 30.00%), (d= D/A= Dividend Portion or Payout= 30.00%), (i= I/S= Interest Portion= 1.00%), (s= [S/S']-1= Sales Growth= 5.00%), (D= Dividend Paid= 2,370 millions USD), and (v= V/S= Variable Portion= 19.00%), are known, then it's (f= Fixed Portion planned) is:

$$f = (\$[1-v-i]-D/\{d[1-t]\})/\{\$'[1+s]\}$$
$$= (51{,}294[1-0.1900-0.0100] - 2{,}370$$
$$/\{0.3000[1-0.3000]\}$$
$$/\{48{,}851[1+0.0500]\}$$
$$= 58.00\%$$

Law-1738:
If ($= Sales or Revenues= 51,294 millions USD), ($'= Sales of Past Year= 48,851 millions USD), (t= T/B= Tax Rate= 30.00%), (d= D/A= Dividend Portion or Payout= 30.00%), (i= I/S= Interest Portion= 1.00%), (f= F/S= Fixed Portion= 58.00%), (D= Dividend Paid= 2,370 millions USD), and (v= V/S= Variable Portion= 19.00%), are known, then it's (s= Sales Growth planned) is:

$$s = (\$[1-v-i]-D/\{d[1-t]\})/[\$'f] - 1$$
$$= (51{,}294[1-0.1900-0.0100] - 2{,}370$$
$$/\{0.3000[1-0.3000]\}$$
$$/[48{,}851*0.5800] - 1$$
$$= 5.00\%$$

Law-1739:

If ($= Sales or Revenues= 51,294 millions USD), ($'= Sales of Past Year= 48,851 millions USD), (s= [S/S']-1= Sales Growth= 5.00%), (d= D/A= Dividend Portion or Payout= 30.00%), (i= I/S= Interest Portion= 1.00%),(f= F/S= Fixed Portion= 58.00%), (D= Dividend Paid= 2,370 millions USD), and (v= V/S= Variable Portion= 19.00%), are known, then it's (t= Tax Rate planned) is:

t= 1-[D/d]/{$-$v-$'f[1+s]-$i]
 = 1-D/(d{$[1-v-i]-$'f[1+s]})
 = 1-2,370/(0.3000{51,294[1-0.1900
   -0.0100]-48,851*0.5800
   [1+0.0500]})
 = 30.00%

Law-1740:

If ($= Sales or Revenues= 51,294 millions USD), ($'= Sales of Past Year= 48,851 millions USD), (s= [S/S']-1= Sales Growth= 5.00%), (t= T/B= Tax Rate= 30.00%), (i= I/S= Interest Portion= 1.00%), (f= F/S= Fixed Portion= 58.00%), (D= Dividend Paid= 2,370 millions USD), and (v= V/S= Variable Portion= 19.00%), are known, then it's (d=Dividend Portion or Payout planned) is:

d= D/([1-t]]{$-$v-$'f[1+s]-$i]
 = D/([1-t]{$[1-v-i]-$'f[1+s]})
 = 2,370/([1-0.3000] {51,294[1-0.1900
   -0.0100] - 48,851*0.5800
   [1+0.0500]})
 = 30.00%

Law-1741:

If ($ = Sales or Revenues = 51,294 millions USD), ($' = Sales of Past Year = 48,851 millions USD), (s = [S/S']-1 = Sales Growth = 5.00%), (t = T/B = Tax Rate = 30.00%), (i = I/S = Interest Portion = 1.00%), (f = F/S = Fixed Portion = 58.00%), (d = D/A = Dividend Portion or Payout = 30.00%), and (v = V/S = Variable Portion = 19.00%), are known, then it's (D = Dividend Paid planned) is:

$$D = d[1-t]\{\$-\$v-\$'f[1+s]-\$'i[1+s]\}$$
$$= d([1-t]\{\$[1-v]-\$'[1+s][f+i]\})$$
$$= 0.3000[1-0.3000]\{51,294[1-0.1900]$$
$$\quad -48,851[1+0.0500]$$
$$\quad [0.5800+0.0100]\})$$
$$= 2,370 \text{ millions USD}$$

Law-1742:

If (D = Dividend Paid = 2,370 millions USD), ($' = Sales of Past Year = 48,851 millions USD), (s = [S/S']-1 = Sales Growth = 5.00%), (t = T/B = Tax Rate = 30.00%), (i = I/S = Interest Portion = 1.00%), (f = F/S = Fixed Portion = 58.00%), (d = D/A = Dividend Portion or Payout = 30.00%), and (v = V/S = Variable Portion = 19.00%), are known, then it's ($ = Sales or Revenues planned) is:

$$\$ = (\$'i[1+s][f+i]+D/\{d[1-t]\})/[1-v]$$
$$= (48,851[1+0.0500][0.5800+0.0100]$$
$$\quad +2,370/\{0.3000[1-0.3000]\}$$
$$\quad /[1-0.1900]$$
$$= 51,294 \text{ millions USD}$$

Law-1743:

If (**D**= Dividend Paid= 2,370 millions USD), (**$'**= Sales of Past Year= 48,851 millions USD), (**s**= [S/S']-1= Sales Growth= 5.00%), (**t**= T/B= Tax Rate= 30.00%), (**i**= I/S= Interest Portion= 1.00%), (**f**= F/S= Fixed Portion= 58.00%), (**d**= D/A=Dividend Portion or Payout= 30.00%), and (**$**= Sales or Revenues= 51,294 millions USD), are known, then it's (**v**= Variable Portion planned) is:

$$v = 1-(\$'i[1+s][f+i]+D/\{d[1-t]\})/\$$$
$$= 1-(48,851[1+0.0500][0.5800+0.0100]$$
$$+2,370/\{0.3000[1-0.3000]\}$$
$$/51,294$$
$$= 19.00\%$$

Law-1744:

If (**D**= Dividend Paid= 2,370 millions USD), (**v**= V/S= Variable Portion= 19.00%), (**s**= [S/S']-1= Sales Growth= 5.00%), (**t**= T/B= Tax Rate= 30.00%), (**i**= I/S= Interest Portion= 1.00%), (**f**= F/S= Fixed Portion= 58.00%), (**d**= D/A= Dividend Portion or Payout= 30.00%), and (**$**= Sales or Revenues= 51,294 millions USD), are known, then it's (**$'**= Sales Past) must be:

$$\$' = (\$[1-v]-D/\{d[1-t]\})/\{[1+s][f+i]\}$$
$$= (51,294[1-0.1900] -2,370/\{0.3000$$
$$[1-0.3000]\})/[1+0.0500]$$
$$[0.5800+0.0100]\}$$
$$= 48,851 \text{ millions USD}$$

Law-1745:

If (**D**= Dividend Paid= 2,370 millions USD), (**v**= V/S= Variable Portion= 19.00%), (**$'**= Sales of Past Year= 48,851 millions USD), (**t**= T/B= Tax Rate= 30.00%), (**i**= I/S= Interest Portion= 1.00%), (**f**= F/S= Fixed Portion= 58.00%), (**d**= D/A= Dividend Portion or Payout= 30.00%), and (**$**= Sales or Revenues= 51,294 millions USD), are known, then it's (**s**= Sales Growth planned) is:

$$s = (\$[1-v] - D/\{d[1-t]\})/\{\$'[f+i]\} - 1$$
$$= (51,294[1-0.1900] - 2,370/\{0.3000[1-0.3000]\})/\{48,851[0.5800+0.0100]\} - 1$$
$$= 5.00\%$$

Law-1746:

If (**D**= Dividend Paid= 2,370 millions USD), (**v**= V/S= Variable Portion= 19.00%), (**$'**= Sales of Past Year= 48,851 millions USD), (**t**= T/B= Tax Rate= 30.00%), (**i**= I/S= Interest Portion= 1.00%), (**s**= [S/S']-1= Sales Growth= 5.00%), (**d**= D/A= Dividend Portion or Payout= 30.00%), and (**$**= Sales or Revenues= 51,294 millions USD), are known, then it's (**f**= Fixed Portion planned) is:

$$f = (\$[1-v] - D/\{d[1-t]\})/\{\$'[1+s]\} - i$$
$$= (51,294[1-0.1900] - 2,370/\{0.3000[1-0.3000]\})/\{48,851[1+0.0500]\} - 0.0100$$
$$= 58.00\%$$

Law-1747:

If (**D**= Dividend Paid= 2,370 millions USD), (**v**= V/S= Variable Portion= 19.00%), (**$'**= Sales of Past Year= 48,851 millions USD), (**t**= T/B= Tax Rate= 30.00%), (**f**= F/S= Fixed Portion= 58.00%), (**s**= [S/S']-1= Sales Growth= 5.00%), (**d**= D/A= Dividend Portion or Payout= 30.00%), and (**$**= Sales or Revenues= 51,294 millions USD), are known, then it's (**i**= Interest Portion planned) is:

$$i = (\$[1-v] - D/\{d[1-t]\}) / \{\$'[1+s]\} - f$$
$$= (51{,}294[1-0.1900] - 2{,}370/\{0.3000[1-0.3000]\}) / \{48{,}851[1+0.0500]\} - 0.5800$$
$$= 1.00\%$$

Law-1748:

If (**D**= Dividend Paid= 2,370 millions USD), (**v**= V/S= Variable Portion= 19.00%), (**$'**= Sales of Past Year= 48,851 millions USD), (**i**= I/S= Interest Portion= 1.00%), (**f**= F/S= Fixed Portion= 58.00%), (**s**= [S/S']-1= Sales Growth= 5.00%), (**d**= D/A= Dividend Portion or Payout= 30.00%), and (**$**= Sales or Revenues= 51,294 millions USD), are known, then it's (**t**= Tax Rate planned) is:

$$t = 1 - [D/d]\{\$ - \$v - \$'f[1+s] - \$'i[1+s]\}$$
$$= 1 - D/\{d\{\$[1-v] - \$'[1+s][f+i]\}\}$$
$$= 1 - 2{,}370/\{0.3000\{(51{,}294[1-0.1900] - 48{,}851[1+0.0500][0.5800+0.0100]\}\}$$
$$= 30.00\%$$

Law-1749:

If (**D**= Dividend Paid= 2,370 millions USD), (**v**= V/S= Variable Portion= 19.00%), (**$'**= Sales of Past Year= 48,851 millions USD), (**i**= I/S= Interest Portion= 1.00%), (**f**= F/S= Fixed Portion= 58.00%), (**s**= [S/S']-1= Sales Growth= 5.00%), (**t**= T/B= Tax Rate= 30.00%), and (**$**= Sales or Revenues= 51,294 millions USD), are known, then it's (**d**= Dividend Portion or Payout planned) is:

$$d = D/([1-t]\{\$-\$v-\$'f[1+s]-\$'i[1+s]\})$$
$$= D/\{[1-t]\{\$[1-v]-\$'[1+s][f+i]\}\}$$
$$= 2,370/([1-0.3000]\{(51,294[1-0.1900]$$
$$-48,851[1+0.0500]$$
$$[0.5800+0.0100]\}$$
$$= 30.00\%$$

Law-1750:

If (**d**= D/A= Dividend Portion or Payout= 30.00%), (**v**= V/S= Variable Portion= 19.00%), (**$'**= Sales of Past Year= 48,851 millions USD), (**I**= Interest Expense= 513 millions USD), (**F**= Fixed Cost= 29,750 millions USD), (**s**= [S/S']-1= Sales Growth= 5.00%), (**t**= T/B= Tax Rate= 30.00%), and (**$**= Sales or Revenues= 51,294 millions USD), are known, then it's (**D**= Dividend Paid planned) is:

$$D = d[1-t]\{\$-F-I-\$'v[1+s]\}$$
$$= 0.3000[1-0.3000]\{(51,294-29,750-513$$
$$-48,851*0.1900[1+0.0500]\}$$
$$= 2,370 \text{ millions USD}$$

Law-1751:
If (d= D/A=Dividend Portion or Payout= 30.00%), (v= V/S= Variable Portion= 19.00%), (S'= Sales of Past Year= 48,851 millions USD), (I= Interest Expense= 513 millions USD), (F= Fixed Cost= 29,750 millions USD), (s= [S/S']-1= Sales Growth= 5.00%), (t= T/B= Tax Rate= 30.00%), and (D= Dividend Paid= 2,370 millions USD), are known, then it's (S= Sales or Revenues planned) is:

$S = F+I+S'v[1+s]+D/\{d[1-t]\}$
  = 29,750+513+48,851*0.1900
    [1+0.0500]+2,370
    /{0.3000[1-0.3000]}
  = 51,294 millions USD

Law-1752:
If (d= D/A= Dividend Portion or Payout= 30.00%), (v= V/S= Variable Portion= 19.00%), (S= Sales or Revenues= 51,294 millions USD), (I= Interest Expense= 513 millions USD), (F= Fixed Cost= 29,750 millions USD), (s= [S/S']-1= Sales Growth= 5.00%), (t= T/B= Tax Rate= 30.00%), and (D= Dividend Paid= 2,370 millions USD), are known, then it's (S'= Sales Past) must be:

$S' = (S-F-I-D/\{d[1-t]\})/\{v[1+s]\}$
   = (51,294-29,750-513-2,370/{0.3000
     [1-0.3000]})/{ 0.1900
     [1+0.0500]}
   = 48,851 millions USD

Law-1753:

If (d= D/A=Dividend Portion or Payout= 30.00%), ($\$'$= Sales of Past Year= 48,851 millions USD), ($\$$= Sales or Revenues= 51,294 millions USD), (I= Interest Expense= 513 millions USD), (F= Fixed Cost= 29,750 millions USD), (s= [S/S']-1= Sales Growth= 5.00%), (t= T/B= Tax Rate= 30.00%), and (D= Dividend Paid= 2,370 millions USD), are known, then it's (v= Variable Portion planned) is:

v= ($\$$-F-I-D/{d[1-t]}/{$\$'$[1+$s$]}
 = (51,294-29,750-513-2,370/{0.3000
  [1-0.3000]})
  /{48,851[1+0.0500]}
 = 19.00%

Law-1754:

If (d= D/A=Dividend Portion or Payout= 30.00%), ($\$'$= Sales of Past Year= 48,851 millions USD), ($\$$= Sales or Revenues= 51,294 millions USD), (I= Interest Expense= 513 millions USD), (F= Fixed Cost= 29,750 millions USD), (v= V/S= Variable Portion= 19.00%), (t= T/B= Tax Rate= 30.00%), and (D= Dividend Paid= 2,370 millions USD), are known, then it's (s= Sales Growth planned) is:

s= ($\$$-F-I-D/{d[1-t]}/{$\$'$[1+$s$]}
 = (51,294-29,750-513-2,370/{ 0.3000
  [1-0.3000]})/[48,851*0.1900]-1
 = 5.00%

Law-1755:

If (d= D/A= Dividend Portion or Payout= 30.00%), (S'= Sales of Past Year= 48,851 millions USD), (S= Sales or Revenues= 51,294 millions USD), (I= Interest Expense= 513 millions USD), (s= [S/S']-1= Sales Growth= 5.00%), (v= V/S= Variable Portion= 19.00%), (t= T/B= Tax Rate= 30.00%), and (D= Dividend Paid= 2,370 millions USD), are known, then it's (F= Fixed Cost planned) is:

$F = (S-I-S'v[1+s]-D/\{d[1-t]\}$
  = (51,294-513-48,851*0.1900
  [1+0.0500]-2,370
  /{0.3000[1-0.3000]})
  = 29,750 millions USD

Law-1756:

If (d= D/A=Dividend Portion or Payout= 30.00%), (S'= Sales of Past Year= 48,851 millions USD), (S= Sales or Revenues= 51,294 millions USD), (F= Fixed Cost= 29,750 millions USD), (s= [S/S']-1= Sales Growth= 5.00%), (v= V/S= Variable Portion= 19.00%), (t= T/B= Tax Rate= 30.00%), and (D= Dividend Paid= 2,370 millions USD), are known, then it's (I= Interest Expense planned) is:

$I = (S-F-S'v[1+s]-D/\{d[1-t]\}$
  = (51,294-29,750-48,851*0.1900
  [1+0.0500]-2,370
  /{0.3000[1-0.3000]})
  = 513 millions USD

Law-1757:

If (**d**= D/A= Dividend Portion or Payout= 30.00%), (**$'**= Sales of Past Year= 48,851 millions USD), (**$**= Sales or Revenues= 51,294 millions USD), (**F**= Fixed Cost= 29,750 millions USD), (**s**= [S/S']-1= Sales Growth= 5.00%), (**v**= V/S= Variable Portion= 19.00%), (**I**= Interest Expense= 513 millions USD), and (**D**= Dividend Paid= 2,370 millions USD), are known, then it's (**t**= Tax Rate planned) is:

$$t = 1 - D/\{d\{\$-F-I-\$'v[1+s]\}\}$$
$$= 1 - 2{,}370/\,0.3000\{51{,}294 - 29{,}750 - 513 - 48{,}851 * 0.1900[1 + 0.0500]\}$$
$$= 30.00\%$$

Law-1758:

If (**t**= T/B= Tax Rate= 30.00%), (**$'**= Sales of Past Year= 48,851 millions USD), (**$**= Sales or Revenues= 51,294 millions USD), (**F**= Fixed Cost= 29,750 millions USD), (**s**= [S/S']-1= Sales Growth= 5.00%), (**v**= V/S= Variable Portion= 19.00%), (**I**= Interest Expense= 513 millions USD), and (**D**= Dividend Paid= 2,370 millions USD), are known, then it's (**d**= Dividend Portion planned) is:

$$d = D/([1-t]\{\$-F-I-\$'v[1+s]\})$$
$$= 2{,}370/([1 - 0.3000]\{51{,}294 - 29{,}750 - 513 - 48{,}851 * 0.1900[1 + 0.0500]\})$$
$$= 30.00\%$$

Law-1759:

If (**t**= T/B= Tax Rate= 30.00%), (**$'**= Sales of Past Year= 48,851 millions USD), (**$**= Sales or Revenues= 51,294 millions USD), (**F**= Fixed Cost= 29,750 millions USD), (**s**= [S/S']-1= Sales Growth= 5.00%), (**v**= V/S= Variable Portion= 19.00%), (**i**= I/S= Interest Portion= 1.00%), and (**d**= D/A= Dividend Portion or Payout= 30.00%), are known, then it's (**D**= Dividend Paid planned) is:

$$D = d[1-t]\{\$-\$'v[1+s]-F-\$i\}$$
$$= d[1-t]\{\$[1-i]-\$'v[1+s]-F\}$$
$$= 0.3000[1-0.3000]\{51{,}294[1-0.0100]$$
$$-48{,}851*0.1900[1+0.0500]$$
$$-29{,}750\}$$
$$= 2{,}370 \text{ millions USD}$$

Law-1760:

If (**t**= T/B= Tax Rate= 30.00%), (**$'**= Sales of Past Year= 48,851 millions USD), (**D**= Dividend Paid= 2,370 millions USD), (**F**= Fixed Cost= 29,750 millions USD), (**s**= [S/S']-1= Sales Growth= 5.00%), (**v**= V/S= Variable Portion= 19.00%), (**i**= I/S= Interest Portion= 1.00%), and (**d**= D/A= Dividend Portion or Payout= 30.00%), are known, then it's (**$**= Sales or Revenues planned) is:

$$\$ = (\$'v[1+s]+F+D/d[1-t]/[1-i]$$
$$= (48{,}851*0.1900[1+0.0500]+29{,}750$$
$$+2{,}370/\{0.3000[1-0.3000]\}$$
$$/[1-0.0100]$$
$$= 51{,}294 \text{ millions USD}$$

Law-1761:

If (**t**= T/B= Tax Rate= 30.00%), (**$'**= Sales of Past Year= 48,851 millions USD), (**$**= Sales or Revenues= 51,294 millions USD), (**D**= Dividend Paid= 2,370 millions USD), (**F**= Fixed Cost= 29,750 millions USD), (**s**= [S/S']-1= Sales Growth= 5.00%), (**v**= V/S= Variable Portion= 19.00%), and (**d**= D/A= Dividend Portion or Payout= 30.00%), are known, then it's (**i**= Interest Portion planned) is:

$$i = 1-(\$'v[1+s]+F+D/d[1-t])/\$$$
$$= (48,851*0.1900[1+0.0500]+29,750$$
$$+2,370/\{0.3000[1-0.3000]\})$$
$$/51,294$$
$$= 1.00\%$$

Law-1762:

If (**t**= T/B= Tax Rate= 30.00%), (**i**= I/S= Interest Portion= 1.00%), (**$**= Sales or Revenues= 51,294 millions USD), (**D**= Dividend Paid= 2,370 millions USD), (**F**= Fixed Cost= 29,750 millions USD), (**s**= [S/S']-1= Sales Growth= 5.00%), (**v**= V/S= Variable Portion= 19.00%), and (**d**= D/A=Dividend Portion or Payout= 30.00%), are known, then it's (**$'**= Sales Past) must be:

$$\$' = (\$[1-i]-F-D/d[1-t]/\{v[1+s]\}$$
$$= (51,294[1-0.0100]-29,750-2,370$$
$$/\{0.3000[1-0.3000]\})$$
$$/\{0.1900[1+0.0500]\}$$
$$= 48,851 \text{ millions USD}$$

Law-1763:

If (t= T/B= Tax Rate= 30.00%), (i= I/S= Interest Portion= 1.00%), ($\$$= Sales or Revenues= 51,294 millions USD), (D= Dividend Paid= 2,370 millions USD), (F= Fixed Cost= 29,750 millions USD), (s= [S/S']-1= Sales Growth= 5.00%), ($\$'$= Sales of Past Year= 48,851 millions USD), and (d= D/A= Dividend Portion or Payout= 30.00%), are known, then it's (v= Variable Portion planned) is:

$$v = (\$[1-i]-F-D/d[1-t])/\{\$'[1+s]\}$$
$$= (51,294[1-0.0100]-29,750-2,370$$
$$/\{0.3000[1-0.3000]\})$$
$$/\{48,851[1+0.0500]\}$$
$$= \underline{19.00\%}$$

Law-1764:

If (t= T/B= Tax Rate= 30.00%), (i= I/S= Interest Portion= 1.00%), ($\$$= Sales or Revenues= 51,294 millions USD), (D= Dividend Paid= 2,370 millions USD), (F= Fixed Cost= 29,750 millions USD), (v= V/S= Variable Portion= 19.00%), ($\$'$= Sales of Past Year= 48,851 millions USD), and (d= D/A= Dividend Portion or Payout= 30.00%), are known, then it's (s= Sales Growth planned) is:

$$s = (\$[1-i]-F-D/d[1-t])/[\$'v]-1$$
$$= (51,294[1-0.0100]-29,750-2,370$$
$$/\{0.3000[1-0.3000]\})$$
$$/[48,851*0.1900]-1$$
$$= \underline{5.00\%}$$

Law-1765:
If (**t**= T/B= Tax Rate= 30.00%), (**i**= I/S= Interest Portion= 1.00%), (**$**= Sales or Revenues= 51,294 millions USD), (**D**= Dividend Paid= 2,370 millions USD), (**s**= [S/S']-1= Sales Growth= 5.00%), (**v**= V/S= Variable Portion= 19.00%), (**$'**= Sales of Past Year= 48,851 millions USD), and (**d**= D/A= Dividend Portion or Payout= 30.00%), are known, then it's (**F**= Fixed Cost planned) is:

F= **$**[1-**i**]-**$'v**[1+**s**]-**D**/{**d**[1-**t**]}
= 51,294[1-0.0100]-48,851*0.1900
  [1+0.0500]-2,370
  /{0.3000[1- 0.3000]}
= 29,750 millions USD

Law-1766:
If (**F**= Fixed Cost= 29,750 millions USD), (**i**= I/S= Interest Portion= 1.00%), (**$**= Sales or Revenues= 51,294 millions USD), (**D**= Dividend Paid= 2,370 millions USD), (**s**= [S/S']-1= Sales Growth= 5.00%), (**v**= V/S= Variable Portion= 19.00%), (**$'**= Sales of Past Year= 48,851 millions USD), and (**d**= D/A= Dividend Portion or Payout= 30.00%), are known, then it's (**t**= Tax Rate planned) is:

t= 1-[**D**/**d**]/{**$**-**$'v**[1+**s**]-**F**-**$i**}
= 1-**D**/(**d**{**$**[1-**i**]-**$'v**[1+**s**]-**F**})
= 1-2,370/(0.3000{51,294[1-0.0100]
  -48,851*0.1900[1+0.0500]
  -29,750})
= 30.00%

Law-1767:

If (F= Fixed Cost= 29,750 millions USD), (i= I/S= Interest Portion= 1.00%), (S= Sales or Revenues= 51,294 millions USD), (D= Dividend Paid= 2,370 millions USD), (s= [S/S']-1= Sales Growth= 5.00%), (v= V/S= Variable Portion= 19.00%), (S'= Sales of Past Year= 48,851 millions USD), and (t= T/B= Tax Rate= 30.00%), are known, then it's (d= Dividend Portion or Payout planned) is:

$d = D/([1-t]\{S-S'v[1+s]-F-Si\}$
$ = 1-D/([1-t]\{S[1-i]-S'v[1+s]-F\})$
$ = 2,370/([1-0.3000]\{51,294[1-0.0100]$
$ -48,851*0.1900[1+0.0500]$
$ -29,750\})$
$ = \underline{30.00\%}$

Law-1768:

If (F= Fixed Cost= 29,750 millions USD), (i= I/S= Interest Portion= 1.00%), (S= Sales or Revenues= 51,294 millions USD), (d= D/A= Dividend Portion or Payout= 30.00%), (s= [S/S']-1= Sales Growth= 5.00%), (v= V/S= Variable Portion= 19.00%), (S'= Sales of Past Year= 48,851 millions USD), and (t= T/B= Tax Rate= 30.00%), are known, then it's (D= Dividend Paid planned) is:

$D = d[1-t]\{S-S'v[1+s]-F-Si\}$
$ = d[1-t]\{S[1-i]-S'v[1+s]-F\})$
$ = 0.3000[1-0.3000]\{51,294[1-0.0100]$
$ -48,851*0.1900[1+0.0500]$
$ -29,750\})$
$ = \underline{30.00\%}$

Law-1769:

If (**F**= Fixed Cost= 29,750 millions USD), (**i**= I/S= Interest Portion= 1.00%), (**D**= Dividend Paid= 2,370 millions USD), (**d**= D/A= Dividend Portion or Payout= 30.00%), (**s**= [S/S']-1= Sales Growth= 5.00%), (**v**= V/S= Variable Portion= 19.00%), (**$'**= Sales of Past Year= 48,851 millions USD), and (**t**= T/B= Tax Rate= 30.00%), are known, then it's (**$**= Sales or Revenues planned) is:

$$\$ = F + \$'[v+i][1+s] + D/\{d[1-t]\}$$
$$= 29{,}750 + 48{,}851[0.1900+0.0100]$$
$$[1+0.0500] + 2{,}370$$
$$/\{0.3000[1-0.3000]\}$$
$$= 51{,}294 \text{ millions USD}$$

Law-1770:

If (**F**= Fixed Cost= 29,750 millions USD), (**i**= I/S= Interest Portion= 1.00%), (**D**= Dividend Paid= 2,370 millions USD), (**d**= D/A= Dividend Portion or Payout= 30.00%), (**s**= [S/S']-1= Sales Growth= 5.00%), (**v**= V/S= Variable Portion= 19.00%), (**$**= Sales or Revenues= 51,294 millions USD), and (**t**= T/B= Tax Rate= 30.00%), are known, then it's (**$'**= Sales Past) must be:

$$\$' = (\$ - F - D/\{d[1-t]\})/\{[v+i][1+s]\}$$
$$= (51{,}294 - 29{,}750 - 2{,}370/\{0.3000$$
$$[1-0.3000]\})/\{[0.1900+0.0100]$$
$$\{1+0.0500\}\}$$
$$= 48{,}851 \text{ millions USD}$$

Law-1771:

If (F= Fixed Cost= 29,750 millions USD), (i= I/S= Interest Portion= 1.00%), (D= Dividend Paid= 2,370 millions USD), (d= D/A= Dividend Portion or Payout= 30.00%), (s= [S/S']-1= Sales Growth= 5.00%), ($\$'$= Sales of Past Year= 48,851 millions USD), ($\$$= Sales or Revenues= 51,294 millions USD), and (t= T/B= Tax Rate= 30.00%), are known, then it's (v= Variable Portion planned) is:

v = ($\$$-F-D/{d[1-t]})/{$\$'$[1+$s$]}-$i$
   = (51,294-29,750-2,370/{0.3000
      [1-0.3000]})/{48,851
      [1+0.0500]}-0.0100
   = 19.00%

Law-1772:

If (F= Fixed Cost= 29,750 millions USD), (v= V/S= Variable Portion= 19.00%), (D= Dividend Paid= 2,370 millions USD), (d= D/A= Dividend Portion or Payout= 30.00%), (s= [S/S']-1= Sales Growth= 5.00%), ($\$'$= Sales of Past Year= 48,851 millions USD), ($\$$= Sales or Revenues= 51,294 millions USD), and (t= T/B= Tax Rate= 30.00%), are known, then it's (i= Interest Portion) is:

i = ($\$$-F-D/{d[1-t]})/{$\$'$[1+$s$]}-$v$
   = (51,294-29,750-2,370/{0.3000
      [1-0.3000]})/{48,851
      [1+0.0500]}-0.5800
   = 19.00%

Law-1773:
If (F= Fixed Cost= 29,750 millions USD), (v= V/S= Variable Portion= 19.00%), (D= Dividend Paid= 2,370 millions USD), (d= D/A= Dividend Portion or Payout= 30.00%), (i= I/S= Interest Portion= 1.00%), ($\$'$= Sales of Past Year= 48,851 millions USD), ($\$$= Sales or Revenues= 51,294 millions USD), and (t= T/B= Tax Rate= 30.00%), are known, then it's (s= Sales Growth planned) is:

$$s = (\$-F-D/\{d[1-t]\})/\{\$'[v+i]\}-1$$
$$= (51,294-29,750-2,370/\{0.3000$$
$$[1-0.3000]\})/\{48,851$$
$$[0.1900+0.0100]\}-1$$
$$= 5.00\%$$

Law-1774:
If (s= [S/S']-1= Sales Growth= 5.00%), (v= V/S= Variable Portion= 19.00%), (D= Dividend Paid= 2,370 millions USD), (d= D/A= Dividend Portion or Payout= 30.00%), (i= I/S= Interest Portion= 1.00%), ($\$'$= Sales of Past Year= 48,851 millions USD), ($\$$= Sales or Revenues= 51,294 millions USD), and (t= T/B= Tax Rate= 30.00%), are known, then it's (F= Fixed Cost planned) is:

$$F = \$-\$'[v+i][1+s]-D/d\{[1-t]\}$$
$$= 51,294-48,851[0.1900+0.0100]$$
$$[1+0.0500]\}-2,370$$
$$/\{0.3000[1-0.3000]\}$$
$$= 29,750 \text{ millions USD}$$

Law-1775:

If (s= [S/S']-1= Sales Growth= 5.00%), (v= V/S= Variable Portion= 19.00%), (D= Dividend Paid= 2,370 millions USD), (d= D/A=Dividend Portion or Payout= 30.00%), (i= I/S= Interest Portion= 1.00%), ($\$'$= Sales of Past Year= 48,851 millions USD), ($\$$= Sales or Revenues= 51,294 millions USD), and (F= Fixed Cost= 29,750 millions USD), are known, then it's (t= Tax Rate planned) is:

t= 1-[D/d]/{$\$$-$\$'v$[1+$s$]-$F$-$\$'i$[1+$s$]}
= 1-D/(d{$\$$-$\$'$[$v$+$i$][1+$s$]-$F$})
= 1-2,370/(0.3000{51,294-48,851
 [0.1900+0.0100]- 29,750})
= 30.00%

Law-1776:

If (s= [S/S']-1= Sales Growth= 5.00%), (v= V/S= Variable Portion= 19.00%), (D= Dividend Paid= 2,370 millions USD), (t= T/B= Tax Rate= 30.00%), (i= I/S= Interest Portion= 1.00%), ($\$'$= Sales of Past Year= 48,851 millions USD), ($\$$= Sales or Revenues= 51,294 millions USD), and (F= Fixed Cost= 29,750 millions USD), are known, then it's (d= Dividend Portion or Payout planned) is:

d= D/[1-t]{$\$$-$\$'v$[1+$s$]-$F$-$\$'i$[1+$s$]}
= D/([1-t]{$\$$-$\$'$[$v$+$i$][1+$s$]-$F$})
= 2,370/([1-0.3000]{51,294-48,851
 [0.1900+0.0100]- 29,750})
= 30.00%

Law-1777:

If (s= [S/S']-1= Sales Growth= 5.00%), (v= V/S= Variable Portion= 19.00%),(d= D/A=Dividend Portion or Payout= 30.00%), (t= T/B= Tax Rate= 30.00%), (I= Interest Expense= 513 millions USD), (S'= Sales of Past Year= 48,851 millions USD), (S= Sales or Revenues= 51,294 millions USD), and (f= F/S= Fixed Portion= 58.00%), are known, then it's (D= Dividend Paid planned) is:

$$D = d[1-t]\{S-S'v[1+s]\}-Sf-I\}$$
$$= d[1-t]\{S[1-f]-I-S'v[1+s]\}$$
$$= 0.3000[1-0.3000]\{51,294[1-0.5800]$$
$$-513-48,851*0.1900[1+0.0500]\}$$
$$= 2,370 \text{ millions USD}$$

Law-1778:

If (s= [S/S']-1= Sales Growth= 5.00%), (v= V/S= Variable Portion= 19.00%), (d= D/A= Dividend Portion or Payout= 30.00%), (t= T/B= Tax Rate= 30.00%), (I= Interest Expense= 513 millions USD), (S'= Sales of Past Year= 48,851 millions USD), (D= Dividend Paid= 2,370 millions USD), and (f= F/S= Fixed Portion= 58.00%), are known, then it's (S=Sales or Revenues planned) is:

$$S = (S'v[1+s]+I+D/\{d[1-t]\}/[1-f]$$
$$= (48,851*0.1900[1+0.0500]\}+513$$
$$+2,370/\{0.3000[1-0.3000]\}$$
$$/[1-0.5800]$$
$$= 51,294 \text{ millions USD}$$

Law-1779:

If (v= V/S= Variable Portion= <u>19.00</u>%), (d= D/A= Dividend Portion orPayout= <u>30.00</u>%), (s= [S/S']-1= Sales Growth= <u>5.00</u>%), (t= T/B= Tax Rate= <u>30.00</u>%), (I= Interest Expense= <u>513</u> millions USD), (S'= Sales of Past Year= <u>48,851</u> millions USD), (D= Dividend Paid= <u>2,370</u> millions USD), and (S= Sales or Revenues= <u>51,294</u> millions USD), are known, then it's (f= Fixed Portion planned) is:

$$f = 1-(S'v[1+s]+I+D/\{d[1-t]\})/S$$
$$= 1-(48{,}851*0.1900[1+0.0500]\}+513$$
$$+2{,}370/\{0.3000[1-0.3000]\}$$
$$/51{,}294$$
$$= \underline{58.00}\%$$

Law-1780:

If (s= [S/S']-1= Sales Growth= <u>5.00</u>%), (v= V/S= Variable Portion= <u>19.00</u>%), (d= D/A=Dividend Portion orPayout= <u>30.00</u>%), (t= T/B= Tax Rate= <u>30.00</u>%), (I= Interest Expense= <u>513</u> millions USD), (f= F/S= Fixed Portion= <u>58.00</u>%), (D= Dividend Paid= <u>2,370</u> millions USD), and (S= Sales or Revenues= <u>51,294</u> millions USD), are known, then it's (S'= Sales Past) must be:

$$S' = (S[1-f]-I-D/\{d[1-t]\})/\{v[1+s]\}$$
$$= (51{,}294[1-0.5800]-513-2{,}370$$
$$/\{0.3000[1-0.3000]\}$$
$$/\{0.1900[1+0.0500]\}$$
$$= \underline{51{,}294} \text{ millions USD}$$

Law-1781:
If (s= [S/S']-1= Sales Growth= 5.00%), (S'= Sales of Past Year= 48,851 millions USD), (d= D/A= Dividend Portion or Payout= 30.00%), (t= T/B= Tax Rate= 30.00%), (I= Interest Expense= 513 millions USD), (f= F/S= Fixed Portion= 58.00%), (D= Dividend Paid= 2,370 millions USD), and (S= Sales or Revenues= 51,294 millions USD), are known, then it's (v= Variable Portion planned) is:

v= (S[1-f]-I-D/{d[1-t]})/{S'[1+s]}
= (51,294[1-0.5800]-513 -2,370
/{0.3000[1-0.3000]}
/{48,851[1+0.0500]})
= 19.00%

Law-1782:
If (v= V/S= Variable Portion= 19.00%), (S'= Sales of Past Year= 48,851 millions USD), (d= D/A= Dividend Portion or Payout= 30.00%), (t= T/B= Tax Rate= 30.00%), (I= Interest Expense= 513 millions USD), (f= F/S= Fixed Portion= 58.00%), (D= Dividend Paid= 2,370 millions USD), and (S= Sales or Revenues= 51,294 millions USD), are known, then it's (s= Sales Growth planned) is:

s= (S[1-f]-I-D/{d[1-t]})/[$S'v$]-1
= (51,294[1-0.5800]-513-2,370
/{0.3000[1-0.3000]}
/[48,851*0.1900])-1
= 5.00%

Law-1783:

If (v= V/S= Variable Portion= 19.00%), ($\$'$= Sales of Past Year= 48,851 millions USD), (d= D/A= Dividend Portion or Payout= 30.00%), (t= T/B= Tax Rate= 30.00%), (s= [S/S']-1= Sales Growth= 5.00%), (f= F/S= Fixed Portion= 58.00%), (D= Dividend Paid= 2,370 millions USD), and ($\$$= Sales or Revenues= 51,294 millions USD), are known, then it's (I= Interest Expense planned) is:

$I = (\$[1-f]-\$'v[1+s]-D/\{d[1-t]\}$
$= (51,294[1-0.5800]-48,851*0.1900]$
$\quad [1+0.0500]-2,370/\{0.3000$
$\quad [1-0.3000]\}$
$= 513$ millions USD

Law-1784:

If (v= V/S= Variable Portion= 19.00%), ($\$'$= Sales of Past Year= 48,851 millions USD), (d= D/A= Dividend Portion or Payout= 30.00%), (I= Interest Expense= 513 millions USD), (s= [S/S']-1= Sales Growth= 5.00%), (f= F/S= Fixed Portion= 58.00%), (D= Dividend Paid= 2,370 millions USD), and ($\$$= Sales or Revenues= 51,294 millions USD), are known, then it's (t= Tax Rate planned) is:

$t = 1-[D/d]\{\$-\$'v[1+s]-\$f-I\}$
$= 1-D/(d(\$[1-f]-\$'v[1+s]-I\})$
$= 1-2,370/(0.3000\{51,294[1-0.5800]$
$\quad -48,851*0.1900][1+0.0500]-513\})$
$= 30.00\%$

Law-1785:

If (v= V/S= Variable Portion= 19.00%), (S'= Sales of Past Year= 48,851 millions USD), (D= Dividend Paid= 2,370 millions USD), (I= Interest Expense= 513 millions USD), (s= [S/S']-1= Sales Growth= 5.00%), (f= F/S= Fixed Portion= 58.00%), (t= T/B= Tax Rate= 30.00%), and (S= Sales or Revenues= 51,294 millions USD), are known, then it's (d= Dividend Portion or Payout planned) is:

d= D/(d[1-t]{S-S'v[1+s]-Sf-I}
  = D/(d[1-t](S[1-f]-S'v[1+s]-I)
  = 2,370/ (0.3000[1-0.3000]{51,294
        [1-0.5800]- 48,851*0.1900]
        [1+0.0500]-513})
  = 30.00%

Law-1786:

If (v= V/S= Variable Portion= 19.00%, (S'= Sales of Past Year= 48,851 millions USD), (d= D/A= Dividend Portion or Payout= 30.00%), (i= I/S= Interest Portion= 1.00%), (s= [S/S']-1= Sales Growth= 5.00%), (f= F/S= Fixed Portion= 58.00%), (t= T/B= Tax Rate= 30.00%), and (S= Sales or Revenues= 51,294 millions USD), are known, then it's (D= Dividend Paid planned) is:

D= d[1-t]{S-S'v[1+s]-Sf-Si}
  = d[1-t](S[1-f-i]-S'v[1+s]})
  = 0.3000[1-0.3000]{51,294[1-0.5800
        -0.0100]- 48,851*0.1900]
        [1+0.0500] })
  = 30.00%

Law-1787:

If (v= V/S= Variable Portion= 19.00%), (S'= Sales of Past Year= 48,851 millions USD), (d= D/A= Dividend Portion or Payout= 30.00%), (i= I/S= Interest Portion= 1.00%), (s= [S/S']-1= Sales Growth= 5.00%), (f= F/S= Fixed Portion= 58.00%), (t= T/B= Tax Rate= 30.00%), and (D= Dividend Paid= 2,370 millions USD), are known, then it's (S= Sales or Revenues planned) is:

$S = S'v[1+s] + D/\{d[1-t]\}/[1-f-i]$
   $= 48,851*0.1900][1+0.0500] + 2,370$
     $/\{0.3000[1-0.3000]\}$
     $/[1-0.5800-0.0100]$
   $= 51,294$ millions USD

Law-1788:

If (v= V/S= Variable Portion= 19.00%), (S'= Sales of Past Year= 48,851 millions USD), (d= D/A= Dividend Portion or Payout= 30.00%), (i= I/S= Interest Portion= 1.00%), (s= [S/S']-1= Sales Growth= 5.00%), (S= Sales or Revenues= 51,294 millions USD), (t= T/B= Tax Rate= 30.00%), and (D= Dividend Paid= 2,370 millions USD), are known, then it's (f= Fixed Portion planned) is:

$f = 1 - i - (S'v[1+s] + D/\{d[1-t]\})/S$
  $= 1 - 0.0100 - 48,851*0.1900][1+0.0500]$
    $+ 2,370/\{0.3000[1-0.3000]\}$
    $/51,294$
  $= 58.00\%$

Law-1789:

If (v= V/S= Variable Portion= 19.00%), ($\$'$= Sales of Past Year= 48,851 millions USD), (d= D/A= Dividend Portion or Payout= 30.00%), (f= F/S= Fixed Portion= 58.00%), (s= [S/S']-1= Sales Growth= 5.00%), ($\$$= Sales or Revenues= 51,294 millions USD), (t= T/B= Tax Rate= 30.00%), and (D= Dividend Paid= 2,370 millions USD), are known, then it's (i= Interest Portion planned) is:

$$i = 1-f-(\$'v[1+s]+D/\{d[1-t]\})/\$$$
$$= 1-0.5800- 48,851*0.1900][1+0.0500]$$
$$+2,370/\{0.3000[1-0.3000]\}$$
$$/51,294$$
$$= 1.00\%$$

Law-1790:

If (v= V/S= Variable Portion= 19.00%), (i= I/S= Interest Portion= 1.00%), (d= D/A= Dividend Portion or Payout= 30.00%), (f= F/S= Fixed Portion= 58.00%), (s= [S/S']-1= Sales Growth= 5.00%), ($\$$= Sales or Revenues= 51,294 millions USD), (t= T/B= Tax Rate= 30.00%), and (D= Dividend Paid= 2,370 millions USD), are known, then it's ($\$'$= Sales Past) must be:

$$\$' = (\$[1-v-i]-D/\{d[1-t]\})/\{v+[1+s]\}$$
$$= (51,294[1-0.5800- 0.0100]- 2,370$$
$$/\{0.3000[1-0.3000]\})$$
$$/\{0.1900][1+0.0500]\}$$
$$= 48,851 \text{ millions USD}$$

Law-1791:

If ($\$'$= Sales of Past Year= 48,851 millions USD), (i= I/S= Interest Portion= 1.00%), (d= D/A= Dividend Portion or Payout= 30.00%), (f= F/S= Fixed Portion= 58.00%), (s= [S/S']-1= Sales Growth= 5.00%), ($\$$= Sales or Revenues= 51,294 millions USD), (t= T/B= Tax Rate= 30.00%), and (D= Dividend Paid= 2,370 millions USD), are known, then it's (v= Variable Portion planned) is:

$v = (\$[1-f-i]-D/\{d[1-t]\})/\{\$'[1+s]\}$
  = (51,294[1-0.5800- 0.0100]- 2,370
      /{0.3000[1-0.3000]})
      /{48,851[1+0.0500]}
  = 19.00%

Law-1792:

If ($\$'$= Sales of Past Year= 48,851 millions USD), (i= I/S= Interest Portion= 1.00%), (d= D/A=Dividend Portion or Payout= 30.00%), (f= F/S= Fixed Portion= 58.00%), (v= V/S= Variable Portion= 19.00%), ($\$$= Sales or Revenues= 51,294 millions USD), (t= T/B= Tax Rate= 30.00%), and (D= Dividend Paid= 2,370 millions USD), are known, then it's (s= Sales Growth planned) is:

$s = (\$[1-f-i]-D/\{d[1-t]\})/[\$'v]-1$
  = (51,294[1-0.5800- 0.0100]- 2,370
      /{0.3000[1-0.3000]})
      /[48,851*0.1900]-1
  = 5.00%

Law-1793:

If ($\$'$= Sales of Past Year= 48,851 millions USD), (i= I/S= Interest Portion= 1.00%), (d= D/A= Dividend Portion or Payout= 30.00%), (f= F/S= Fixed Portion= 58.00%), (v= V/S= Variable Portion= 19.00%), ($\$$= Sales or Revenues= 51,294 millions USD), (s= [S/S']-1= Sales Growth= 5.00%), and (D= Dividend Paid= 2,370 millions USD), are known, then it's (t= Tax Rate planned) is:

t = 1-[D/d]/({$\$-\$'v[1+s]$}-$\$f$-$\$i$}
   = 1-D/(d{$\$[1-f-i]-\$'v[1+s]$})
   = 1-2,370/(0.3000{51,294[1-0.5800
       -0.0100]-48,851*0.1900]
       [1+0.0500]})
   = 30.00%

Law-1794:

If ($\$'$= Sales of Past Year= 48,851 millions USD), (i= I/S= Interest Portion= 1.00%), (t= T/B= Tax Rate= 30.00%), (f= F/S= Fixed Portion= 58.00%), (v= V/S= Variable Portion= 19.00%), ($\$$= Sales or Revenues= 51,294 millions USD), (s= [S/S']-1= Sales Growth= 5.00%), and (D= Dividend Paid= 2,370 millions USD), are known, then it's (d= Dividend Portion planned) is:

d = D/([1-t]{$\$-\$'v[1+s]$}-$\$f$-$\$i$}
   = D/([1-t]{$\$[1-f-i]-\$'v[1+s]$})
   = 2,370/([1-0.3000]{51,294[1-0.5800
       -0.0100]-48,851*0.1900]
       [1+0.0500]})
   = 30.00%

Law-1795:

If ($\$'$= Sales of Past Year= 48,851 millions USD), (i= I/S= Interest Portion= 1.00%), (t= T/B= Tax Rate= 30.00%), (f= F/S= Fixed Portion= 58.00%), (v= V/S= Variable Portion= 19.00%), ($\$$= Sales or Revenues= 51,294 millions USD), (s= [S/S']-1= Sales Growth= 5.00%), and (d= D/A= Dividend Portion or Payout= 30.00%), are known, then it's (D= Dividend Paid planned) is:

$$D = d[1-t]\{\$-\$'v[1+s]-\$f-\$'i[1+s]\}$$
$$= d[1-t]\{\$[1-f]-\$'[1+s][v+i]\}$$
$$= 0.3000\,[1-0.3000]\{51{,}294[1-0.5800]$$
$$-48{,}851[1+0.0500]$$
$$[0.1900+0.0100]\})$$
$$= 2{,}370 \text{ millions USD}$$

Law-1796:

If ($\$'$= Sales of Past Year= 48,851 millions USD), (i= I/S= Interest Portion= 1.00%), (t= T/B= Tax Rate= 30.00%), ($\$$= Sales or Revenues= 51,294 millions USD), (v= V/S= Variable Portion= 19.00%), (D= Dividend Paid= 2,370 millions USD), (s= [S/S']-1= Sales Growth= 5.00%), and (d= D/A= Dividend Portion or Payout= 30.00%), are known, then it's ($\$$= Sales or Revenues planned) is:

$$\$ = (\$'[1+s][v+i]\}+D/\{d[1-t]\})/[1-f]$$
$$= (48{,}851[1+0.0500][0.1900+0.0100]\})$$
$$+2{,}370/\{0.3000\,[1-0.3000]\}$$
$$/[1-0.5800]$$
$$= 51{,}294 \text{ millions USD}$$

Law-1797:
If ($’= Sales of Past Year= 48,851 millions USD), (i= I/S= Interest Portion= 1.00%), (t= T/B= Tax Rate= 30.00%), ($= Sales or Revenues= 51,294 millions USD), (v= V/S= Variable Portion= 19.00%), (D= Dividend Paid= 2,370 millions USD), (s= [S/S']-1= Sales Growth= 5.00%), and (d= D/A= Dividend Portion or Payout= 30.00%), are known, then it's (f= Fixed Portion planned) is:

f= 1- ($'[1+s][v+i]}+D/{d[1-t]})/$
 = (48,851[1+0.0500] [0.1900+0.0100]})
    +2,370/{0.3000 [1-0.3000]}
    /51,294
 = 58.00%

Law-1798:
If (f= F/S= Fixed Portion= 58.00%), (i= I/S= Interest Portion= 1.00%), (t= T/B= Tax Rate= 30.00%), ($= Sales or Revenues= 51,294 millions USD), (v= V/S= Variable Portion= 19.00%), (D= Dividend Paid= 2,370 millions USD), (s= [S/S']-1= Sales Growth= 5.00%), and (d= D/A= Dividend Portion or Payout= 30.00%), are known, then it's ($'= Sales Past) must be:

$'= ($[1-f]-D/{d[1-t]})/{[1+s][v+i]}
 = (51,294[1-0.5800] -2,370/{0.3000
    [1-0.3000]})/{[1+0.0500]
    [0.1900+0.0100]}
 = 48,851 millions USD

Law-1799:

If (**f**= F/S= Fixed Portion= 58.00%), (**i**= I/S= Interest Portion= 1.00%), (**t**= T/B= Tax Rate= 30.00%), (**$**= Sales or Revenues= 51,294 millions USD), (**v**= V/S= Variable Portion= 19.00%), (**D**= Dividend Paid= 2,370 millions USD), (**$'**= Sales of Past Year= 48,851 millions USD), and (**d**= D/A= Dividend Portion or Payout= 30.00%), are known, then it's (**s**= Sales Growth planned) is:

s= (**\$**[1-**f**]-**D**{**d**[1-**t**]})/{**\$'**[**v**+**i**]}-1
= (51,294[1-0.5800] -2,370/{0.3000
[1-0.3000]})/{48,851
[0.1900+0.0100]}-1
= 5.00%

Law-1800:

If (**f**= F/S= Fixed Portion= 58.00%), (**i**= I/S= Interest Portion= 1.00%), (**t**= T/B= Tax Rate= 30.00%), (**$**= Sales or Revenues= 51,294 millions USD), (**s**= [S/S']-1= Sales Growth= 5.00%), (**D**= Dividend Paid= 2,370 millions USD), (**$'**= Sales of Past Year= 48,851 millions USD), and (**d**= D/A= Dividend Portion or Payout= 30.00%), are known, then it's (**v**= Variable Portion planned) is:

v= (**\$**[1-**f**]-**D**{**d**[1-**t**]})/{**\$'**[1+**s**]}-**i**
= (51,294[1-0.5800]-2,370/{0.3000
[1-0.3000]})/ {48,851[1
+0.0500]}-0.0100
= 19.00%

Law-1801:

If (**f**= F/S=Fixed Portion= 58.00%), (**v**= V/S= Variable Portion= 19.00%), (**t**= T/B= Tax Rate= 30.00%), (**$**= Sales or Revenues= 51,294 millions USD), (**s**= [S/S']-1= Sales Growth= 5.00%), (**D**= Dividend Paid= 2,370 millions USD), (**$'**= Sales of Past Year= 48,851 millions USD), and (**d**= D/A= Dividend Portion or Payout= 30.00%), are known, then it's (**i**= Interest Portion planned) is:

$$i = (\$[1-f] - D\{d[1-t]\})/\{\$'[1+s]\} - v$$
$$= (51{,}294[1-0.5800] - 2{,}370/\{0.3000[1-0.3000]\})/\{48{,}851[1+0.0500]\} - 0.1900$$
$$= 1.00\%$$

Law-1802:

If (**f**= F/S=Fixed Portion= 58.00%), (**v**= V/S= Variable Portion= 19.00%), (**i**= I/S= Interest Portion= 1.00%), (**$**= Sales or Revenues= 51,294 millions USD), (**s**= [S/S']-1= Sales Growth= 5.00%), (**D**= Dividend Paid= 2,370 millions USD), (**$'**= Sales of Past Year= 48,851 millions USD), and (**d**= D/A= Dividend Portion or Payout= 30.00%), are known, then it's (**t**= Tax Rate planned) is:

$$t = 1 - [D/d]\{\$ - \$'v[1+s]\} - \$f\} - \$'[1+s]\}$$
$$= 1 - D/(d\{\$[1-f] - \$'[1+s][v+i]\})$$
$$= 1 - 2{,}370/(0.3000\,\{51{,}294[1-0.5800] - 48{,}851[1+0.0500][0.1900+0.0100]\})$$
$$= 30.00\%$$

Law-1803:

If (f= F/S=Fixed Portion= 58.00%), (v= V/S= Variable Portion= 19.00%), (i= I/S= Interest Portion= 1.00%), ($\$$= Sales or Revenues= 51,294 millions USD), (s= [S/S']-1= Sales Growth= 5.00%), (D= Dividend Paid= 2,370 millions USD), ($\$'$= Sales of Past Year= 48,851 millions USD), and (t= T/B= Tax Rate= 30.00%), are known, then it's (d= Dividend Portion planned) is:

$$d = D/([1-t]\{\$-\$'v[1+s]-\$f-\$'i[1+s]\}$$
$$= D/([1-t]\{\$[1-f]-\$'[1+s][v+i]\})$$
$$= 2,370/([1-0.3000]\{51,294[1-0.5800]$$
$$-48,851[1+0.0500]$$
$$[0.1900+0.0100]\})$$
$$= \underline{30.00\%}$$

Law-1804:

If (f= F/S=Fixed Portion= 58.00%), (v= V/S= Variable Portion= 19.00%), (I= Interest Expense= 513 millions USD), ($\$$= Sales or Revenues= 51,294 millions USD), (s= [S/S']-1= Sales Growth= 5.00%), (d= D/A=Dividend Portion or Payout= 30.00%), ($\$'$= Sales of Past Year= 48,851 millions USD), and (t= T/B= Tax Rate= 30.00%), are known, then it's (D= Dividend Paid planned) is:

$$D = d[1-t]\{\$-\$'v[1+s]-\$'f[1+s]-I\}$$
$$= d[1-t]\{\$-I-\$'[1+s][v+f]\})$$
$$= 0.3000[1-0.3000]\{51,294-513$$
$$-48,851[1+0.0500]$$
$$[0.1900+0.5800]\})$$
$$= \underline{2,370} \text{ millions USD}$$

Law-1805:

If (f= F/S= Fixed Portion= 58.00%), (v= V/S= Variable Portion= 19.00%), (I= Interest Expense= 513 millions USD), (D= Dividend Paid= 2,370 millions USD), (s= [S/S']-1= Sales Growth= 5.00%), (d= D/A=Dividend Portion or Payout= 30.00%), (S'= Sales of Past Year= 48,851 millions USD), and (t= T/B= Tax Rate= 30.00%), are known, then it's (S= Sales or Revenues planned) is:

$$S = I + S'[1+s][v+f] + D/\{d[1-t]\}$$
$$= 513 + 48,851[1+0.0500][0.1900 + 0.5800] + 2,370/\{0.3000[1-0.3000]\}$$
$$= 51,294 \text{ millions USD}$$

Law-1806:

If (f= F/S= Fixed Portion= 58.00%), (v= V/S= Variable Portion= 19.00%), (I= Interest Expense= 513 millions USD), (D= Dividend Paid= 2,370 millions USD), (s= [S/S']-1= Sales Growth= 5.00%), (d= D/A= Dividend Portion or Payout= 30.00%), (S= Sales or Revenues= 51,294 millions USD), and (t= T/B= Tax Rate= 30.00%), are known, then it's (S'= Sales Past) must be:

$$S' = (S - I - D/\{d[1-t]\}) / \{[1+s][v+f]\}$$
$$= (51,294 - 513 - 2,370/\{0.3000[1-0.3000]\}) / \{[1+0.0500][0.1900+0.5800]\}$$
$$= 48,851 \text{ millions USD}$$

Law-1807:

If (**f**= F/S= Fixed Portion= 58.00%), (**v**= V/S= Variable Portion= 19.00%), (**I**= Interest Expense= 513 millions USD), (**D**= Dividend Paid= 2,370 millions USD), (**$'**= Sales of Past Year= 48,851 millions USD), (**d**= D/A= Dividend Portion or Payout= 30.00%), (**$**= Sales or Revenues= 51,294 millions USD), and (**t**= T/B= Tax Rate= 30.00%), are known, then it's (**s**= Sales Growth planned) is:

$$s = (\$-I-D/\{d[1-t]\})/\{\$'[v+f]\}-1$$
$$= (51,294-513-2,370/\{0.3000$$
$$[1-0.3000]\})/\{48,851$$
$$[0.1900+0.5800]\}-1$$
$$= 5.00\%$$

Law-1808:

If (**f**= F/S= Fixed Portion= 58.00%), (**s**= [S/S']-1= Sales Growth= 5.00%), (**I**= Interest Expense= 513 millions USD), (**D**= Dividend Paid= 2,370 millions USD), (**$'**= Sales of Past Year= 48,851 millions USD), (**d**= D/A= Dividend Portion or Payout= 30.00%), (**$**= Sales or Revenues= 51,294 millions USD), and (**t**= T/B= Tax Rate= 30.00%), are known, then it's (**v**= Variable Portion planned) is:

$$v = (\$-I-D/\{d[1-t]\})/\{\$'[1+s]\}-f$$
$$= (51,294-513-2,370/\{0.3000$$
$$[1-0.3000]\})/\{48,851$$
$$[1+0.0500]\}-0.5800$$
$$= 19.00\%$$

<u>Law-1809</u>:

If (v= V/S= Variable Portion= <u>19.00</u>%), (s= [S/S']-1= Sales Growth= <u>5.00</u>%), (I= Interest Expense= <u>513</u> millions USD), (D= Dividend Paid= <u>2,370</u> millions USD), ($\$'$= Sales of Past Year= <u>48,851</u> millions USD), (d= D/A= Dividend Portion or Payout= <u>30.00</u>%), ($\$$= Sales or Revenues= <u>51,294</u> millions USD), and (t= T/B= Tax Rate= <u>30.00</u>%), are known, then it's (f= Fixed Portion planned) is:

$f = (\$-I-D/\{d[1-t]\})/\{\$'[1+s]\}-v$
$= (51,294-513-2,370/\{0.3000$
$[1-0.3000]\})/\{48,851$
$[1+0.0500]\}-0.1900$
$= \underline{58.00}\%$

<u>Law-1810</u>:

If (v= V/S= Variable Portion= <u>19.00</u>%), (s= [S/S']-1= Sales Growth= <u>5.00</u>%), (f= F/S= Fixed Portion= <u>58.00</u>%), (D= Dividend Paid= <u>2,370</u> millions USD), ($\$'$= Sales of Past Year= <u>48,851</u> millions USD), (d= D/A= Dividend Portion or Payout= <u>30.00</u>%), ($\$$= Sales or Revenues= <u>51,294</u> millions USD), and (t= T/B= Tax Rate= <u>30.00</u>%), are known, then it's (I= Interest Expense planned) is:

$I = (\$-\$'[1+s][v+f]-D/\{d[1-t]\}$
$= (51,294-48,851[1+0.0500][0.1900$
$+0.5800]-2,370$
$/\{0.3000[1-0.3000]\}$
$= \underline{513}$ millions USD

Law-1811:

If (v= V/S= Variable Portion= 19.00%), (s= [S/S']-1= Sales Growth= 5.00%), (f= F/S= Fixed Portion= 58.00%), (D= Dividend Paid= 2,370 millions USD), (S'= Sales of Past Year= 48,851 millions USD), (d= D/A= Dividend Portion or Payout= 30.00%), (S= Sales or Revenues= 51,294 millions USD), and (I= Interest Expense= 513 millions USD), are known, then it's (t= Tax Rate planned) is:

$$t = 1-[D/d]\{S-S'v[1+s]-S'f[1+s]-I\}$$
$$= 1-[D/d]\{S-I-S'[1+s][v+f]\})$$
$$= 1-2,370/(0.3000\{51,294-513-48,851$$
$$[1+0.0500][0.1900+0.5800]\})$$
$$= 30.00\%$$

Law-1812:

If (v= V/S= Variable Portion= 19.00%), (s= [S/S']-1= Sales Growth= 5.00%), (f= F/S= Fixed Portion= 58.00%), (D= Dividend Paid= 2,370 millions USD), (S'= Sales of Past Year= 48,851 millions USD), (t= T/B= Tax Rate= 30.00%), (S= Sales or Revenues= 51,294 millions USD), and (I= Interest Expense= 513 millions USD), are known, then it's (d= Dividend Portion planned) is:

$$d = D/([1-t]\{S-S'v[1+s]-S'f[1+s]-I\}$$
$$= D/([1-t]\{S-I-S'[1+s][v+f]\})$$
$$= 2,370/([1-0.3000]\{51,294-513$$
$$-48,851[1+0.0500]$$
$$[0.1900+0.5800]\})$$
$$= 30.00\%$$

Law-1813:

If (v= V/S= Variable Portion= 19.00%), (s= [S/S']-1= Sales Growth= 5.00%), (f= F/S= Fixed Portion= 58.00%), (d= D/A= Dividend Portion or Payout= 30.00%), ($\$'$= Sales of Past Year= 48,851 millions USD), (t= T/B= Tax Rate= 30.00%), ($\$$= Sales or Revenues= 51,294 millions USD), and (i= I/S= Interest Portion= 1.00%), are known, then it's (D= Dividend Paid planned) is:

$$D = d[1-t]\{\$-\$'v[1+s]-\$'f[1+s]-\$i\}$$
$$= d[1-t]\{\$[1-i]-\$'[1+s][v+f]\})$$
$$= 0.3000([1-0.3000]\{51,294[1-0.0100]$$
$$-48,851[1+0.0500]$$
$$[0.1900 +0.5800]\})$$
$$= 2,370 \text{ millions USD}$$

Law-1814:

If (v= V/S= Variable Portion= 19.00%), (s= [S/S']-1= Sales Growth= 5.00%), (f= F/S= Fixed Portion= 58.00%), (d= D/A= Dividend Portion or Payout= 30.00%), ($\$'$= Sales of Past Year= 48,851 millions USD), (t= T/B= Tax Rate= 30.00%), (D= Dividend Paid= 2,370 millions USD), and (i= I/S= Interest Portion= 1.00%), are known, then it's ($\$$=Sales or Revenues planned) is:

$$\$ = (\$'[1+s][v+f]+D/\{d[1-t]\}/[1-i]$$
$$= 48,851[1+0.0500][0.1900 +0.5800]$$
$$+2,370/\{0.3000([1-0.3000]\}$$
$$/[1-0.0100]$$
$$= 51,294 \text{ millions USD}$$

Law-1815:

If (v= V/S= Variable Portion= 19.00%), (s= [S/S']-1= Sales Growth= 5.00%), (f= F/S= Fixed Portion= 58.00%), (d= D/A= Dividend Portion or Payout= 30.00%), ($\$'$= Sales of Past Year= 48,851 millions USD), (t= T/B= Tax Rate= 30.00%), (D= Dividend Paid= 2,370 millions USD), and ($\$$= Sales or Revenues= 51,294 millions USD), are known, then it's (i= Interest Portion planned) is:

i= 1-($\$'$[1+$s$][$v$+$f$]+$D$/{$d$[1-$t$]})/$\$$
 = 1-(48,851[1+0.0500][0.1900 +0.5800]
   +2,370/{0.3000([1-0.3000]})
   /51,294
 = 1.00%

Law-1816:

If (v= V/S= Variable Portion= 19.00%), (s= [S/S']-1= Sales Growth= 5.00%), (f= F/S= Fixed Portion= 58.00%), (d= D/A= Dividend Portion or Payout= 30.00%), (i= I/S= Interest Portion= 1.00%), (t= T/B= Tax Rate= 30.00%), (D= Dividend Paid= 2,370 millions USD), and ($\$$= Sales or Revenues= 51,294 millions USD), are known, then it's ($\$'$= Sales Past) must be:

$\$'$= ($\$$[1-i]-D/{d[1-t]})/{[1+s][v+f]}
 = (51,294[1-0.0100] +2,370/{0.3000
   [1-0.3000]})/{[1+0.0500]
   [0.1900 +0.5800]}
 = 48,851 millions USD

Law-1817:

If (v= V/S= Variable Portion= 19.00%), ($\$'$= Sales of Past Year= 48,851 millions USD), (f= F/S= Fixed Portion= 58.00%), (d= D/A= Dividend Portion or Payout= 30.00%), (i= I/S= Interest Portion= 1.00%), (t= T/B= Tax Rate= 30.00%), (D= Dividend Paid= 2,370 millions USD), and ($\$$= Sales or Revenues= 51,294 millions USD), are known, then it's (s= Sales Growth planned) is:

s = ($\$[1-i]-D/\{d[1-t]\})/\{[\$'[v+f]\}-1$
= (51,294[1-0.0100] -2,370/{0.3000 [1-0.3000]})/{48,851[0.1900 +0.5800]}-1
= 5.00%

Law-1818:

If (s= [S/S']-1= Sales Growth= 5.00%), ($\$'$= Sales of Past Year= 48,851 millions USD), (f= F/S= Fixed Portion= 58.00%), (d= D/A= Dividend Portion or Payout= 30.00%), (i= I/S= Interest Portion= 1.00%), (t= T/B= Tax Rate= 30.00%), (D= Dividend Paid= 2,370 millions USD), and ($\$$= Sales or Revenues= 51,294 millions USD), are known, then it's (v= Variable Portion planned) is:

v = ($\$[1-f-i]-D/\{d[1-t]\})/\{[\$'[1+s]\}$
= (51,294[1-0.5800-0.0100] -2,370 /{0.3000([1-0.3000]})/{48,851 [1 +0.0500]}
= 19.00%

Law-1819:

If ($=[S/S']-1=$ Sales Growth= $\underline{5.00}$%), ($\$'=$ Sales of Past Year= $\underline{48,851}$ millions USD), ($v= V/S=$ Variable Portion= $\underline{19.00}$%), ($d= D/A=$ Dividend Portion or Payout= $\underline{30.00}$%), ($i= I/S=$ Interest Portion= $\underline{1.00}$%), ($t= T/B=$ Tax Rate= $\underline{30.00}$%), ($D=$ Dividend Paid= $\underline{2,370}$ millions USD), and ($\$=$ Sales or Revenues= $\underline{51,294}$ millions USD), are known, then it's ($f=$ Fixed Portion planned) is:

$$f = (\$[1-v-i]-D/\{d[1-t]\})/\{\$'[1+s]\}$$
$$= (51,294[1-0.1900-0.0100] -2,370$$
$$/\{0.3000([1-0.3000]\})/\{48,851$$
$$[1+0.0500]\}$$
$$= \underline{58.00\%}$$

Law-1820:

If ($=[S/S']-1=$ Sales Growth= $\underline{5.00}$%), ($\$'=$ Sales of Past Year= $\underline{48,851}$ millions USD), ($v= V/S=$ Variable Portion= $\underline{19.00}$%), ($d= D/A=$ Dividend Portion or Payout= $\underline{30.00}$%), ($i= I/S=$ Interest Portion= $\underline{1.00}$%), ($f= F/S=$ Fixed Portion= $\underline{58.00}$%), ($D=$ Dividend Paid= $\underline{2,370}$ millions USD), and ($\$=$ Sales or Revenues= $\underline{51,294}$ millions USD), are known, then it's ($t=$ Tax Rate planned) is:

$$t = 1-[D/d]\{\$-\$'v[1+s]-\$'f[1+s]-\$i\}$$
$$= 1-D/(d\{\$[1-i]-\$'[1+s][v+f]\})$$
$$= 1-2,370/(0.3000\{51,294[1-0.0100]$$
$$-48,851[1+0.0500]$$
$$[0.1900+0.5800]\})$$
$$= \underline{30.00\%}$$

Law-1821:

If (s= [S/S']-1= Sales Growth= 5.00%), ($\$'$= Sales of Past Year= 48,851 millions USD), (v= V/S= Variable Portion= 19.00%), (t= T/B= Tax Rate= 30.00%), (i= I/S= Interest Portion= 1.00%), (f= F/S= Fixed Portion= 58.00%), (D= Dividend Paid= 2,370 millions USD), and ($\$$= Sales or Revenues= 51,294 millions USD), are known, then it's (d= Dividend Portion planned) is:

d = D/([1-t]{$\$$-$\$'v$[1+$s$]-$\$'f$[1+$s$]-$\$i$}
  = D/(d{$\$$[1-i]-$\$'$[1+$s$][$v$+$f$]})
  = 1-2,370/(0.3000{51,294[1-0.0100]
   -48,851[1 +0.0500]
   [0.1900+0.5800]})
  = 30.00%

Law-1822:

If (s= [S/S']-1= Sales Growth= 5.00%), ($\$'$= Sales of Past Year= 48,851 millions USD), (v= V/S= Variable Portion= 19.00%), (t= T/B= Tax Rate= 30.00%), (i= I/S= Interest Portion= 1.00%), (f= F/S= Fixed Portion= 58.00%), (d= D/A= Dividend Portion or Payout= 30.00%), and ($\$$= Sales or Revenues= 51,294 millions USD), are known, then it's (D= Dividend Paid planned) is:

D = d[1-t]{$\$$-$\$'v$[1+$s$]-$\$'f$[1+$s$]-$\$i$}
  = d[1-t]{$\$$[1-i]-$\$'$[1+$s$][$v$+$f$]})
  = 0.3000[1-0.3000]{51,294[1-0.0100]
   -48,851[1 +0.0500]
   [0.1900+0.5800]})
  = 2,370 millions USD

Law-1823:

If (s= [S/S']-1= Sales Growth= 5.00%), (S'= Sales of Past Year= 48,851 millions USD), (v= V/S= Variable Portion= 19.00%), (t= T/B= Tax Rate= 30.00%), (i= I/S= Interest Portion= 1.00%), (f= F/S= Fixed Portion= 58.00%), (d= D/A= Dividend Portion or Payout= 30.00%), and (D= Dividend Paid= 2,370 millions USD), are known, then it's (S= Sales or Revenues planned) is:

$$S = S'[1+s][v+f+i] + D/\{d[1-t]\}$$
$$= 48,851[1+0.0500][0.1900+0.5800 + 0.0100] + 2,370 / \{0.3000[1-0.3000]\}$$
$$= \underline{51,294} \text{ millions USD}$$

Law-1824:

If (s= [S/S']-1= Sales Growth= 5.00%), (S= Sales or Revenues= 51,294 millions USD), (v= V/S= Variable Portion= 19.00%), (t= T/B= Tax Rate= 30.00%), (i= I/S= Interest Portion= 1.00%), (f= F/S= Fixed Portion= 58.00%), (d= D/A= Dividend Portion or Payout= 30.00%), and (D= Dividend Paid= 2,370 millions USD), are known, then it's (S'= Sales Past) must be:

$$S' = (S - D/\{d[1-t]\}) / \{[1+s][v+f+i]\}$$
$$= (51,294 - 2,370/\{0.3000[1-0.3000]\}) / \{[1+0.0500][0.1900 + 0.5800+0.0100]\}$$
$$= \underline{48,851} \text{ millions USD}$$

Law-1825:
If ($\$'$= Sales of Past Year= 48,851 millions USD), ($\$$= Sales or Revenues= 51,294 millions USD), (v= V/S= Variable Portion= 19.00%), (t= T/B= Tax Rate= 30.00%), (i= I/S= Interest Portion= 1.00%), (f= F/S=Fixed Portion= 58.00%), (d= D/A= Dividend Portion or Payout= 30.00%), and (D= Dividend Paid= 2,370 millions USD), are known, then it's (s= Sales Growth planned) is:

s= ($\$$-D/{d[1-t]})/{$\$'$[$v$+$f$+$i$]}-1
   = (51,294-2,370/{0.3000[1-0.3000]})
     /{48,851[0.1900+0.5800
     +0.0100] }-1
   = 5.00%

Law-1826:
If ($\$'$= Sales of Past Year= 48,851 millions USD), ($\$$= Sales or Revenues= 51,294 millions USD), (s= [S/S']-1= Sales Growth= 5.00%), (t= T/B= Tax Rate= 30.00%), (i= I/S= Interest Portion= 1.00%), (f= F/S= Fixed Portion= 58.00%), (d= D/A= Dividend Portion or Payout= 30.00%), and (D= Dividend Paid= 2,370 millions USD), are known, then it's (v=Variable Portion planned) is:

v= ($\$$-D/{d[1-t]})/{$\$'$[1+$s$]})-$f$-$i$
   = (51,294-2,370/{0.3000[1-0.3000]})
     /{ 48,851[1+0.0500]})-0.5800
     -0.0100
   = 19.00%

Law-1827:

If ($= Sales of Past Year= 48,851 millions USD), ($= Sales or Revenues= 51,294 millions USD), (s= [S/S']-1= Sales Growth= 5.00%), (t= T/B= Tax Rate= 30.00%), (i= I/S= Interest Portion= 1.00%), (v= V/S= Variable Portion= 19.00%), (d= D/A= Dividend Portion or Payout= 30.00%), and (D= Dividend Paid= 2,370 millions USD), are known, then it's (f=Fixed Portion planned) is:

$$f = (\$-D/\{d[1-t]\})/\{\$'[1+s]\}-v-i$$
$$= (51,294-2,370/\{0.3000[1-0.3000]\})$$
$$/\{48,851[1+0.0500]\}-0.1900$$
$$-0.0100$$
$$= 58.00\%$$

Law-1828:

If ($'= Sales of Past Year= 48,851 millions USD), ($= Sales or Revenues= 51,294 millions USD), (s= [S/S']-1= Sales Growth= 5.00%), (t= T/B= Tax Rate= 30.00%), (v= V/S= Variable Portion= 19.00%), (f= F/S= Fixed Portion= 58.00%), (d= D/A= Dividend Portion or Payout= 30.00%), and (D= Dividend Paid= 2,370 millions USD), are known, then it's (i= Interest Portion planned) is:

$$i = (\$-D/\{d[1-t]\})/\{\$'[1+s]\}-v-f$$
$$= 1-0.5800-0.0100-(51,294-2,370$$
$$/\{0.3000[1-0.3000]\})/\{48,851$$
$$[1+0.0500]\} -0.1900-0.5800$$
$$= 1.00\%$$

Law-1829:

If ($'$= Sales of Past Year= 48,851 millions USD), ($$= Sales or Revenues= 51,294 millions USD), (s= [S/S']-1= Sales Growth= 5.00%), (v= V/S= Variable Portion= 19.00%), (i= I/S= Interest Portion= 1.00%), (f= F/S= Fixed Portion= 58.00%), (d= D/A= Dividend Portion or Payout= 30.00%), and (D= Dividend Paid= 2,370 millions USD), are known, then it's (t= Tax Rate planned) is:

t= 1-[D/d]/{$$-$'v$[1+s]-$'f$[1+s]-$'i$[1+s]}
  = 1-D/(d{$$-$'$[1+s][v+f+i]})
  = 1-2,370/(0.3000{51,294-48,851
    [1+0.0500][0.1900+0.5800
    +0.0100})
  = 30.00%

Law-1830:

If ($'$= Sales of Past Year= 48,851 millions USD), ($$= Sales or Revenues= 51,294 millions USD), (s= [S/S']-1= Sales Growth= 5.00%), (i= I/S= Interest Portion= 1.00%),(v= V/S= Variable Portion= 19.00%), (f= F/S= Fixed Portion= 58.00%), (t= T/B= Tax Rate= 30.00%), and (D= Dividend Paid= 2,370 millions USD), are known, then it's (d= Dividend Portion or Payout planned) is:

d= D/([1-t]{$$-$'v$[1+s]-$'f$[1+s]-$'i$[1+s]}
  = D/([1-t]{$$-$'$[1+s][v+f+i]})
  = 1-2,370/(0.3000{51,294-48,851
    [1+0.0500][0.1900+0.5800
    +0.0100})
  = 30.00%

Law-1831:

If ($'= Sales of Past Year= 48,851 millions USD), (s= [S/S']-1= Sales Growth= 5.00%), (I= Interest Expense= 513 millions USD), (V= Variable Cost= 9,746 millions USD), (F= Fixed Cost= 29,750 millions USD), (t= T/B= Tax Rate= 30.00%), and (d= D/A= Dividend Portion or Payout= 30.00%), are known, then it's (D= Dividend Paid planned) is:

$$D = d[1-t]\{\$'[1+s]-V-F-I\}$$
$$= 0.3000[1-0.3000]\{48,851[1+0.0500]$$
$$-9,746-29,750-513\}$$
$$= 2,370 \text{ millions USD}$$

Law-1832:

If (D= Dividend Paid= 2,370 millions USD), (s= [S/S']-1= Sales Growth= 5.00%), (I= Interest Expense= 513 millions USD), (V= Variable Cost= 9,746 millions USD), (F= Fixed Cost= 29,750 millions USD), (t= T/B= Tax Rate= 30.00%), and (d= D/A= Dividend Portion or Payout= 30.00%), are known, then it's ($'= Sales Past) must be:

$$\$' = (V+F+I+D/\{d[1-t]\})/[1+s]$$
$$= (9,746+29,750+513+2,370/\{0.3000$$
$$[1-0.3000\})/[1+0.0500]$$
$$= 48,851 \text{ millions USD}$$

Law-1833:
If (**D**= Dividend Paid= 2,370 millions USD), (**I**= Interest Expense= 513 millions USD), (**V**= Variable Cost= 9,746 millions USD), (**S'**= Sales of Past Year= 48,851 millions USD), (**F**= Fixed Cost= 29,750 millions USD), (**t**= T/B= Tax Rate= 30.00%), and (**d**= D/A= Dividend Portion or Payout= 30.00%), are known, then it's (**s**= Sales Growth planned) is:

$$s = (V+F+I+D/\{d[1-t]\})/S' - 1$$
$$= (9,746+29,750+513+2,370/\{0.3000[1-0.3000]\})/48,851 - 1$$
$$= 5.00\%$$

Law-1834:
If (**D**= Dividend Paid= 2,370 millions USD), (**I**= Interest Expense= 513 millions USD), (**s**= [S/S']-1= Sales Growth= 5.00%), (**F**= Fixed Cost= 29,750 millions USD), (**S'**= Sales of Past Year= 48,851 millions USD), (**t**= T/B= Tax Rate= 30.00%), and (**d**= D/A= Dividend Portion or Payout= 30.00%), are known, then it's (**V**=Variable Cost planned) is:

$$V = S'[1+s] - F - I - D/\{d[1-t]\}$$
$$= 48,851[1+0.0500] - 29,750 - 513 - 2,370/\{0.3000[1-0.3000]\}$$
$$= 9,756 \text{ millions USD}$$

Law-1835:

If (**D**= Dividend Paid= 2,370 millions USD), (**I**= Interest Expense= 513 millions USD), (**s**= [S/S']-1= Sales Growth= 5.00%), (**V**= Variable Cost= 9,746 millions USD), (**S'**= Sales of Past Year= 48,851 millions USD), (**t**= T/B= Tax Rate= 30.00%), and (**d**= D/A= Dividend Portion or Payout= 30.00%), are known, then it's (**F**= Fixed Cost planned) is:

$$F = S'[1+s] - V - I - D/\{d[1-t]\}$$
$$= 48,851[1+0.0500] - 9,746 - 513 - 2,370 / \{0.3000[1-0.3000]\}$$
$$= \underline{29,750} \text{ millions USD}$$

Law-1836:

If (**D**= Dividend Paid= 2,370 millions USD), (**F**= Fixed Cost= 29,750 millions USD), (**s**= [S/S']-1= Sales Growth= 5.00%), (**V**= Variable Cost= 9,746 millions USD), (**S'**= Sales of Past Year= 48,851 millions USD), (**t**= T/B= Tax Rate= 30.00%), and (**d**= D/A= Dividend Portion or Payout= 30.00%), are known, then it's (**I**= Fixed Cost planned) is:

$$I = S'[1+s] - V - F - D/\{d[1-t]\}$$
$$= 48,851[1+0.0500] - 9,746 - 29,750 - 2,370 / \{0.3000[1-0.3000]\}$$
$$= \underline{513} \text{ millions USD}$$

Law-1837:
If (**D**= Dividend Paid= 2,370 millions USD), (**F**= Fixed Cost= 29,750 millions USD), (**s**= [S/S']-1= Sales Growth= 5.00%), (**V**= Variable Cost= 9,746 millions USD), (**S'**= Sales of Past Year= 48,851 millions USD), (**I**= Interest Expense= 513 millions USD), and (**d**= D/A= Dividend Portion or Payout= 30.00%), are known, then it's (**t**= Tax Rate planned) is:

$$t = 1 - D/(d\{S'[1+s]-V-F-I\})$$
$$= 1 - 2{,}370/\{0.3000\{48{,}851[1+0.0500]\\ -9{,}746 - 29{,}750 - 513\}\})$$
$$= 30.00\%$$

Law-1838:
If (**D**= Dividend Paid= 2,370 millions USD), (**F**= Fixed Cost= 29,750 millions USD), (**s**= [S/S']-1= Sales Growth= 5.00%), (**V**= Variable Cost= 9,746 millions USD), (**S'**= Sales of Past Year= 48,851 millions USD), (**I**= Interest Expense= 513 millions USD), and (**t**= T/B= Tax Rate= 30.00%), are known, then it's (**d**= Dividend Portion or Payout planned) is:

$$d = D/([1-t]\{S'[1+s]-V-F-I\})$$
$$= 2{,}370/\{[1-0.3000]\{48{,}851[1+0.0500]\\ -9{,}746 - 29{,}750 - 513\}\})$$
$$= 30.00\%$$

Law-1839:
If (**d**= D/A= Dividend Portion or Payout= 30.00%), (**F**= Fixed Cost= 29,750 millions USD), (**s**= [S/S']-1= Sales Growth= 5.00%), (**S**= Sales or Revenues= 51,294 millions USD), (**V**= Variable Cost= 9,746 millions USD), (**S'**= Sales of Past Year= 48,851 millions USD), (**i**= I/S= Interest Portion= 1.00%), and (**t**= T/B= Tax Rate= 30.00%), are known, then it's (**D**= Dividend Paid planned) is:

$$D = d[1-t]\{S'[1+s]-V-F-Si\}$$
$$= 0.3000[1-0.3000]\{48,851[1+0.0500]$$
$$-9,746-29,750-51,294*0.0100\})$$
$$= 2,370 \text{ millions USD}$$

Law-1840:
If (**d**= D/A=Dividend Portion or Payout= 30.00%), (**F**= Fixed Cost= 29,750 millions USD), (**s**= [S/S']-1= Sales Growth= 5.00%), (**S**= Sales or Revenues= 51,294 millions USD), (**V**= Variable Cost= 9,746 millions USD), (**D**= Dividend Paid= 2,370 millions USD), (**i**= I/S= Interest Portion= 1.00%), and (**t**= T/B= Tax Rate= 30.00%), are known, then it's (**S'**= Sales Past) must be:

$$S' = (D/d[1-t]+V+F+Si\})/[1+s]$$
$$= (2,370/\{ 0.3000[1-0.3000]+9,746$$
$$+29,750+51,294*0.0100\})$$
$$/ [1+0.0500]$$
$$= 48,851 \text{ millions USD}$$

Law-1841:

If (d= D/A=Dividend Portion or Payout= 30.00%), (F= Fixed Cost= 29,750 millions USD), (S'= Sales of Past Year= 48,851 millions USD), (S= Sales or Revenues= 51,294 millions USD), (V= Variable Cost= 9,746 millions USD), (D= Dividend Paid= 2,370 millions USD), (i= I/S= Interest Portion= 1.00%), and (t= T/B= Tax Rate= 30.00%), are known, then it's (s= Sales Growth planned) is:

$$s = (D/d[1-t]+V+F+Si\})/S'-1$$
$$= (2,370/\{0.3000[1-0.3000]+9,746 +29,750+51,294*0.0100\}) /48,851-1$$
$$= 5.00\%$$

Law-1842:

If (d= D/A=Dividend Portion or Payout= 30.00%), (F= Fixed Cost= 29,750 millions USD), (S'= Sales of Past Year= 48,851 millions USD), (S= Sales or Revenues= 51,294 millions USD), (s= [S/S']-1= Sales Growth= 5.00%), (D= Dividend Paid= 2,370 millions USD), (i= I/S= Interest Portion= 1.00%), and (t= T/B= Tax Rate= 30.00%), are known, then it's (V= Variable Cost planned) is:

$$V = S'[1+s]-F-Si-D/\{d[1-t]\}$$
$$= 48,851[1+0.0500] - 29,750-51,294 *0.0100 -2,370/\{0.3000 [1-0.3000]\}$$
$$= 9,746 \text{ millions USD}$$

Law-1843:

If (**d**= D/A= Dividend Portion or Payout= 30.00%), (**V**= Variable Cost= 9,746 millions USD), (**S'**= Sales of Past Year= 48,851 millions USD), (**S**= Sales or Revenues= 51,294 millions USD), (**s**= [S/S']-1= Sales Growth= 5.00%), (**D**= Dividend Paid= 2,370 millions USD), (**i**= I/S= Interest Portion= 1.00%), and (**t**= T/B= Tax Rate= 30.00%), are known, then it's (**F**= Fixed Cost planned) is:

$$F = S'[1+s]-V-Si-D/\{d[1-t]\}$$
$$= 48,851[1+0.0500]-9,746-51,294$$
$$*0.0100 -2,370/\{0.3000$$
$$[1-0.3000]$$
$$= 29,750 \text{ millions USD}$$

Law-1844:

If (**d**= D/A= Dividend Portion or Payout= 30.00%), (**V**= Variable Cost= 9,746 millions USD), (**S'**= Sales of Past Year= 48,851 millions USD), (**F**= Fixed Cost= 29,750 millions USD), (**s**= [S/S']-1= Sales Growth= 5.00%), (**D**= Dividend Paid= 2,370 millions USD), (**i**= I/S= Interest Portion= 1.00%), and (**t**= T/B= Tax Rate= 30.00%), are known, then it's (**S**= Sales or Revenues planned) is:

$$S = S'[1+s]-V-F-D/\{d[1-t]\}/i$$
$$= 48,851[1+0.0500]-9,746-29,750$$
$$-2,370/\{0.3000[1-0.3000]\}$$
$$/0.0100$$
$$= 51,294 \text{ millions USD}$$

Law-1845:

If (**d**= D/A=Dividend Portion or Payout= 30.00%), (**V**= Variable Cost= 9,746 millions USD), (**$'**= Sales of Past Year= 48,851 millions USD), (**F**= Fixed Cost= 29,750 millions USD), (**s**= [S/S']-1= Sales Growth= 5.00%), (**D**= Dividend Paid= 2,370 millions USD), (**$**= Sales or Revenues= 51,294 millions USD), and (**t**= T/B= Tax Rate= 30.00%), are known, then it's (**i**= Interest Portion planned) is:

$i = \$'[1+s]-V-F-D/\{d[1-t]\}/\$$
$= 48,851[1+0.0500]-9,746-29,750$
$\quad -2,370/\{0.3000[1-0.3000]\}$
$\quad /51,294$
$= \underline{1.00\%}$

Law-1846:

If (**d**= D/A= Dividend Portion or Payout= 30.00%), (**V**= Variable Cost= 9,746 millions USD), (**$'**= Sales of Past Year= 48,851 millions USD), (**F**= Fixed Cost= 29,750 millions USD), (**s**= [S/S']-1= Sales Growth= 5.00%), (**D**= Dividend Paid= 2,370 millions USD), (**$**= Sales or Revenues= 51,294 millions USD), and (**i**= I/S= Interest Portion= 1.00%), are known, then it's (**t**= Tax Rate planned) is:

$t = 1-D/(d\{\$'[1+s]-V-F-\$i\}$
$= 1-2,370/(0.3000\{48,851[1+0.0500]$
$\quad -9,746-29,750 -51,294*0.0100\}$
$= \underline{30.00\%}$

Law-1847:

If (t= T/B= Tax Rate= 30.00%), (V= Variable Cost= 9,746 millions USD), (S'= Sales of Past Year= 48,851 millions USD), (F= Fixed Cost= 29,750 millions USD), (s= [S/S']-1= Sales Growth= 5.00%), (D= Dividend Paid= 2,370 millions USD), (S= Sales or Revenues= 51,294 millions USD), and (i= I/S= Interest Portion= 1.00%), are known, then it's (d= Dividend Portion or Payout planned) is:

d= D/([1-t]{S'[1+s]-V-F-Si}
 = 2,370/([1- 0.3000]{48,851[1+0.0500]
  -9,746-29,750 -51,294*0.0100}
 = 30.00%

Law-1848:

If (t= T/B= Tax Rate= 30.00%), (V= Variable Cost= 9,746 millions USD), (S'= Sales of Past Year= 48,851 millions USD), (F= Fixed Cost= 29,750 millions USD), (s= [S/S']-1= Sales Growth= 5.00%), (d= D/A=Dividend Portion or Payout= 30.00%), (S= Sales or Revenues= 51,294 millions USD), and (i= I/S= Interest Portion= 1.00%), are known, then it's (D= Dividend Paid planned) is:

D= d[1-t]{S'[1+s]-V-F-S'i[1+s]}
 = 0.3000[1- 0.3000]{48,851[1+0.0500]
  -9,746-29,750 - 48,851*0.0100
  [1+0.0500]}
 = 2,370 millions USD

Law-1849:

If (**t**= T/B= Tax Rate= 30.00%), (**V**= Variable Cost= 9,746 millions USD),(**D**= Dividend Paid= 2,370 millions USD), (**F**= Fixed Cost= 29,750 millions USD), (**s**= [S/S']-1= Sales Growth= 5.00%), (**d**= D/A=Dividend Portion or Payout= 30.00%), (**$**= Sales or Revenues= 51,294 millions USD), and (**i**= I/S= Interest Portion= 1.00%), are known, then it's (**$'**= Sales Past) must be:

$'= (**V**+**F**+**D**/{**d**[1-**t**]})/{[1+**s**][1-**i**]}
　= (9,746+29,750 +2,370/{ 0.3000
　　　　[1-0.3000]})/{[1+0.0500]
　　　　[1-0.0100]}
　= 48,851 millions USD

Law-1850:

If (**t**= T/B= Tax Rate= 30.00%), (**V**= Variable Cost= 9,746 millions USD), (**D**= Dividend Paid= 2,370 millions USD), (**F**= Fixed Cost= 29,750 millions USD), (**$'**= Sales of Past Year= 48,851 millions USD), (**d**=D/A= Dividend Portion or Payout= 30.00%), (**$**= Sales or Revenues= 51,294 millions USD), and (**i**= I/S= Interest Portion= 1.00%), are known, then it's (**s**=Sales Growth planned) is:

s= (**V**+**F**+**D**/{**d**[1-**t**]})/{**$'**[1-**i**]}-1
　= (9,746+29,750+2,370/{0.3000
　　　　[1-0.3000]})/{48,851
　　　　[1-0.0100]}-1
　= 5.00%

Law-1851:

If (**t**= T/B= Tax Rate= 30.00%), (**V**= Variable Cost= 9,746 millions USD), (**D**= Dividend Paid= 2,370 millions USD), (**F**= Fixed Cost= 29,750 millions USD), (**$'**= Sales of Past Year= 48,851 millions USD), (**d**=D/A =Dividend Portion or Payout= 30.00%), (**$**= Sales or Revenues= 51,294 millions USD), and (**s**= [S/S']-1= Sales Growth= 5.00%), are known, then it's (**i**= Interest Portion planned) is:

$$i = 1 - (V+F+D/\{d[1-t]\})/\{\$'[1+s]\}$$
$$= -(9,746+29,750 +2,370/\{0.3000[1- 0.3000]\})/\{48,851[1+ 0.0500]\}$$
$$= 1.00\%$$

Law-1852:

If (**t**= T/B= Tax Rate= 30.00%), (**i**= I/S= Interest Portion= 1.00%), (**D**= Dividend Paid= 2,370 millions USD), (**F**= Fixed Cost= 29,750 millions USD), (**$'**= Sales of Past Year= 48,851 millions USD), (**d**= D/A= Dividend Portion or Payout= 30.00%), (**$**= Sales or Revenues= 51,294 millions USD), and (**s**= [S/S']-1= Sales Growth= 5.00%), are known, then it's (**V**=Variable Cost planned) is:

$$V = \$'[1+s][1-i]-F-D/\{d[1-t]\}$$
$$= 48,851[1+ 0.0500][1-0.0100]-29,750 -2,370/\{ 0.3000[1- 0.3000]\}$$
$$= 9,746 \text{ millions USD}$$

Law-1853:

If (t= T/B= Tax Rate= 30.00%), (i= I/S= Interest Portion= 1.00%), (D= Dividend Paid= 2,370 millions USD), (V= Variable Cost= 9,746 millions USD), (S'= Sales of Past Year= 48,851 millions USD), (d= D/A= Dividend Portion or Payout= 30.00%), (S= Sales or Revenues= 51,294 millions USD), and (s= [S/S']-1= Sales Growth= 5.00%), are known, then it's (F=Fixed Cost planned) is:

$F = S'[1+s][1-i] - V - D/\{d[1-t]\}$
  = 48,851[1+ 0.0500][1-0.0100] -9,746
    -2,370/{ 0.3000[1-0.3000]}
  = 29,750 millions USD

Law-1854:

If (F= Fixed Cost= 29,750 millions USD), (i= I/S= Interest Portion= 1.00%), (D= Dividend Paid= 2,370 millions USD), (V= Variable Cost= 9,746 millions USD), (S'= Sales of Past Year= 48,851 millions USD), (d=D/A=Dividend Portion or Payout= 30.00%), (S= Sales or Revenues= 51,294 millions USD), and (s= [S/S']-1= Sales Growth= 5.00%), are known, then it's (t=Tax Rate planned) is:

$t = 1 - [D/d]/\{S'[1+s] - V - F - S'i[1+s]\}$
  = 1 - D/(d{S'[1+s][1-i] - V - F})
  = 1 - 2,370/(0.3000{48,851[1+ 0.0500]
    [1-0.0100]- 9,746 -29,750})
  = 30.00%

Law-1855:

If (F= Fixed Cost= 29,750 millions USD), (i= I/S= Interest Portion= 1.00%), (D= Dividend Paid= 2,370 millions USD), (V= Variable Cost= 9,746 millions USD), ($\$'$= Sales of Past Year= 48,851 millions USD), (t= T/B= Tax Rate= 30.00%), ($\$$= Sales or Revenues= 51,294 millions USD), and (s= [S/S']-1= Sales Growth= 5.00%), are known, then it's (d= Dividend Portion or Payout planned) is:

d= D/([1-t]{$\$'$[1+$s$]-$V$-$F$-$\$'i$[1+$s$]})
  = D/([1-t]{$\$'$[1+$s$][1-$i$]-$V$-$F$})
  = 2,370/([1-0.3000]{48,851[1+ 0.0500][1-0.0100]- 9,746-29,750})
  = 30.00%

Law-1856:

If (f= F/S=Fixed Portion= 58.00%), (I= Interest Expense= 513 millions USD), (d= D/A= Dividend Portion or Payout= 30.00%), (V= Variable Cost= 9,746 millions USD), ($\$'$= Sales of Past Year= 48,851 millions USD), (t= T/B= Tax Rate= 30.00%), ($\$$= Sales or Revenues= 51,294 millions USD), and (s= [S/S']-1= Sales Growth= 5.00%), are known, then it's (D= Dividend Paid planned) is:

D= d[1-t]{$\$'$[1+$s$]-$V$-$F$-$\$'i$[1+$s$]})
  = d[1-t]{$\$'$[1+$s$][1-$i$]-$V$-$F$})
  = 0.3000[1-0.3000]{48,851[1+ 0.0500][1-0.0100]- 9,746 -29,750})
  = 2,370 millions USD

Law-1857:
  If (**f**= F/S=Fixed Portion= 58.00%), (**I**= Interest Expense= 513 millions USD), (**d**= D/A= Dividend Portion or Payout= 30.00%), (**V**= Variable Cost= 9,746 millions USD), (**D**= Dividend Paid= 2,370 millions USD), (**t**= T/B= Tax Rate= 30.00%), (**$**= Sales or Revenues= 51,294 millions USD), and (**s**= [S/S']-1= Sales Growth= 5.00%), are known, then it's (**$'**= Sales Past) must be:
  $' = (**V**+**$f**+**I**+**D**/{**d**[1-**t**]})/[1+**s**]
    = 9,746 +51,294*0.5800+513+2,370
      /{0.3000[1-0.3000]})/ [1+0.0500]
    = 48,851 millions USD

Law-1858:
  If (**f**= F/S=Fixed Portion= 58.00%), (**I**= Interest Expense= 513 millions USD), (**d**= D/A= Dividend Portion or Payout= 30.00%), (**V**= Variable Cost= 9,746 millions USD), (**D**= Dividend Paid= 2,370 millions USD), (**t**= T/B= Tax Rate= 30.00%), (**$**= Sales or Revenues= 51,294 millions USD), and (**$'**= Sales of Past Year= 48,851 millions USD), are known, then it's (**s**= Sales Growth planned) is:
  s= (**V**+**$f**+**I**+**D**/{**d**[1-**t**]})/$'-1
    = 9,746 +51,294*0.5800+513+2,370
      /{0.3000[1-0.3000]})/ 48,851-1
    = 5.00%

Law-1859:
If (**f**= F/S=Fixed Portion= 58.00%), (**I**= Interest Expense= 513 millions USD), (**d**= D/A= Dividend Portion or Payout= 30.00%), (**s**= [S/S']-1= Sales Growth= 5.00%), (**D**= Dividend Paid= 2,370 millions USD), (**t**= T/B= Tax Rate= 30.00%), (**$**= Sales or Revenues= 51,294 millions USD), and (**$'**= Sales of Past Year= 48,851 millions USD), are known, then it's (**V**= Variable Cost planned) is:

$V = (\$'[1+s] - \$f - I - D/\{d[1-t]\})$
$= 48,851[1+0.0500] - 51,294*0.5800 - 513 - 2,370/\{0.3000[1-0.3000]\})$
$= 9,746$ millions USD

Law-1860:
If (**f**= F/S=Fixed Portion= 58.00%), (**I**= Interest Expense= 513 millions USD), (**d**= D/A= Dividend Portion or Payout= 30.00%), (**V**= Variable Cost= 9,746 millions USD), (**D**= Dividend Paid= 2,370 millions USD), (**t**= T/B= Tax Rate= 30.00%), (**s**= [S/S']-1= Sales Growth= 5.00%), and (**$'**= Sales of Past Year= 48,851 millions USD), are known, then it's (**$**= Sales or Revenues planned) is:

$\$ = (\$'[1+s] - V - I - D/\{d[1-t]\})/f$
$= (48,851[1+0.0500] - 9,746 - 513 - 2,370 /\{0.3000[1-0.3000]\})/ 0.5800$
$= 51,294$ millions USD

Law-1861:
If ($= Sales or Revenues= 51,294 millions USD), (I= Interest Expense= 513 millions USD), (d= D/A= Dividend Portion or Payout= 30.00%), (V= Variable Cost= 9,746 millions USD), (D= Dividend Paid= 2,370 millions USD), (t= T/B= Tax Rate= 30.00%), (s= [S/S']-1= Sales Growth= 5.00%), and ($'= Sales of Past Year= 48,851 millions USD), are known, then it's (f= Fixed Portion planned) is:

f= ($'[1+s]-V-I-D/{d[1-t]})/$
 = (48,851[1+0.0500]-9,746-513-2,370
      /{0.3000[1-0.3000]})/51,294
 = 58.00%

Law-1862:
If ($= Sales or Revenues= 51,294 millions USD), (f= F/S=Fixed Portion= 58.00%), (d= D/A= Dividend Portion or Payout= 30.00%), (V= Variable Cost= 9,746 millions USD), (D= Dividend Paid= 2,370 millions USD), (t= T/B= Tax Rate= 30.00%), (s= [S/S']-1= Sales Growth= 5.00%), and ($'= Sales of Past Year= 48,851 millions USD), are known, then it's (I= Interest Expense planned) is:

I= ($'[1+s]-V-$f-D/{d[1-t]})
 = (48,851[1+0.0500]-9,746-51,294
      *0.5800-2,370
      /{0.3000[1-0.3000]})
 = 513 millions USD

Law-1863:
If ($= Sales or Revenues= 51,294 millions USD), (f= F/S=Fixed Portion= 58.00%), (d= D/A= Dividend Portion or Payout= 30.00%), (V= Variable Cost= 9,746 millions USD), (D= Dividend Paid= 2,370 millions USD), (I= Interest Expense= 513 millions USD), (s= [S/S']-1= Sales Growth= 5.00%), and ($'= Sales of Past Year= 48,851 millions USD), are known, then it's (t= Tax Rate planned) is:

t= 1-D/(d{$'[1+s]-V-$f-I})
= 1-2,370/(0.3000{48,851[1+0.0500]
-9,746-51,294*0.5800-513})
=30.00%

Law-1864:
If ($= Sales or Revenues= 51,294 millions USD), (f= F/S=Fixed Portion= 58.00%), (t= T/B= Tax Rate= 30.00%), (V= Variable Cost= 9,746 millions USD), (D= Dividend Paid= 2,370 millions USD), (I= Interest Expense= 513 millions USD), (s= [S/S']-1= Sales Growth= 5.00%), and ($'= Sales of Past Year= 48,851 millions USD), are known, then it's (d= Dividend Portion planned) is:

d= D/([1-t]{$'[1+s]-V-$f-I})
= 2,370/([1-0.3000]{48,851[1+0.0500]
-9,746-51,294*0.5800-513})
=30.00%

Law-1865:
If ($\$$= Sales or Revenues= 51,294 millions USD), (f= F/S=Fixed Portion= 58.00%), (t= T/B= Tax Rate= 30.00%), (V= Variable Cost= 9,746 millions USD), (d= D/A= Dividend Portion or Payout= 30.00%), (i= I/S= Interest Portion= 1.00%), (s= [S/S']-1= Sales Growth= 5.00%), and ($\$'$= Sales of Past Year= 48,851 millions USD), are known, then it's (D= Dividend Paid planned) is:

$D = d[1-t]\{\$'[1+s]-V-\$f-\$i\})$
$ = d[1-t]\{\$'[1+s]-V-\$[f+i]\})$
$ = 0.3000[1-0.3000]\{48,851[1+0.0500]$
$\phantom{D = 0.3000[1-0.3000]\{}-9,746-51,294[0.5800+0.0100]\})$
$ = \underline{2,370}$ millions USD

Law-1866:
If ($\$$= Sales or Revenues= 51,294 millions USD), (f= F/S=Fixed Portion= 58.00%), (t= T/B= Tax Rate= 30.00%), (V= Variable Cost= 9,746 millions USD), (d= D/A= Dividend Portion or Payout= 30.00%), (i= I/S= Interest Portion= 1.00%), (s= [S/S']-1= Sales Growth= 5.00%), and (D= Dividend Paid= 2,370 millions USD), are known, then it's ($\$'$= Sales Past) must be:

$\$' = \{V+\$[f+i]+D/\{d[1-t]\}/[1+s]$
$ = \{9,746+51,294[0.5800+0.0100]+2,370$
$\phantom{\$' = \{}/\{0.3000[1-0.3000]\}/[1+0.0500]$
$ = \underline{48,851}$ millions USD

Law-1867:

If ($= Sales or Revenues= 51,294 millions USD), (f= F/S=Fixed Portion= 58.00%), (t= T/B= Tax Rate= 30.00%), (V= Variable Cost= 9,746 millions USD), (d= D/A= Dividend Portion or Payout= 30.00%), (i= I/S= Interest Portion= 1.00%), ($'= Sales of Past Year= 48,851 millions USD), and (D= Dividend Paid= 2,370 millions USD), are known, then it's (s= Sales Growth planned) is:

$s = \{V + \$[f+i] + D/\{d[1-t]\}\}/\$' - 1$
$= \{9,746 + 51,294[0.5800 + 0.0100] + 2,370 /\{0.3000[1-0.3000]\}\}/48,851 - 1$
$= \underline{5.00\%}$

Law-1868:

If ($= Sales or Revenues= 51,294 millions USD), (f= F/S=Fixed Portion= 58.00%), (t= T/B= Tax Rate= 30.00%), (s= [S/S']-1= Sales Growth= 5.00%), (d= D/A=Dividend Portion or Payout= 30.00%), (i= I/S= Interest Portion= 1.00%), ($'= Sales of Past Year= 48,851 millions USD), and (D= Dividend Paid= 2,370 millions USD), are known, then it's (V= Variable Cost planned) is:

$V = \{\$'[1+s] - \$[f+i] - D/\{d[1-t]\}\}$
$= \{48,851[1+0.0500] - 51,294[0.5800 + 0.0100] - 2,370 /\{0.3000[1-0.3000]\}\}$
$= \underline{9,746}$ millions USD

Law-1869:
If (**V**= Variable Cost= 9,746 millions USD), (**f**= F/S=Fixed Portion= 58.00%), (**t**= T/B= Tax Rate= 30.00%), (**s**= [S/S']-1= Sales Growth= 5.00%), (**d**= D/A=Dividend Portion or Payout= 30.00%), (**i**= I/S= Interest Portion= 1.00%), (**$'**= Sales of Past Year= 48,851 millions USD), and (**D**= Dividend Paid= 2,370 millions USD), are known, then it's (**$**= Sales or Revenues planned) is:

$= ($'[1+s]-**V**-**D**/{**d**[1-**t**]})/[**f**+**i**]
  = (48,851[1+0.0500]-9,746-2,370
       /{0.3000[1-0.3000]})
       /[0.5800+0.0100]
  =51,294 millions USD

Law-1870:
If (**V**= Variable Cost= 9,746 millions USD), (**$**= Sales or Revenues= 51,294 millions USD), (**t**= T/B= Tax Rate= 30.00%), (**s**= [S/S']-1= Sales Growth= 5.00%), (**d**= D/A=Dividend Portion or Payout= 30.00%), (**i**= I/S= Interest Portion= 1.00%), (**$'**= Sales of Past Year= 48,851 millions USD), and (**D**= Dividend Paid= 2,370 millions USD), are known, then it's (**f**= Fixed Portion planned) is:

f= ($'[1+s]-**V**-**D**/{**d**[1-**t**]})/$-**i**
  = (48,851[1+0.0500]-9,746-2,370
       /{0.3000[1-0.3000]})/51,294
       -0.0100
  =58.00%

Law-1871:

If (V= Variable Cost= 9,746 millions USD), (S= Sales or Revenues= 51,294 millions USD), (t= T/B= Tax Rate= 30.00%), (s= [S/S']-1= Sales Growth= 5.00%), (d= D/A= Dividend Portion or Payout= 30.00%), (f= F/S=Fixed Portion= 58.00%), (S'= Sales of Past Year= 48,851 millions USD), and (D= Dividend Paid= 2,370 millions USD), are known, then it's (i= Interest Portion planned) is:

i= (S[1+s]-V-D/{d[1-t]})/S-f
= (48,851[1+0.0500]-9,746-2,370
/{0.3000[1-0.3000]})/51,294
-0.5800
= 1.00%

Law-1872:

If (V= Variable Cost= 9,746 millions USD), (S= Sales or Revenues= 51,294 millions USD), (i= I/S= Interest Portion= 1.00%), (s= [S/S']-1= Sales Growth= 5.00%), (d= D/A= Dividend Portion or Payout= 30.00%), (f= F/S=Fixed Portion= 58.00%), (S'= Sales of Past Year= 48,851 millions USD), and (D= Dividend Paid= 2,370 millions USD), are known, then it's (t= Tax Rate planned) is:

t= 1-[D/d]/{S'[1+s]-V-$S$$f$-$S$$i$}
= 1-D/(d{S'[1+s]-V-S[f+i]})
= 1-2,370/(0.3000{ 48,851[1+0.0500]
-9,746- 51,294
[0.5800+0.0100]})
= 30.00%

Law-1873:
If (V= Variable Cost= 9,746 millions USD), (S= Sales or Revenues= 51,294 millions USD), (i= I/S= Interest Portion= 1.00%), (s= [S/S']-1= Sales Growth= 5.00%), (t= T/B= Tax Rate= 30.00%), (f= F/S=Fixed Portion= 58.00%), (S'= Sales of Past Year= 48,851 millions USD), and (D= Dividend Paid= 2,370 millions USD), are known, then it's (d= Dividend Portion or Payout planned) is:

$d = D/(1-t)\{S'[1+s]-V-Sf-Si\}$
 $= D/([1-t]\{S'[1+s]-V-S[f+i]\})$
 $= 2,370/([1-0.3000]\{48,851[1+0.0500]$
  $-9,746-51,294[0.5800+0.0100]\})$
$=30.00\%$

Law-1874:
If (V= Variable Cost= 9,746 millions USD), (S= Sales or Revenues= 51,294 millions USD), (i= I/S= Interest Portion= 1.00%), (s= [S/S']-1= Sales Growth= 5.00%), (t= T/B= Tax Rate= 30.00%), (f= F/S=Fixed Portion= 58.00%), (S'= Sales of Past Year= 48,851 millions USD), and (d= D/A=Dividend Portion or Payout= 30.00%), are known, then it's (D= Dividend Paid planned) is:

$D = d[1-t]\{S'[1+s]-V-Sf-S'i[1+s]\}$
 $= d[1-t]\{S'[1+s][1-i]-V-Sf\})$
 $= 0.3000[1-0.3000]\{48,851[1+0.0500]$
  $[1-0.0100]-9,746$
  $-51,294*0.5800\}$
$=2,370$ millions USD

Law-1875:

If (**V**= Variable Cost= 9,746 millions USD), (**S**= Sales or Revenues= 51,294 millions USD), (**i**= I/S= Interest Portion= 1.00%), (**s**= [S/S']-1= Sales Growth= 5.00%), (**t**= T/B= Tax Rate= 30.00%), (**f**= F/S= Fixed Portion= 58.00%), (**D**= Dividend Paid= 2,370 millions USD), and (**d**= D/A= Dividend Portion or Payout= 30.00%), are known, then it's (**S'**= Sales Past) must be:

$$S' = (V+Sf+D/\{d[1-t]\})/\{[1+s][1-i]\}$$
$$= (9,746+ 51,294*0.5800+2,370$$
$$/\{0.3000[1- 0.3000]\})$$
$$/\{[1+0.0500][1-0.0100]\}$$
$$=48,851 \text{ millions USD}$$

Law-1876:

If (**V**= Variable Cost= 9,746 millions USD), (**S**= Sales or Revenues= 51,294 millions USD), (**i**= I/S= Interest Portion= 1.00%), (**S'**= Sales of Past Year= 48,851 millions USD), (**t**= T/B= Tax Rate= 30.00%), (**f**= F/S=Fixed Portion= 58.00%), (**D**= Dividend Paid= 2,370 millions USD), and (**d**= D/A=Dividend Portion or Payout= 30.00%), are known, then it's (**s**= Sales Growth planned) is:

$$s = (V+Sf+D/\{d[1-t]\})/\{S'[1-i]\}-1$$
$$= (9,746+ 51,294*0.5800+2,370$$
$$/\{0.3000[1- 0.3000]\})$$
$$/\{48,851[1-0.0100]\}-1$$
$$=5.00\%$$

Law-1877:

If (**V**= Variable Cost= 9,746 millions USD), (**$**= Sales or Revenues= 51,294 millions USD), (**s**= [S/S']-1= Sales Growth= 5.00%), (**$'**= Sales of Past Year= 48,851 millions USD), (**t**= T/B= Tax Rate= 30.00%), (**f**= F/S=Fixed Portion= 58.00%), (**D**= Dividend Paid= 2,370 millions USD), and (**d**= D/A=Dividend Portion or Payout= 30.00%), are known, then it's (**i**= Interest Portion planned) is:

i= 1-(**V**+**$f**+**D**/{**d**[1-**t**]})/{**$'**[1+**s**]}
= 1-(9,746+ 51,294*0.5800+2,370
/{0.3000[1- 0.3000]})
/{48,851[1+0.0500]}
=1.00%

Law-1878:

If (**D**= Dividend Paid= 2,370 millions USD), (**$**= Sales or Revenues= 51,294 millions USD), (**i**= I/S= Interest Portion= 1.00%), (**s**= [S/S']-1= Sales Growth= 5.00%), (**t**= T/B= Tax Rate= 30.00%), (**f**= F/S= Fixed Portion= 58.00%), (**$'**= Sales of Past Year= 48,851 millions USD), and (**d**= D/A=Dividend Portion or Payout= 30.00%), are known, then it's (**V**= Variable Cost planned) is:

V= **$'**[1+**s**][1-**i**]-**$f**-**D**/{**d**[1-**t**]}
= 48,851[1+0.0500][1-0.0100]
-51,294*0.5800
-2,370/{0.3000[1- 0.3000]}
=9,746 millions USD

Law-1879:

If (**D**= Dividend Paid= 2,370 millions USD), (**V**= Variable Cost= 9,746 millions USD), (**i**= I/S= Interest Portion= 1.00%), (**s**= [S/S']-1= Sales Growth= 5.00%), (**t**= T/B= Tax Rate= 30.00%), (**f**= F/S= Fixed Portion= 58.00%), (**S'**= Sales of Past Year= 48,851 millions USD), and (**d**= D/A=Dividend Portion or Payout= 30.00%), are known, then it's (**S**= Sales or Revenues planned) is:

$S = (S'[1+s][1-i]-V-D/\{d[1-t]\})/f$
$= (48,851[1+0.0500][1-0.0100]-9,746$
$\quad -2,370/\{0.3000[1-0.3000]\})$
$\quad /0.5800$
$= \underline{51,294}$ millions USD

Law-1880:

If (**D**= Dividend Paid= 2,370 millions USD), (**V**= Variable Cost= 9,746 millions USD), (**i**= I/S= Interest Portion= 1.00%), (**s**= [S/S']-1= Sales Growth= 5.00%), (**t**= T/B= Tax Rate= 30.00%), (**S**= Sales or Revenues= 51,294 millions USD), (**S'**= Sales of Past Year= 48,851 millions USD), and (**d**= D/A= Dividend Portion or Payout= 30.00%), are known, then it's (**f**= Fixed Portion planned) is:

$f = (S'[1+s][1-i]-V-D/\{d[1-t]\})/S$
$= (48,851[1+0.0500][1-0.0100]- 9,746$
$\quad -2,370/\{0.3000[1-0.3000]\})$
$\quad /51,294$
$= \underline{58.00\%}$

Law-1881:
If (**D**= Dividend Paid= 2,370 millions USD), (**V**= Variable Cost= 9,746 millions USD), (**i**= I/S= Interest Portion= 1.00%), (**s**= [S/S']-1= Sales Growth= 5.00%), (**f**= F/S=Fixed Portion= 58.00%), (**$**= Sales or Revenues= 51,294 millions USD), (**$'**= Sales of Past Year= 48,851 millions USD), and (**d**= D/A= Dividend Portion or Payout= 30.00%), are known, then it's (**t**=Tax Rate planned) is:

$$t = 1-[D/d]\{\$i[1+s]-V-\$f-\$'i[1+s]\}$$
$$= 1-D/(d\{\$'[1+s][1-i]-V-\$f\})$$
$$= 1-2,370/(0.3000\{48,851[1+0.0500]$$
$$[1-0.0100]-9,746$$
$$-51,294*0.5800\})$$
$$=30.00\%$$

Law-1882:
If (**D**= Dividend Paid= 2,370 millions USD), (**V**= Variable Cost= 9,746 millions USD), (**i**= I/S= Interest Portion= 1.00%), (**s**= [S/S']-1= Sales Growth= 5.00%), (**f**= F/S= Fixed Portion= 58.00%), (**$**= Sales or Revenues= 51,294 millions USD), (**$'**= Sales of Past Year= 48,851 millions USD), and (**t**= T/B= Tax Rate= 30.00%), are known, then it's (**d**= Dividend Portion planned) is:

$$d = D/([1-t]]\{\$[1+s]-V-\$f-\$'i[1+s]\}$$
$$= D/([1-t]\{\$'[1+s][1-i]-V-\$f\})$$
$$= 2,370([1-0.3000]\{48,851[1+0.0500]$$
$$[1-0.0100]-9,746$$
$$-51,294*0.5800\})$$
$$=30.00\%$$

Law-1883:
If (**d**= D/A=Dividend Portion or Payout= 30.00%), (**V**= Variable Cost= 9,746 millions USD), (**i**= I/S= Interest Portion= 1.00%), (**s**= [S/S']-1= Sales Growth= 5.00%), (**f**= F/S=Fixed Portion= 58.00%), (**$**= Sales or Revenues= 51,294 millions USD), (**$'**= Sales of Past Year= 48,851 millions USD), and (**t**= T/B= Tax Rate= 30.00%), are known, then it's (**D**= Dividend Paid planned) is:

$$\begin{aligned}\mathbf{D} &= \mathbf{d}[1\text{-}\mathbf{t}]]\{\$[1+\mathbf{s}]\text{-}\mathbf{V}\text{-}\$'\mathbf{f}[1+\mathbf{s}]\text{-}\$'\mathbf{i}[1+\mathbf{s}]\} \\ &= \mathbf{d}[1\text{-}\mathbf{t}]\{\$'[1+\mathbf{s}][1\text{-}\mathbf{f}\text{-}\mathbf{i}]\text{-}\mathbf{V}\}) \\ &= 0.3000[1\text{-}0.3000]\{48,851[1+0.0500] \\ &\qquad [1\text{-}0.5800\text{-}0.0100]\text{-} 9,746\}) \\ &= \underline{2,370} \text{ millions USD}\end{aligned}$$

Law-1884:
If (**d**= D/A=Dividend Portion or Payout= 30.00%), (**V**= Variable Cost= 9,746 millions USD), (**i**= I/S= Interest Portion= 1.00%), (**s**= [S/S']-1= Sales Growth= 5.00%), (**f**= F/S= Fixed Portion= 58.00%), (**$**= Sales or Revenues= 51,294 millions USD), (**D**= Dividend Paid= 2,370 millions USD), and (**t**= T/B= Tax Rate= 30.00%), are known, then it's (**$'**= Sales Past) must be:

$$\begin{aligned}\$' &= (\mathbf{V}+\mathbf{D}/\{\mathbf{d}[1\text{-}\mathbf{t}]\})/\{[1+\mathbf{s}][1\text{-}\mathbf{f}\text{-}\mathbf{i}]\} \\ &= (9,746+2,370/\{\ 0.3000[1\text{-}0.3000]\} \\ &\qquad /\{[1+0.0500][1\text{-}0.5800\text{-}0.0100]\} \\ &= \underline{48,851} \text{ millions USD}\end{aligned}$$

Law-1885:
If (d= D/A= Dividend Portion or Payout= 30.00%), (V= Variable Cost= 9,746 millions USD), (i= I/S= Interest Portion= 1.00%), (S'= Sales of Past Year= 48,851 millions USD), (f= F/S= Fixed Portion= 58.00%), (S= Sales or Revenues= 51,294 millions USD), (D= Dividend Paid= 2,370 millions USD), and (t= T/B= Tax Rate= 30.00%), are known, then it's (s= Sales Growth planned) is:

$$s = (V+D/\{d[1-t]\})/\{S'[1-f-i]\}-1$$
$$= (9,746+2,370/\{0.3000[1-0.3000]\})/\{48,851[1-0.5800-0.0100]\}-1$$
$$= 5.00\%$$

Law-1886:
If (d= D/A= Dividend Portion or Payout= 30.00%), (V= Variable Cost= 9,746 millions USD), (i= I/S= Interest Portion= 1.00%), (S'= Sales of Past Year= 48,851 millions USD), (s= [S/S']-1= Sales Growth= 5.00%), (S= Sales or Revenues= 51,294 millions USD), (D= Dividend Paid= 2,370 millions USD), and (t= T/B= Tax Rate= 30.00%), are known, then it's (f= Fixed Portion planned) is:

$$f = 1-i-(V+D/\{d[1-t]\})/\{S'[1+s]\}$$
$$= 1-0.0100-(9,746+2,370/\{0.3000[1-0.3000]\})/\{48,851[1+0.0500]\}$$
$$= 58.00\%$$

Law-1887:

If (d= D/A= Dividend Portion or Payout= 30.00%), (V= Variable Cost= 9,746 millions USD), (f= F/S=Fixed Portion= 58.00%), (S'= Sales of Past Year= 48,851 millions USD), (s= [S/S']-1= Sales Growth= 5.00%), (S= Sales or Revenues= 51,294 millions USD), (D= Dividend Paid= 2,370 millions USD), and (t= T/B= Tax Rate= 30.00%), are known, then it's (i= Interest Portion planned) is:

$i = 1 - f - (V + D/\{d[1-t]\})/\{S'[1+s]\}$
$= 1 - 0.5800 - (9,746 + 2,370/\{0.3000$
$[1-0.3000]\})/\{48,851[1+0.0500]\}$
$= 58.00\%$

Law-1888:

If (d= D/A= Dividend Portion or Payout= 30.00%), (i= I/S= Interest Portion= 1.00%), (f= F/S=Fixed Portion= 58.00%), (S'= Sales of Past Year= 48,851 millions USD), (s= [S/S']-1= Sales Growth= 5.00%), (S= Sales or Revenues= 51,294 millions USD), (D= Dividend Paid= 2,370 millions USD), and (t= T/B= Tax Rate= 30.00%), are known, then it's (V= Variable Cost planned) is:

$V = S'[1+s][1-f-i] - D/\{d[1-t]\}$
$= 48,851[1+0.0500][1-0.5800-0.0100]$
$-2,370/\{0.3000[1-0.3000]\}$
$= 9,746$ millions USD

Law-1889:

If (**d**= D/A=Dividend Portion or Payout= 30.00%), (**i**= I/S= Interest Portion= 1.00%), (**f**= F/S= Fixed Portion= 58.00%), (**$'**= Sales of Past Year= 48,851 millions USD), (**s**= [S/S']-1= Sales Growth= 5.00%), (**$**= Sales or Revenues= 51,294 millions USD), (**D**= Dividend Paid= 2,370 millions USD), and (**V**= Variable Cost= 9,746 millions USD), are known, then it's (**t**= Tax Rate planned) is:

$$t = 1-[D/d]/\{\$'[1+s]-V-\$'f[1+s]-\$'i[1+s]\}$$
$$= 1-D/(d\{\$'[1+s][1-f-i]-V\})$$
$$= 1-2,370/(0.3000\{48,851[1+0.0500][1-0.5800-0.0100]-9,746]\})$$
$$= 30.00\%$$

Law-1890:

If (**t**= T/B= Tax Rate= 30.00%), (**i**= I/S= Interest Portion= 1.00%), (**f**= F/S=Fixed Portion= 58.00%), (**$'**= Sales of Past Year= 48,851 millions USD), (**s**= [S/S']-1= Sales Growth= 5.00%), (**$**= Sales or Revenues= 51,294 millions USD), (**D**= Dividend Paid= 2,370 millions USD), and (**V**= Variable Cost= 9,746 millions USD), are known, then it's (**d**= Dividend Portion or Payout planned) is:

$$d = D/([1-t]\{\$'[1+s]-V-\$'f[1+s]-\$'i[1+s]\})$$
$$= D/(d\{\$'[1+s][1-f-i]-V\})$$
$$= 2,370/([1-0.3000]\{48,851[1+0.0500][1-0.5800-0.0100]-9,746]\})$$
$$= 30.00\%$$

Law-1891:
If (**t**= T/B= Tax Rate= 30.00%), (**I**= Interest Expense= 513 millions USD), (**F**= Fixed Cost= 29,750 millions USD), (**$'**= Sales of Past Year= 48,851 millions USD), (**s**= [S/S']-1= Sales Growth= 5.00%), (**$**= Sales or Revenues= 51,294 millions USD), (**d**= D/A= Dividend Portion or Payout= 30.00%), and (**v**= V/S= Variable Portion= 19.00%), are known, then it's (**D**= Dividend Paid planned) is:

$$\mathbf{D= d[1-t]\{\$'[1+s]-\$v-F-I\}}$$
$$= 0.3000[1-0.3000]\{48,851[1+0.0500]$$
$$-51,294*0.1900-29,750-513]\}$$
$$= 2,370 \text{ millions USD}$$

Law-1892:
If (**t**= T/B= Tax Rate= 30.00%), (**I**= Interest Expense= 513 millions USD), (**F**= Fixed Cost= 29,750 millions USD), (**D**= Dividend Paid= 2,370 millions USD), (**s**= [S/S']-1= Sales Growth= 5.00%), (**$**= Sales or Revenues= 51,294 millions USD), (**d**= D/A= Dividend Portion or Payout= 30.00%), and (**v**= V/S= Variable Portion= 19.00%), are known, then it's (**$'**= Sales Past) must be:

$$\mathbf{\$'= (\$v+F+I+D/\{d[1-t]\})/[1+s]}$$
$$= (51,294*0.1900 +29,750+513 +2,370$$
$$/\{0.3000[1-0.3000]\})/[1+0.0500]$$
$$= 48,851 \text{ millions USD}$$

Law-1893:
   If (**t**= T/B= Tax Rate= 30.00%), (**I**= Interest Expense= 513 millions USD), (**F**= Fixed Cost= 29,750 millions USD), (**D**= Dividend Paid= 2,370 millions USD), (**$'**= Sales of Past Year= 48,851 millions USD), (**$**= Sales or Revenues= 51,294 millions USD), (**d**= D/A= Dividend Portion or Payout= 30.00%), and (**v**= V/S= Variable Portion= 19.00%), are known, then it's (**s**= Sales Growth planned) is:
   $s = (\$v + F + I + D/\{d[1-t]\})/\$' - 1$
   $= (51,294*0.1900 + 29,750 + 513 + 2,370 /\{0.3000[1-0.3000]\})/48,851 - 1$
   $= 5.00\%$

Law-1894:
   If (**t**= T/B= Tax Rate= 30.00%), (**I**= Interest Expense= 513 millions USD), (**F**= Fixed Cost= 29,750 millions USD), (**D**= Dividend Paid= 2,370 millions USD), (**$'**= Sales of Past Year= 48,851 millions USD), (**s**= [S/S']-1= Sales Growth= 5.00%), (**d**= D/A= Dividend Portion or Payout= 30.00%), and (**v**= V/S= Variable Portion= 19.00%), are known, then it's (**$**= Sales or Revenues planned) is:
   $\$ = (\$'[1+s] - F - I - D/\{d[1-t]\})/v$
   $= (48,851[1+0.0500] - 29,750 - 513 - 2,370/\{0.3000[1-0.3000]\}) /0.1900$
   $= 51,294$ millions USD

Law-1895:

If (t= T/B= Tax Rate= 30.00%), (I= Interest Expense= 513 millions USD), (F= Fixed Cost= 29,750 millions USD), (D= Dividend Paid= 2,370 millions USD), ($\$'$= Sales of Past Year= 48,851 millions USD), (s= [S/S']-1= Sales Growth= 5.00%), (d= D/A= Dividend Portion or Payout= 30.00%), and ($\$$= Sales or Revenues= 51,294 millions USD), are known, then it's (v= Variable Portion planned) is:

v= ($\$'$[1+$s$]-$F$-$I$-$D$/{$d$[1-$t$]})/$\$$
 = (48,851[1+0.0500] -29,750-513
   -2,370/{0.3000[1-0.3000]})
   /51,294
 = 19.00%

Law-1896:

If (t= T/B= Tax Rate= 30.00%), (I= Interest Expense= 513 millions USD), (v= V/S= Variable Portion= 19.00%), (D= Dividend Paid= 2,370 millions USD), ($\$'$= Sales of Past Year= 48,851 millions USD), (s= [S/S']-1= Sales Growth= 5.00%), (d= D/A= Dividend Portion or Payout= 30.00%), and ($\$$= Sales or Revenues= 51,294 millions USD), are known, then it's (F= Fixed Cost planned) is:

F= ($\$'$[1+$s$]-$\$v$-$I$-$D$/{$d$[1-$t$]})
 = (48,851[1+0.0500] -51,294*0.1900
   -513 -2,370/{0.3000[1-0.3000]}
 = 29,750 millions USD

Law-1897:

If (t= T/B= Tax Rate= 30.00%), (F= Fixed Cost= 29,750 millions USD), (v= V/S= Variable Portion= 19.00%), (D= Dividend Paid= 2,370 millions USD), ($\$'$= Sales of Past Year= 48,851 millions USD), (s= [S/S']-1= Sales Growth= 5.00%), (d= D/A= Dividend Portion or Payout= 30.00%), and ($\$$= Sales or Revenues= 51,294 millions USD), are known, then it's (I= Interest Expense planned) is:

I= ($\$'$[1+$s$]-$\$v$-$F$-$D$/{$d$[1-$t$]})
   = (48,851[1+0.0500] -51,294*0.1900
       -29,750 -2,370/ {0.3000
       [1-0.3000]})
   = 513 millions USD

Law-1898:

If (I= Interest Expense= 513 millions USD), (F= Fixed Cost= 29,750 millions USD), (v= V/S= Variable Portion= 19.00%), (D= Dividend Paid= 2,370 millions USD), ($\$'$= Sales of Past Year= 48,851 millions USD), (s= [S/S']-1= Sales Growth= 5.00%), (d= D/A= Dividend Portion or Payout= 30.00%), and ($\$$= Sales or Revenues= 51,294 millions USD), are known, then it's (t= Tax Rate planned) is:

t= 1-D/(d{$\$'$[1+$s$]-$\$v$-$F$-$I$})
   = 1-2,370/(0.3000{48,851[1+0.0500]
       -51,294*0.1900-29,750 -513})
   = 30.00%

Law-1899:

If (**I**= Interest Expense= 513 millions USD), (**F**= Fixed Cost= 29,750 millions USD), (**v**= V/S= Variable Portion= 19.00%), (**D**= Dividend Paid= 2,370 millions USD), (**$'**= Sales of Past Year= 48,851 millions USD), (**s**= [S/S']-1= Sales Growth= 5.00%), (**t**= T/B= Tax Rate= 30.00%), and (**$**= Sales or Revenues= 51,294 millions USD), are known, then it's (**d**= Dividend Portion or Payout planned) is:

$$d = D/([1-t]\{\$'[1+s]-\$v-F-I\})$$
$$= 2,370/([1-0.3000]\{48,851[1+0.0500]$$
$$-51,294*0.1900-29,750-513\})$$
$$= \underline{30.00}\%$$

Law-1900:

If (**i**= I/S= Interest Portion= 1.00%), (**F**= Fixed Cost= 29,750 millions USD), (**v**= V/S= Variable Portion= 19.00%), (**d**= D/A= Dividend Portion or Payout= 30.00%), (**$'**= Sales of Past Year= 48,851 millions USD), (**s**= [S/S']-1= Sales Growth= 5.00%), (**t**= T/B= Tax Rate= 30.00%), and (**$**= Sales or Revenues= 51,294 millions USD), are known, then it's (**D**= Dividend Paid planned) is:

$$D = d[1-t]\{\$'[1+s]-\$v-F-\$i\})$$
$$= d[1-t]\{\$'[1+s]-\$[v+i]-F\})$$
$$= 0.3000[1-0.3000]\{48,851[1+0.0500]$$
$$-51,294[0.1900+0.0100]-29,750\}$$
$$= \underline{2,370} \text{ millions USD}$$

Law-1901:

If (i= I/S= Interest Portion= 1.00%), (F= Fixed Cost= 29,750 millions USD), (v= V/S= Variable Portion= 19.00%), (d= D/A= Dividend Portion or Payout= 30.00%), (D= Dividend Paid= 2,370 millions USD), (s= [S/S']-1= Sales Growth= 5.00%), (t= T/B= Tax Rate= 30.00%), and ($\$$= Sales or Revenues= 51,294 millions USD), are known, then it's ($\$'$= Sales Past) must be:

$$\$' = (\$[v+i]+F+D/\{d[1-t]\})/[1+s]$$
$$= (51,294[0.1900+0.0100]+29,750$$
$$+2,370/\{0.3000[1-0.3000]\})$$
$$/[1+0.0500]$$
$$= 48,851 \text{ millions USD}$$

Law-1902:

If (i= I/S= Interest Portion= 1.00%), (F= Fixed Cost= 29,750 millions USD), (v= V/S= Variable Portion= 19.00%), (d= D/A= Dividend Portion or Payout= 30.00%), (D= Dividend Paid= 2,370 millions USD), ($\$'$= Sales of Past Year= 48,851 millions USD), (t= T/B= Tax Rate= 30.00%), and ($\$$= Sales or Revenues= 51,294 millions USD), are known, then it's (s= Sales Growth planned) must be:

$$s = (\$[v+i]+F+D/\{d[1-t]\})/\$'-1$$
$$= (51,294[0.1900+0.0100]+29,750$$
$$+2,370/\{0.3000[1-0.3000]\})$$
$$/48,851-1$$
$$= 5.00\%$$

Law-1903:

If (i= I/S= Interest Portion= 1.00%), (F= Fixed Cost= 29,750 millions USD), (v= V/S= Variable Portion= 19.00%), (d= D/A= Dividend Portion or Payout= 30.00%), (D= Dividend Paid= 2,370 millions USD), ($\$'$= Sales of Past Year= 48,851 millions USD), (t= T/B= Tax Rate= 30.00%), and (s= [S/S']-1= Sales Growth= 5.00%), are known, then it's ($\$$= Sales or Revenues planned) is:

$$\$ = (\$'[1+s]\text{-}F\text{-}D/\{d[1\text{-}t]\})/[v+i]$$
$$= (48,851[1+0.0500]\text{-}29,750\text{-}2,370$$
$$/\{0.3000[1\text{-}0.3000]\})$$
$$/[0.1900+0.0100]$$
$$= 51,294 \text{ millions USD}$$

Law-1904:

If (i= I/S= Interest Portion= 1.00%), (F= Fixed Cost= 29,750 millions USD), ($\$$= Sales or Revenues= 51,294 millions USD), (d= D/A= Dividend Portion or Payout= 30.00%), (D= Dividend Paid= 2,370 millions USD), ($\$'$= Sales of Past Year= 48,851 millions USD), (t= T/B= Tax Rate= 30.00%), and (s= [S/S']-1= Sales Growth= 5.00%), are known, then it's (v= Variable Portion planned) is:

$$v = (\$'[1+s]\text{-}F\text{-}D/\{d[1\text{-}t]\})/\$\text{-}i$$
$$= (48,851[1+0.0500]\text{-}29,750\text{-}2,370$$
$$/\{0.3000[1\text{-}0.3000]\})/51,294$$
$$-0.0100$$
$$= 19.00\%$$

Law-1905:
If (v= V/S= Variable Portion= 19.00%), (**F**= Fixed Cost= 29,750 millions USD), ($= Sales or Revenues= 51,294 millions USD), (**d**= D/A= Dividend Portion or Payout= 30.00%), (**D**= Dividend Paid= 2,370 millions USD), ($'= Sales of Past Year= 48,851 millions USD), (**t**= T/B= Tax Rate= 30.00%), and (**s**= [S/S']-1= Sales Growth= 5.00%), are known, then it's (**i**= Interest Portion planned) is:

$$i= (\$'[1+s]-F-D/\{d[1-t]\})/\$-v$$
$$= (48,851[1+0.0500]-29,750-2,370$$
$$/\{0.3000[1-0.3000]\})/51,294$$
$$-0.1900$$
$$= 1.00\%$$

Law-1906:
If (v= V/S= Variable Portion= 19.00%), (**i**= I/S= Interest Portion= 1.00%), ($= Sales or Revenues= 51,294 millions USD), (**d**= D/A= Dividend Portion or Payout= 30.00%), (**D**= Dividend Paid= 2,370 millions USD), ($'= Sales of Past Year= 48,851 millions USD), (**t**= T/B= Tax Rate= 30.00%), and (**s**= [S/S']-1= Sales Growth= 5.00%), are known, then it's (**F**= Fixed Cost planned) is:

$$F= \$'[1+s]-\$[v+i]-D/\{d[1-t]\}$$
$$= (48,851[1+0.0500]- 51,294[0.1900$$
$$+0.0100]-2,370/\{0.3000$$
$$[1-0.3000]\})$$
$$= 29,750 \text{ millions USD}$$

Law-1907:

If (v= V/S= Variable Portion= 19.00%), (i= I/S= Interest Portion= 1.00%), ($\$$= Sales or Revenues= 51,294 millions USD), (d= D/A= Dividend Portion or Payout= 30.00%), (D= Dividend Paid= 2,370 millions USD), ($\$'$= Sales of Past Year= 48,851 millions USD), (F= Fixed Cost= 29,750 millions USD), and (s= [S/S']-1= Sales Growth= 5.00%), are known, then it's (t= Tax Rate planned) is:

$$t = 1-[D/d]/\{\$'[1+s]-\$v-F-\$i\}$$
$$= 1-D/(d\{\$'[1+s]-\$[v+i]-F\})$$
$$= 1-2,370/(0.3000\{48,851[1+0.0500]$$
$$-51,294[0.1900 +0.0100]$$
$$-29,750\})$$
$$= \underline{30.00\%}$$

Law-1908:

If (v= V/S= Variable Portion= 19.00%), (i= I/S= Interest Portion= 1.00%), ($\$$= Sales or Revenues= 51,294 millions USD), (t= T/B= Tax Rate= 30.00%), (D= Dividend Paid= 2,370 millions USD), ($\$'$= Sales of Past Year= 48,851 millions USD), (F= Fixed Cost= 29,750 millions USD), and (s= [S/S']-1= Sales Growth= 5.00%), are known, then it's (d= Dividend Portion or Payout planned) is:

$$d = D/([1-t]\{\$'[1+s]-\$v-F-\$i\}$$
$$= D/([1-t]\{\$'[1+s]-\$[v+i]-F\})$$
$$= 2,370/([1-0.3000]\{48,851[1+0.0500]$$
$$-51,294[0.1900 +0.0100]$$
$$-29,750\})$$
$$= \underline{30.00\%}$$

Law-1909:
If (v= V/S= Variable Portion= 19.00%), (i= I/S= Interest Portion= 1.00%), ($\$$= Sales or Revenues= 51,294 millions USD), (t= T/B= Tax Rate= 30.00%), (d=D/A= Dividend Portion or Payout= 30.00%), ($\$$'= Sales of Past Year= 48,851 millions USD), (F= Fixed Cost= 29,750 millions USD), and (s= [S/S']-1= Sales Growth= 5.00%), are known, then it's (D= Dividend Paid planned) is:

$D = d[1-t]\{\$'[1+s]-\$v-F-\$i\}$
 $= d[1-t]\{\$'[1+s]-\$[v+i]-F\}$)
 $= 0.3000[1-0.3000]\{48,851[1+0.0500]$
  $-51,294[0.1900 +0.0100]$
  $-29,750\}$
 $= \underline{2,370}$ millions USD

Law-1910:
If (v= V/S= Variable Portion= 19.00%), (i= I/S= Interest Portion= 1.00%), ($\$$= Sales or Revenues= 51,294 millions USD), (t= T/B= Tax Rate= 30.00%), (d=D/A= Dividend Portion or Payout= 30.00%), (D= Dividend Paid= 2,370 millions USD), (F= Fixed Cost= 29,750 millions USD), and (s= [S/S']-1= Sales Growth= 5.00%), are known, then it's ($\$$'= Sales Past) must be:

$\$' = (\$v+F+D/\{d[1-t]\})/\{[1+s][1-i]\}$
 $= (51,294*0.1900 + 29,750+2,370$
  $/\{0.3000[1-0.3000]\})$
  $/\{[1+0.0500][1-0.0100]\}$
 $= \underline{48,851}$ millions USD

Law-1911:

If (v= V/S= Variable Portion= 19.00%), (i= I/S= Interest Portion= 1.00%), ($\$$= Sales or Revenues= 51,294 millions USD), (t= T/B= Tax Rate= 30.00%), (d=D/A= Dividend Portion or Payout= 30.00%), (D= Dividend Paid= 2,370 millions USD), (F= Fixed Cost= 29,750 millions USD), and ($\$'$= Sales of Past Year= 48,851 millions USD), are known, then it's (s= Sales Growth planned) is:

$s = (\$v + F + D/\{d[1-t]\})/\{\$'[1-i]\} - 1$
  $= (51,294*0.1900 + 29,750 + 2,370$
    $/\{0.3000[1-0.3000]\})$
    $/\{48,851[1-0.0100]\} - 1$
  $= 5.00\%$

Law-1912:

If (v= V/S= Variable Portion= 19.00%), (s= [S/S']-1= Sales Growth= 5.00%), ($\$$= Sales or Revenues= 51,294 millions USD), (t= T/B= Tax Rate= 30.00%), (d=D/A= Dividend Portion or Payout= 30.00%), (D= Dividend Paid= 2,370 millions USD), (F= Fixed Cost= 29,750 millions USD), and ($\$'$= Sales of Past Year= 48,851 millions USD), are known, then it's (i= Interest Portion planned) is:

$i = 1 - (\$v + F + D/\{d[1-t]\})/\{\$'[1+s]\}$
  $= 1 - (51,294*0.1900 + 29,750 + 2,370$
    $/\{0.3000[1-0.3000]\})$
    $/\{48,851[1+0.0500]\})$
  $= 1.00\%$

Law-1913:
If (**v**= V/S= Variable Portion= 19.00%), (**s**= [S/S']-1= Sales Growth= 5.00%), (**i**= I/S= Interest Portion= 1.00%), (**t**= T/B= Tax Rate= 30.00%), **d**= D/A= Dividend Portion or Payout= 30.00%), (**D**= Dividend Paid= 2,370 millions USD), (**F**= Fixed Cost= 29,750 millions USD), and (**S'**= Sales of Past Year= 48,851 millions USD), are known, then it's (**S**=Sales or Revenues planned) is:

$$S = S'[1+s][1-i] - F - D/\{d[1-t]\})/v$$
$$= 48,851[1+0.0500][1-0.0100] - 29,750$$
$$\quad -2,370/\{0.3000[1-0.3000]\})$$
$$\quad /0.1900$$
$$= 51,294 \text{ millions USD}$$

Law-1914:
If (**S**= Sales or Revenues= 51,294 millions USD), (**s**= [S/S']-1= Sales Growth= 5.00%), (**i**= I/S= Interest Portion= 1.00%), (**t**= T/B= Tax Rate= 30.00%), **d**= D/A= Dividend Portion or Payout= 30.00%), (**D**= Dividend Paid= 2,370 millions USD), (**F**= Fixed Cost= 29,750 millions USD), and (**S'**= Sales of Past Year= 48,851 millions USD), are known, then it's (**v**= Variable Portion planned) is:

$$v = S'[1+s][1-i] - F - D/\{d[1-t]\})$$
$$= 48,851[1+0.0500][1-0.0100] - 29,750$$
$$\quad -2,370/\{0.3000[1-0.3000]\})$$
$$\quad /51,294$$
$$= 19.00\%$$

Law-1915:

If ($= Sales or Revenues= 51,294 millions USD), (s= [S/S']-1= Sales Growth= 5.00%), (i= I/S= Interest Portion= 1.00%), (t= T/B= Tax Rate= 30.00%), d= D/A= Dividend Portion or Payout= 30.00%), (D= Dividend Paid= 2,370 millions USD), (v= V/S= Variable Portion= 19.00%), and ($'= Sales of Past Year= 48,851 millions USD), are known, then it's (F= Fixed Cost planned) is:

$$F = \$'[1+s][1-i] - \$v - D/\{d[1-t]\}$$
$$= 48,851[1+0.0500][1-0.0100]$$
$$\quad -51,294*0.1900 - 2,370/\{0.3000$$
$$\quad [1-0.3000]\})/51,294$$
$$= 29,750 \text{ millions USD}$$

Law-1916:

If ($= Sales or Revenues= 51,294 millions USD), (s= [S/S']-1= Sales Growth= 5.00%), (i= I/S= Interest Portion= 1.00%), (F= Fixed Cost= 29,750 millions USD), d= D/A= Dividend Portion or Payout= 30.00%), (D= Dividend Paid= 2,370 millions USD), (v= V/S= Variable Portion= 19.00%), and ($'= Sales of Past Year= 48,851 millions USD), are known, then it's (t= Tax Rate planned) is:

$$t = 1 - [D/d]/\$'[1+s] - \$v - F - \$'i[1+s]\}$$
$$= 1 - D/(d\{\$'[1+s][1-i] - \$v - F\})$$
$$= 1 - 2,370/(0.3000\{48,851[1+0.0500]$$
$$\quad [1-0.0100] - 51,294*0.1900$$
$$\quad -29,750\})$$
$$= 30.00\%$$

Law-1917:
If ($= Sales or Revenues= 51,294 millions USD), ($= [S/S']-1= Sales Growth= 5.00%), (i= I/S= Interest Portion= 1.00%), (F= Fixed Cost= 29,750 millions USD), (t= T/B= Tax Rate= 30.00%), (D= Dividend Paid= 2,370 millions USD), (v= V/S= Variable Portion= 19.00%), and ($'= Sales of Past Year= 48,851 millions USD), are known, then it's (d= Dividend Portion planned) is:

d= D/([1-t]/{$'[1+s]-$v-F-$'i[1+s]}
= D/([1-t]($'[1+s][1-i]-$v-F})
= 2,370/([1-0.3000]{48,851[1+0.0500]
[1-0.0100]- 51,294*0.1900
-29,750})
= 30.00%

Law-1918:
If ($= Sales or Revenues= 51,294 millions USD), ($= [S/S']-1= Sales Growth= 5.00%), (I= Interest Expense= 513 millions USD), (f= F/S=Fixed Portion= 58.00%), (t= T/B= Tax Rate= 30.00%), (d= D/A= Dividend Portion or Payout= 30.00%), (v= V/S= Variable Portion= 19.00%), and ($'= Sales of Past Year= 48,851 millions USD), are known, then it's (D= Dividend Paid planned) is:

D= d[1-t]/$'[1+s]-$v-$f-I}
= d[1-t]{$'[1+s]-I-$[v+f]}
= 0.3000[1-0.3000]{48,851[1+0.0500]
-513-51,294[0.1900+0.5800]}
= 2,370 millions USD

Law-1919:

If ($= Sales or Revenues= 51,294 millions USD), (s= [S/S']-1= Sales Growth= 5.00%), (I= Interest Expense= 513 millions USD), (f= F/S= Fixed Portion= 58.00%), (t= T/B= Tax Rate= 30.00%), (d= D/A= Dividend Portion or Payout= 30.00%), (v= V/S= Variable Portion= 19.00%), and (D= Dividend Paid= 2,370 millions USD), are known, then it's ($'=Sales Past) must be:

$$\$' = (\$[v+f]+I+D/\{d[1-t]\})/[1+s]$$
$$= (51,294[0.1900+0.5800]+513+2,370$$
$$/\{0.3000[1-0.3000]\})/[1+0.0500]$$
$$= \underline{48,851} \text{ millions USD}$$

Law-1920:

If ($= Sales or Revenues= 51,294 millions USD), ($'= Sales of Past Year= 48,851 millions USD), (f= F/S= Fixed Portion= 58.00%), (I= Interest Expense= 513 millions USD), (t= T/B= Tax Rate= 30.00%), (d= D/A= Dividend Portion or Payout= 30.00%), (v= V/S= Variable Portion= 19.00%), and (D= Dividend Paid= 2,370 millions USD), are known, then it's (s= Sales Growth planned) is:

$$s = (\$[v+f]+I+D/\{d[1-t]\})/\$'-1$$
$$= (51,294[0.1900+0.5800]+513+2,370$$
$$/\{0.3000[1-0.3000]\})/ 48,851-1$$
$$= \underline{5.00\%}$$

Law-1921:

If ($=[S/S']-1=$ Sales Growth= 5.00%), (I= Interest Expense= 513 millions USD), (f= F/S= Fixed Portion= 58.00%), ($'= Sales of Past Year= 48,851 millions USD), (t= T/B= Tax Rate= 30.00%), (d= D/A= Dividend Portion or Payout= 30.00%), (v= V/S= Variable Portion= 19.00%), and (D= Dividend Paid= 2,370 millions USD), are known, then it's ($= Sales or Revenues planned) is:

$$\$= (\$'[1+\$]-I-D/\{d[1-t]\})/[v+f]$$
$$= (48,851[1+0.0500]-513-2,370$$
$$/\{0.3000[1-0.3000]\})$$
$$/[0.1900+0.5800]$$
$$= 51,294 \text{ millions USD}$$

Law-1922:

If ($=[S/S']-1=$ Sales Growth= 5.00%), (I= Interest Expense= 513 millions USD), ($'= Sales of Past Year= 48,851 millions USD), (f= F/S= Fixed Portion= 58.00%), (t= T/B= Tax Rate= 30.00%), (d= D/A= Dividend Portion or Payout= 30.00%), ($= Sales or Revenues= 51,294 millions USD), and (D= Dividend Paid= 2,370 millions USD), are known, then it's (v= Variable Portion planned) is:

$$v= (\$'[1+\$]-I-D/\{d[1-t]\})/\$-f$$
$$= (48,851[1+0.0500]-513-2,370$$
$$/\{0.3000[1-0.3000]\})/ 51,294$$
$$-0.5800$$
$$= 19.00\%$$

Law-1923:
If (s= [S/S']-1= Sales Growth= 5.00%), (I= Interest Expense= 513 millions USD), (v= V/S= Variable Portion= 19.00%), (S'= Sales of Past Year= 48,851 millions USD), (t= T/B= Tax Rate= 30.00%), (d= D/A= Dividend Portion or Payout= 30.00%), (S= Sales or Revenues= 51,294 millions USD), and (D= Dividend Paid= 2,370 millions USD), are known, then it's (f= Fixed Portion planned) is:

f= (S'[1+s]-I-D/{d[1-t]})/S-v
  = (48,851[1+0.0500]-513-2,370
    /{0.3000[1-0.3000]})/51,294
    -0.1900
  = 58.00%

Law-1924:
If (s= [S/S']-1= Sales Growth= 5.00%), (v= V/S= Variable Portion= 19.00%), (S'= Sales of Past Year= 48,851 millions USD), (f= F/S= Fixed Portion= 58.00%), (t= T/B= Tax Rate= 30.00%), (d= D/A= Dividend Portion or Payout= 30.00%), (S= Sales or Revenues= 51,294 millions USD), and (D= Dividend Paid= 2,370 millions USD), are known, then it's (I= Interest Expense planned) is:

I= (S'[1+s]-S[v+f]-D/{d[1-t]}
  = (48,851[1+0.0500]-51,294[0.1900
    +0.5800]-2,370/{0.3000
    [1-0.3000]}
  = 513 millions USD

Law-1925:

If (s= [S/S']-1= Sales Growth= 5.00%), (v= V/S= Variable Portion= 19.00%), (S'= Sales of Past Year= 48,851 millions USD), (f= F/S= Fixed Portion= 58.00%), (I= Interest Expense= 513 millions USD), (d= D/A= Dividend Portion or Payout= 30.00%), (S= Sales or Revenues= 51,294 millions USD), and (D= Dividend Paid= 2,370 millions USD), are known, then it's (t= Tax Expense planned) is:

t = 1-[D/d]{S'[1+s]-Sv-Sf-I}
  = 1-D/(d{S'[1+s]-S[v+f]-I})
  = 1-2,370/(0.3000{48,851[1+0.0500]
      -51,294[0.1900+0.5800]-513})
  = 30.00%

Law-1926:

If (s= [S/S']-1= Sales Growth= 5.00%), (v= V/S= Variable Portion= 19.00%), (S'= Sales of Past Year= 48,851 millions USD), (f= F/S= Fixed Portion= 58.00%), (I= Interest Expense= 513 millions USD), (t= T/B= Tax Rate= 30.00%), (S= Sales or Revenues= 51,294 millions USD), and (D= Dividend Paid= 2,370 millions USD), are known, then it's (d= Dividend Portion or Payout planned) is:

d = D/[1-t]{S'[1+s]-Sv-Sf-I}
  = D/([1-t]{S'[1+s]-S[v+f]-I})
  = 2,370/([1-0.3000]{48,851[1+0.0500]
      -51,294[0.1900+0.5800]-513})
  = 30.00%

Law-1927:
If (s= [S/S']-1= Sales Growth= 5.00%), (v= V/S= Variable Portion= 19.00%), ($\$'$= Sales of Past Year= 48,851 millions USD), (f= F/S= Fixed Portion= 58.00%), (I= Interest Expense= 513 millions USD), (t= T/B= Tax Rate= 30.00%), ($\$$= Sales or Revenues= 51,294 millions USD), and (d= D/A= Dividend Portion or Payout= 30.00%), are known, then it's (D= Dividend Paid planned) is:

$D= d[1-t]\{\$'[1+s]-\$v-\$f-I\}$
$= d([1-t]\{\$'[1+s]-\$[v+f]-I\}$
$= 0.3000[1-0.3000]\{48,851[1+0.0500]$
$-51,294[0.1900+0.5800]-513\}$
$= 2,370$ millions USD

Law-1928:
If (s= [S/S']-1= Sales Growth= 5.00%), (v= V/S= Variable Portion= 19.00%), (D= Dividend Paid= 2,370 millions USD), (f= F/S= Fixed Portion= 58.00%), (i= I/S= Interest Portion= 1.00%), (t= T/B= Tax Rate= 30.00%), ($\$$= Sales or Revenues= 51,294 millions USD), and (d= D/A= Dividend Portion or Payout= 30.00%), are known, then it's ($\$'$= Sales Past) must be:

$\$'= (\$[v+f+i]+D/\{d[1-t]\})/[1+s]$
$= (51,294[0.1900+0.5800+0.0100]$
$+2,370/\{0.3000[1-0.3000]\})$
$/[1+0.0500]$
$= 48,851$ millions USD

Law-1929:
If ($'= Sales of Past Year= 48,851 millions USD), (ʋ= V/S= Variable Portion= 19.00%), (**D**= Dividend Paid= 2,370 millions USD), (**f**= F/S= Fixed Portion= 58.00%), (**i**= I/S= Interest Portion= 1.00%), (**t**= T/B= Tax Rate= 30.00%), ($= Sales or Revenues= 51,294 millions USD), and (**d**= D/A= Dividend Portion or Payout= 30.00%), are known, then it's (**s**= Sales Growth planned) is:

$$s = (\$[ʋ+f+i]+D/\{d[1-t]\})/\$' - 1$$
$$= (51,294[0.1900+0.5800+0.0100]$$
$$\quad +2,370/\{0.3000[1-0.3000]\})$$
$$\quad /48,851-1$$
$$= 5.00\%$$

Law-1930:
If (**s**= [S/S']-1= Sales Growth= 5.00%), (ʋ= V/S= Variable Portion= 19.00%), (**D**= Dividend Paid= 2,370 millions USD), (**f**= F/S= Fixed Portion= 58.00%), (**i**= I/S= Interest Portion= 1.00%), (**t**= T/B= Tax Rate= 30.00%), ($'= Sales of Past Year= 48,851 millions USD), and (**d**= D/A= Dividend Portion or Payout= 30.00%), are known, then it's ($ = Sales or Revenues planned) is:

$$\$ = (\$'[1+s] - D/\{d[1-t]\})/[ʋ+f+i]$$
$$= (48,851[1+0.0500] - 2,370/\{0.3000$$
$$\quad [1-0.3000]\})/[0.1900$$
$$\quad +0.5800+0.0100]$$
$$= 51,294 \text{ USD}$$

Law-1931:
If (s= [S/S']-1= Sales Growth= 5.00%), ($\$$= Sales or Revenues= 51,294 millions USD), (D= Dividend Paid= 2,370 millions USD), (f= F/S= Fixed Portion= 58.00%), (i= I/S= Interest Portion= 1.00%), (t= T/B= Tax Rate= 30.00%), ($\$'$= Sales of Past Year= 48,851 millions USD), and (d= D/A=Dividend Portion or Payout= 30.00%), are known, then it's (v = Variable Portion planned) is:

v= ($\$'$[1+$s$]-$D$/{$d$[1-$t$]})/$\$$-f-i
= (48,851[1+0.0500] -2,370/{0.3000
[1-0.3000]})/ 51,294-0.5800
-0.0100
= 19.00%

Law-1932:
If (s= [S/S']-1= Sales Growth= 5.00%), ($\$$= Sales or Revenues= 51,294 millions USD), (D= Dividend Paid= 2,370 millions USD), (v= V/S= Variable Portion= 19.00%), (i= I/S= Interest Portion= 1.00%), (t= T/B= Tax Rate= 30.00%), ($\$'$= Sales of Past Year= 48,851 millions USD), and (d= D/A=Dividend Portion or Payout= 30.00%), are known, then it's (f = Fixed Portion planned) is:

f= ($\$'$[1+$s$]=$D$/{$d$[1-$t$]})/$\$$-v-i
= (48,851[1+0.0500] -2,370/{0.3000
[1-0.3000]})/ 51,294-0.1900
-0.0100
= 58.00%

Law-1933:
If (s= [S/S']-1= Sales Growth= 5.00%), (S= Sales or Revenues= 51,294 millions USD), (**D**= Dividend Paid= 2,370 millions USD), (**v**= V/S= Variable Portion= 19.00%), (**f**= F/S= Fixed Portion= 58.00%), (**t**= T/B= Tax Rate= 30.00%), (S'= Sales of Past Year= 48,851 millions USD), and (**d**= D/A=Dividend Portion or Payout= 30.00%), are known, then it's (**i** = Interest Portion planned) is:

$$i = (S'[1+s]-D/\{d[1-t]\})/S-f-v$$
$$= (48,851[1+0.0500] -2,370/\{0.3000[1-0.3000]\})/ 51,294-0.5800 -0.1900$$
$$= 1.00\%$$

Law-1934:
If (s= [S/S']-1= Sales Growth= 5.00%), (S= Sales or Revenues= 51,294 millions USD), (**D**= Dividend Paid= 2,370 millions USD), (**v**= V/S= Variable Portion= 19.00%), (**f**= F/S= Fixed Portion= 58.00%), (**i**= I/S= Interest Portion= 1.00%), (S'= Sales of Past Year= 48,851 millions USD), and (**d**= D/A= Dividend Portion or Payout= 30.00%), are known, then it's (**t** = Tax Rate planned) is:

$$t = 1-D/(d\{S'[1+s]-Sv-Sf-Si\})$$
$$= 1-D/(d\{S'[1+s]-S[v+f+i]\})$$
$$= 1-2,370/(0.3000\{48,851[1+0.0500] -51,294[0.1900+0.5800 +0.0100]\})$$
$$= 30.00\%$$

Law-1935:

If (s= [S/S']-1= Sales Growth= 5.00%), (S= Sales or Revenues= 51,294 millions USD), (D= Dividend Paid= 2,370 millions USD), (v= V/S= Variable Portion= 19.00%), (f= F/S= Fixed Portion= 58.00%), (i= I/S= Interest Portion= 1.00%), (S'= Sales of Past Year= 48,851 millions USD), and (t= T/B= Tax Rate= 30.00%), are known, then it's (d = Dividend Portion or Payout planned) is:

$$d = D/([1-t]\{S'[1+s]-Sv-Sf-Si\})$$
$$= D/([1-t]\{S'[1+s]-S[v+f+i]\})$$
$$= 2{,}370/([1-0.3000]\{48{,}851[1+0.0500]$$
$$-51{,}294[0.1900+0.5800$$
$$+0.0100]\})$$
$$= 30.00\%$$

Law-1936:

If (s= [S/S']-1= Sales Growth= 5.00%), (S= Sales or Revenues= 51,294 millions USD), (d= D/A= Dividend Portion or Payout= 30.00%), (v= V/S= Variable Portion= 19.00%), (f= F/S=Fixed Portion= 58.00%), (i= I/S= Interest Portion= 1.00%), (S'= Sales of Past Year= 48,851 millions USD), and (t= T/B= Tax Rate= 30.00%), are known, then it's (D= Dividend Paid planned) is:

$$D = d[1-t]\{S'[1+s]-Sv-Sf-Si\})$$
$$= d[1-t]\{S'[1+s]-S[v+f+i]\})$$
$$= 0.3000[1-0.3000]\{48{,}851[1+0.0500]$$
$$-51{,}294[0.1900+0.5800$$
$$+0.0100]\})$$
$$= 2{,}370 \text{ millions USD}$$

Law-1937:
If (s= [S/S']-1= Sales Growth= 5.00%), ($\$$= Sales or Revenues= 51,294 millions USD), (d= D/A= Dividend Portion or Payout= 30.00%), (v= V/S= Variable Portion= 19.00%), (f= F/S= Fixed Portion= 58.00%), (i= I/S= Interest Portion= 1.00%), (D= Dividend Paid= 2,370 millions USD), and (t= T/B= Tax Rate= 30.00%), are known, then it's ($\$$'= Sales Past) must be:

$\$$' = ($\$$[v+f]-$\$i$}-D/{$d$[1-$t$]})/{[1+$s$][1-$i$]
　　= (51,294[0.1900+0.5800]+2,370
　　　／{0.3000[1-0.3000]})
　　　／[1+0.0500] [1-0.0100]
　　= 48,851 millions USD

Law-1938:
If ($\$$'= Sales of Past Year= 48,851 millions USD), ($\$$= Sales or Revenues= 51,294 millions USD), (d= D/A= Dividend Portion or Payout= 30.00%), (v= V/S= Variable Portion= 19.00%), (f= F/S= Fixed Portion= 58.00%), (i= I/S= Interest Portion= 1.00%), (D= Dividend Paid= 2,370 millions USD), and (t= T/B= Tax Rate= 30.00%), are known, then it's (s= Sales Growth planned) is:

s= ($\$$[v+f]-$\$i$}+D/{$d$[1-$t$]})/{$\$$'[1-i]-1
　　= (51,294[0.1900+0.5800]+2,370
　　　／{0.3000[1-0.3000]})／ 48,851
　　　[1-0.0100]-1
　　= 5.00%

Law-1939:

If ($\$'$= Sales of Past Year= 48,851 millions USD), ($\$$= Sales or Revenues= 51,294 millions USD), (**d**= D/A=Dividend Portion or Payout= 30.00%), (**v**= V/S= Variable Portion= 19.00%), (**f**= F/S= Fixed Portion= 58.00%), (**s**= [S/S']-1= Sales Growth= 5.00%), (**D**= Dividend Paid= 2,370 millions USD), and (**t**= T/B= Tax Rate= 30.00%), are known, then it's (**i**= Interest Portion planned) is:

$$\mathbf{i} = 1-(\$[\mathbf{v}+\mathbf{f}]-\$\mathbf{i}\}+\mathbf{D}/\{\mathbf{d}[1-\mathbf{t}]\})/\{\$'[1+\mathbf{s}]\}$$
$$= 1-51,294[0.1900+0.5800]+2,370$$
$$/\{0.3000[1-0.3000]\})/\{48,851$$
$$[1+0.0500]\}$$
$$= 1.00\%$$

Law-1940:

If ($\$'$= Sales of Past Year= 48,851 millions USD), (**i**= I/S= Interest Portion= 1.00%), (**d**= D/A= Dividend Portion or Payout= 30.00%), (**v**= V/S= Variable Portion= 19.00%), (**f**= F/S= Fixed Portion= 58.00%), (**s**= [S/S']-1= Sales Growth= 5.00%), (**D**= Dividend Paid= 2,370 millions USD), and (**t**= T/B= Tax Rate= 30.00%), are known, then it's ($\$$ = Sales or Revenues planned) is:

$$\$ = (\$'[1+\mathbf{s}][1-\mathbf{i}]-\mathbf{D}/\{\mathbf{d}[1-\mathbf{t}]\})/[\mathbf{v}+\mathbf{f}]$$
$$= (48,851\ [1+0.0500][1-0.0100]-2,370$$
$$/\{0.3000[1-0.3000]\})$$
$$/\ [0.1900+0.5800]$$
$$= 51,294 \text{ millions USD}$$

Law-1941:
If ($**$'**= Sales of Past Year= 48,851 millions USD), (**i**= I/S= Interest Portion= 1.00%), (**d**= D/A= Dividend Portion or Payout= 30.00%), ($**$**= Sales or Revenues= 51,294 millions USD), (**f**= F/S= Fixed Portion= 58.00%), (**s**= [S/S']-1= Sales Growth= 5.00%), (**D**= Dividend Paid= 2,370 millions USD), and (**t**= T/B= Tax Rate= 30.00%), are known, then it's (**v** = Variable Portion planned) is:

$$v = (\$'[1+s][1-i]-D/\{d[1-t]\})/\$-f$$
$$= (48,851 \ [1+0.0500][1-0.0100]-2,370 \ /\{0.3000[1-0.3000]\})/51,294$$
$$-0.5800$$
$$= 19.00\%$$

Law-1942:
If ($**$'**= Sales of Past Year= 48,851 millions USD), (**i**= I/S= Interest Portion= 1.00%), (**d**= D/A= Dividend Portion or Payout= 30.00%), ($**$**= Sales or Revenues= 51,294 millions USD), (**v**= V/S= Variable Portion= 19.00%), (**s**= [S/S']-1= Sales Growth= 5.00%), (**D**= Dividend Paid= 2,370 millions USD), and (**t**= T/B= Tax Rate= 30.00%), are known, then it's (**f** = Fixed Portion planned) is:

$$f = (\$'[1+s][1-i]-D/\{d[1-t]\})/\$-v$$
$$= (48,851 \ [1+0.0500][1-0.0100]-2,370 \ /\{0.3000[1-0.3000]\})/51,294$$
$$-0.1900$$
$$= 58.00\%$$

Law-1943:

If ($\$'$= Sales of Past Year= 48,851 millions USD), (i= I/S= Interest Portion= 1.00%), (d= D/A= Dividend Portion or Payout= 30.00%), ($\$$= Sales or Revenues= 51,294 millions USD), (v= V/S= Variable Portion= 19.00%), (s= [S/S']-1= Sales Growth= 5.00%), (D= Dividend Paid= 2,370 millions USD), and (f= F/S= Fixed Portion= 58.00%), are known, then it's (t = Taxed Rate planned) is:

t= 1-D/(d\{$\$'$[1+$s$]-$\$v$-$\$f$-$\$'i$[1+$s$]
    = 1-D/(d\{$\$'$[1+$s$][1-$i$]-$\$$[v+f]\})
    = 1-2,370/(0.3000\{48,851[1+0.0500][1-0.0100]- 51,294[0.1900+0.5800]\})
    = 30.00%

Law-1944:

If ($\$'$= Sales of Past Year= 48,851 millions USD), (i= I/S= Interest Portion= 1.00%), (t= T/B= Tax Rate= 30.00%), ($\$$= Sales or Revenues= 51,294 millions USD), (v= V/S= Variable Portion= 19.00%), (s= [S/S']-1= Sales Growth= 5.00%), (D= Dividend Paid= 2,370 millions USD), and (f= F/S=Fixed Portion= 58.00%), are known, then it's (d = Dividend Portion or Payout planned) is:

d= D/(d[1-t]\{$\$'$[1+$s$]-$\$v$-$\$f$-$\$'i$[1+$s$]
    = D/([1-t]\{$\$'$[1+$s$][1-$i$]-$\$$[v+f]\})
    = 2,370/([1-0.3000]\{48,851[1+0.0500][1-0.0100]- 51,294[0.1900+0.5800]\})
    = 30.00%

Law-1945:
   If ($**S'**$= Sales of Past Year= 48,851 millions USD), (**I**= Interest Expense= 513 millions USD), (**t**= T/B= Tax Rate= 30.00%), (**S**= Sales or Revenues= 51,294 millions USD), (**v**= V/S= Variable Portion= 19.00%), (**s**= [S/S']-1= Sales Growth= 5.00%), (**d**= D/A= Dividend Portion or Payout= 30.00%), and (**f**= F/S=Fixed Portion= 58.00%), are known, then it's (**D**= Dividend Paid planned) is:
   $$D = d[1-t]\{S'[1+s]-Sv-S'f[1+s]-I\}$$
   $$= d[1-t]\{S'[1+s][1-f]-Sv-I\})$$
   $$= 0.3000[1-0.3000]\{48,851[1+0.0500][1-0.5800]-51,294*0.1900-513\}$$
   $$= 2,370 \text{ millions USD}$$

Law-1946:
   If (**I**= Interest Expense= 513 millions USD), (**t**= T/B= Tax Rate= 30.00%), (**D**= Dividend Paid= 2,370 millions USD), (**S**= Sales or Revenues= 51,294 millions USD), (**v**= V/S= Variable Portion= 19.00%), (**s**= [S/S']-1= Sales Growth= 5.00%), (**d**= D/A= Dividend Portion or Payout= 30.00%), and (**f**= F/S=Fixed Portion= 58.00%), are known, then it's (**S'**= Sales Past) must be:
   $$S' = (Sv+I+D/\{d[1-t]\}/\{[1+s][1-f]\}$$
   $$= 51,294*0.1900+513+2,370/ 0.3000[1-0.3000]/\{[1+0.0500][1-0.5800]\}$$
   $$= 48,851 \text{ millions USD}$$

Law-1947:

If (**I**= Interest Expense= 513 millions USD), (**t**= T/B= Tax Rate= 30.00%), (**$**= Sales or Revenues= 51,294 millions USD), (**D**= Dividend Paid= 2,370 millions USD), (**v**= V/S= Variable Portion= 19.00%), (**$'**= Sales of Past Year= 48,851 millions USD), (**d**= D/A=Dividend Portion or Payout= 30.00%), and (**f**= F/S=Fixed Portion= 58.00%), are known, then it's (**s**= Sales Growth planned) is:

$s = (\$v+I+D/\{d[1-t]\})/\{\$'[1-f]\} - 1$
$= (51,294*0.1900+513+2,370/0.3000$
$\quad [1-0.3000]\})/\{48,851$
$\quad [1-0.5800]\} - 1$
$= \underline{5.00\%}$

Law-1948:

If (**I**= Interest Expense= 513 millions USD), (**t**= T/B= Tax Rate= 30.00%), (**$**= Sales or Revenues= 51,294 millions USD), (**D**= Dividend Paid= 2,370 millions USD), (**v**= V/S= Variable Portion= 19.00%), (**$'**= Sales of Past Year= 48,851 millions USD), (**d**= D/A= Dividend Portion or Payout= 30.00%), and (**s**= [S/S']-1= Sales Growth= 5.00%), are known, then it's (**f**= Fixed Portion planned) is:

$f = 1 - (\$v+I+D/\{d[1-t]\})/\{\$'[1+s]\}$
$= 1 - (51,294*0.1900+513+2,370/0.3000$
$\quad [1-0.3000]/\{48,851[1+0.0500]\}$
$= \underline{58.00\%}$

Law-1949:

If (**I**= Interest Expense= 513 millions USD), (**t**= T/B= Tax Rate= 30.00%), (**f**= F/S= Fixed Portion= 58.00%), (**D**= Dividend Paid= 2,370 millions USD), (**v**= V/S= Variable Portion= 19.00%), (**$'**= Sales of Past Year= 48,851 millions USD), (**d**= D/A= Dividend Portion or Payout= 30.00%), and (**s**= [S/S']-1= Sales Growth= 5.00%), are known, then it's (**$**= Sales or Revenues planned) is:

$= $'[1+s][1-f]-I-D/{d[1-t]}/v
   = 48,851[1+0.0500][1-0.5800]-513
     -2,370/ 0.3000[1-0.3000]
     / 0.1900
   = 51,294 millions USD

Law-1950:

If (**I**= Interest Expense= 513 millions USD), (**t**= T/B= Tax Rate= 30.00%), (**f**= F/S= Fixed Portion= 58.00%), (**D**= Dividend Paid= 2,370 millions USD), (**$**= Sales or Revenues= 51,294 millions USD), (**$'**= Sales of Past Year= 48,851 millions USD), (**d**= D/A= Dividend Portion or Payout= 30.00%), and (**s**= [S/S']-1= Sales Growth= 5.00%), are known, then it's (**v**= Variable Portion planned) is:

v= $'[1+s][1-f]-I-D/{d[1-t]}/$
   = 48,851[1+0.0500][1-0.5800]-513
     -2,370/ 0.3000[1-0.3000]
     / 51,294
   = 19.00%

Law-1951:

If (v= V/S= Variable Portion= 19.00%), (t= T/B= Tax Rate= 30.00%), (f= F/S= Fixed Portion= 58.00%), (D= Dividend Paid= 2,370 millions USD), (S= Sales or Revenues= 51,294 millions USD), (S'= Sales of Past Year= 48,851 millions USD), (d= D/A= Dividend Portion or Payout= 30.00%), and (s= [S/S']-1= Sales Growth= 5.00%), are known, then it's (I= Interest Expense planned) is:

$$I = S'[1+s][1-f]-Sv-D/\{d[1-t]\}$$
$$= 48,851[1+0.0500][1-0.5800]-51,294$$
$$*0.1900-2,370/\{0.3000$$
$$[1-0.3000]\}$$
$$= 513 \text{ millions USD}$$

Law-1952:

If (v= V/S= Variable Portion= 19.00%), (I= Interest Expense= 513 millions USD), (f= F/S= Fixed Portion= 58.00%), (D= Dividend Paid= 2,370 millions USD), (S= Sales or Revenues= 51,294 millions USD), (S'= Sales of Past Year= 48,851 millions USD), (d= D/A= Dividend Portion or Payout= 30.00%), and (s= [S/S']-1= Sales Growth= 5.00%), are known, then it's (t= Tax Rate planned) is:

$$t = 1-D/(d\{S'[1+s]-Sv-S'f[1+s]-I\}$$
$$= 1-D/(d\{S'[1+s]-[1-f]-Sv-I\})$$
$$= 1-2,370/(0.3000\{48,851[1+0.0500]$$
$$[1-0.5800]-51,294*0.1900-513\})$$
$$= 30.00\%$$

Law-1953:

If (v= V/S= Variable Portion= 19.00%), (I= Interest Expense= 513 millions USD), (f= F/S= Fixed Portion= 58.00%), (D= Dividend Paid= 2,370 millions USD), ($\$$= Sales or Revenues= 51,294 millions USD), ($\$'$= Sales of Past Year= 48,851 millions USD), (t= T/B= Tax Rate= 30.00%), and (s= [S/S']-1= Sales Growth= 5.00%), are known, then it's (d= Dividend Portion or Payout planned) is:

d= D/([1-t]{$\$'$[1+$s$]-$\$v$-$\$'f$[1+$s$]-$I$}
  = D/([1-t]{$\$'$[1+$s$]-[1-$f$]-$\$v$-$I$})
  = 2,370/(0.3000[1-0.3000]{48,851
       [1+0.0500][1-0.5800]-51,294
       *0.1900-513})
  = 30.00%

Law-1954:

If (v= V/S= Variable Portion= 19.00%), (i= I/S= Interest Portion= 1.00%), (f= F/S= Fixed Portion= 58.00%), (d= D/A= Dividend Portion or Payout= 30.00%), ($\$$= Sales or Revenues= 51,294 millions USD), ($\$'$= Sales of Past Year= 48,851 millions USD), (t= T/B= Tax Rate= 30.00%), and (s= [S/S']-1= Sales Growth= 5.00%), are known, then it's (D =Dividend Paid planned) is:

D= d[1-t]{$\$'$[1+$s$]-$\$v$-$\$'f$[1+$s$]-$\$i$}
  = d[1-t]{$\$'$[1+$s$]-[1-$f$]-$\$$[v+i]}
  = 0.3000[1-0.3000]{48,851[1+0.0500]
       [1-0.5800]-51,294[0.1900
       +0.0100}
  = 2,370 millions USD

Law-1955:

If (v= V/S= Variable Portion= 19.00%), (i= I/S= Interest Portion= 1.00%), (f= F/S= Fixed Portion= 58.00%), (s= [S/S']-1= Sales Growth= 5.00%), (d= D/A=Dividend Portion or Payout= 30.00%), (S= Sales or Revenues= 51,294 millions USD), (D= Dividend Paid= 2,370 millions USD), and (t= B/T= Tax Rate= 30.00%), are known, then it's (S'= Sales Past) must be:

$$S' = (S[v+i]+D/\{d[1-t]\})/\{[1+s][1-f]\}$$
$$= (51,294[0.1900+0.0100]+2,370$$
$$/\{0.3000[1-0.3000]\})$$
$$/\{[1+0.0500][1-0.5800]\}$$
$$= 48,851 \text{ millions USD}$$

Law-1956:

If (v= V/S= Variable Portion= 19.00%), (i= I/S= Interest Portion= 1.00%), (f= F/S=Fixed Portion= 58.00%), (S'= Sales of Past Year= 48,851 millions USD), (d= D/A= Dividend Portion or Payout= 30.00%), (S= Sales or Revenues= 51,294 millions USD), (D= Dividend Paid= 2,370 millions USD), and (t= T/B=Tax Rate= 30.00%), are known, then it's (s= Sales Growth planned) is:

$$s = (S[v+i]+D/\{d\}[1-t]\})/\{S'[1-f]\}-1$$
$$= 51,294[0.1900+0.0100]+2,370$$
$$/\{0.3000[1-0.3000]\})$$
$$/\{48,851[1-0.5800]\}-1$$
$$= 5.00\%$$

Law-1957:

If (v= V/S= Variable Portion= 19.00%), (i= I/S= Interest Portion= 1.00%), (s= [S/S']-1= Sales Growth= 5.00%), ($\$'$= Sales of Past Year= 48,851 millions USD), (d= D/A= Dividend Portion or Payout= 30.00%), ($\$$= Sales or Revenues= 51,294 millions USD), (D= Dividend Paid= 2,370 millions USD), and (t= T/B= Tax Rate= 30.00%), are known, then it's (f= Fixed Portion planned) is:

f= 1-($\$[v+i]+D/\{d\}[1-t]\})/\{\$'[1+s]\}$
= 51,294[0.1900+0.0100]+2,370
/{0.3000[1-0.3000]})
/{48,851[1+0.0500]}
= 58.00%

Law-1958:

If (v= V/S= Variable Portion= 19.00%), (i= I/S= Interest Portion= 1.00%), (s= [S/S']-1= Sales Growth= 5.00%), ($\$'$= Sales of Past Year= 48,851 millions USD), (d= D/A= Dividend Portion or Payout= 30.00%), (f= F/S=Fixed Portion= 58.00%), (D= Dividend Paid= 2,370 millions USD), and (t= T/B= Tax Rate= 30.00%), are known, then it's ($\$$= Sales or Revenues planned) is:

$\$$= ($\$'[1+s][1-f]-D/\{d\}[1-t]\})/[v+i]$
= (48,851[1+0.0500][1-0.5800]-2,370
/{0.3000[1-0.3000]})
/[0.1900+0.0100]
= 51,294 millions USD

Law-1959:

If ($= Sales or Revenues= 51,294 millions USD), (i= I/S= Interest Portion= 1.00%), (s= [S/S']-1= Sales Growth= 5.00%), ($'= Sales of Past Year= 48,851 millions USD), (d= D/A= Dividend Portion or Payout= 30.00%), (f= F/S= Fixed Portion= 58.00%), (D= Dividend Paid= 2,370 millions USD), and (t= T/B= Tax Rate= 30.00%), are known, then it's (v= Variable Portion planned) is:

$$v = (\$'[1+s][1-f]-D/\{d\}[1-t]\})/\$-i$$
$$= (48,851[1+0.0500][1-0.5800]-2,370$$
$$/\{0.3000[1-0.3000]\})/51,294$$
$$-0.0100$$
$$= 19.00\%$$

Law-1960:

If ($= Sales or Revenues= 51,294 millions USD), (v= V/S= Variable Portion= 19.00%), (s= [S/S']-1= Sales Growth= 5.00%), ($'= Sales of Past Year= 48,851 millions USD), d= D/A= Dividend Portion or Payout= 30.00%), (f= F/S= Fixed Portion= 58.00%), (D= Dividend Paid= 2,370 millions USD), and (t= T/B= Tax Rate= 30.00%), are known, then it's (v= Variable Portion planned) is:

$$i = (\$'[1+s][1-f]-D/\{d\}[1-t]\})/\$-v$$
$$= (48,851[1+0.0500][1-0.5800]-2,370$$
$$/\{0.3000[1-0.3000]\})/51,294$$
$$-0.1900$$
$$= 1.00\%$$

Law-1961:
If ($= Sales or Revenues= 51,294 millions USD), (v= V/S= Variable Portion= 19.00%), (s= [S/S']-1= Sales Growth= 5.00%), ($'= Sales of Past Year= 48,851 millions USD), (d= D/A= Dividend Portion or Payout= 30.00%), (f= F/S= Fixed Portion= 58.00%), (D= Dividend Paid= 2,370 millions USD), and (i= I/S= Interest Portion= 1.00%), are known, then it's (t= Tax Rate planned) is:

$$t = 1 - D/(d)\{\$'[1+s] - \$v - \$'f[1+s] - \$i\}$$
$$= 1 - D/(d(\$'[1+s][1-f] - \$[v+i]\})$$
$$= 1 - 2,370/(0.3000\{48,851[1+0.0500]$$
$$[1-0.5800] - 51,294$$
$$[0.1900 + 0.0100]\})$$
$$= 30.00\%$$

Law-1962:
If ($= Sales or Revenues= 51,294 millions USD), (v= V/S= Variable Portion= 19.00%), (s= [S/S']-1= Sales Growth= 5.00%), ($'= Sales of Past Year= 48,851 millions USD), (t= T/B= Tax Rate= 30.00%), (f= F/S=Fixed Portion= 58.00%), (D= Dividend Paid= 2,370 millions USD), and (i= I/S= Interest Portion= 1.00%),are known, then it's (d= Dividend Portion or Payout planned) is:

$$d = D/([1-t]\{\$'[1+s] - \$v - \$'f[1+s] - \$i\})$$
$$= D/([1-t]\{\$'[1+s][1-f] - \$[v+i]\})$$
$$= 2,370/([1-0.3000]\{48,851[1+0.0500]$$
$$[1-0.5800] - 51,294$$
$$[0.1900 + 0.0100]\})$$
$$= 30.00\%$$

Law-1963:

If ($= Sales or Revenues= 51,294 millions USD), (v= V/S= Variable Portion= 19.00%), (s= [S/S']-1= Sales Growth= 5.00%), ($'= Sales of Past Year= 48,851 millions USD), (t= T/B= Tax Rate= 30.00%), (f= F/S=Fixed Portion= 58.00%), (d= D/A=Dividend Portion or Payout= 30.00%), and (i= I/S= Interest Portion= 1.00%), are known, then it's (D= Dividend Paid planned) is:

$$D = d[1-t]\{\$'[1+s]-\$v-\$'f[1+s]-\$'i[1+s]\}$$
$$= d[1-t]\{\$'[1+s][1-f-i]-\$v\}$$
$$= 0.3000[1-0.3000]\{48,851[1+0.0500]$$
$$[1-0.5800-0.0100]$$
$$- 51,294*0.1900\}$$
$$= \underline{2,370} \text{ millions USD}$$

Law-1964:

If ($= Sales or Revenues= 51,294 millions USD), (v= V/S= Variable Portion= 19.00%), (s= [S/S']-1= Sales Growth= 5.00%), (D= Dividend Paid= 2,370 millions USD), (t= T/B= Tax Rate= 30.00%), (f= F/S= Fixed Portion= 58.00%), (d= D/A= Dividend Portion or Payout= 30.00%), and (i= I/S= Interest Portion= 1.00%), are known, then it's ($'= Sales Past) must be:

$$\$' = (\$v+D/\{d[1-t]\}/\{[1+s][1-f-i]\}$$
$$= (51,294*0.1900+2,370/ (0.3000$$
$$[1-0.3000]\})/\{ [1+0.0500]$$
$$[1-0.5800-0.0100]\}$$
$$= \underline{48,851} \text{ millions USD}$$

Law-1965:
  If ($= Sales or Revenues= 51,294 millions USD), (v= V/S= Variable Portion= 19.00%), ($'= Sales of Past Year= 48,851 millions USD), (D= Dividend Paid= 2,370 millions USD), (t= T/B= Tax Rate= 30.00%), (f= F/S=Fixed Portion= 58.00%), (d= D/A= Dividend Portion or Payout= 30.00%), and (i= I/S= Interest Portion= 1.00%), are known, then it's (s= Sales Growth planned) is:

  $s = (\$v+D/\{d[1-t]\}/\{\$'[1-f-i]\})-1$
  $= (51{,}294*0.1900+2{,}370/(0.3000$
  $[1-0.3000]\})/\{48{,}851$
  $[1-0.5800-0.0100]\}-1$
  $= 5.00\%$

Law-1966:
  If ($= Sales or Revenues= 51,294 millions USD), (v= V/S= Variable Portion= 19.00%), ($'= Sales of Past Year= 48,851 millions USD), (D= Dividend Paid= 2,370 millions USD), (t= T/B= Tax Rate= 30.00%), (s= [S/S']-1= Sales Growth= 5.00%), (d= D/A= Dividend Portion or Payout= 30.00%), and (i= I/S= Interest Portion= 1.00%), are known, then it's (f= Fixed Portion planned) is:

  $f = 1-i-(\$v+D/\{d[1-t]\})/\{\$'[1+s]\}$
  $= 1-0.0100-(51{,}294*0.1900+2{,}370$
  $/(0.3000[1-0.3000]\})$
  $/\{48{,}851[1+0.0500]\}$
  $= 58.00\%$

Law-1967:

If ($\$$= Sales or Revenues= 51,294 millions USD), (υ= V/S= Variable Portion= 19.00%), ($\$'$= Sales of Past Year= 48,851 millions USD), (**D**= Dividend Paid= 2,370 millions USD), (**t**= T/B= Tax Rate= 30.00%), (**s**= [S/S']-1= Sales Growth= 5.00%), (**d**= D/A= Dividend Portion or Payout= 30.00%), and (**f**= F/S=Fixed Portion= 58.00%), are known, then it's (**i**= Interest Portion planned) is:

$$i = 1-f-(\$\upsilon+D/\{d[1-t]\})/\{\$'[1+s]\}$$
$$= 1-0.5800-(51,294*0.1900+2,370$$
$$/(0.3000[1-0.3000]\})$$
$$/\{48,851[1+0.0500]\}$$
$$= 1.00\%$$

Law-1968:

If (**i**= I/S= Interest Portion= 1.00%), (υ= V/S= Variable Portion= 19.00%), ($\$'$= Sales of Past Year= 48,851 millions USD), (**D**= Dividend Paid= 2,370 millions USD), (**t**= T/B= Tax Rate= 30.00%), (**s**= [S/S']-1= Sales Growth= 5.00%), (**d**= D/A= Dividend Portion or Payout= 30.00%), and (**f**= F/S= Fixed Portion= 58.00%), are known, then it's ($\$$= Sales or Revenues planned) is:

$$\$ = (\$'[1+s][1-f-i]-D/\{d[1-t]\})/\upsilon$$
$$= (48,851[1+0.0500][1-0.5800-0.0100]$$
$$-2,370/ (0.3000[1-0.3000]\})$$
$$/0.1900$$
$$= 51,294 \text{ millions USD}$$

Law-1969:
If (i= I/S= Interest Portion= 1.00%), (S= Sales or Revenues= 51,294 millions USD), (S'= Sales of Past Year= 48,851 millions USD), (D= Dividend Paid= 2,370 millions USD), (t= T/B= Tax Rate= 30.00%), (s= [S/S']-1= Sales Growth= 5.00%), (d= D/A= Dividend Portion or Payout= 30.00%), and (f= F/S=Fixed Portion= 58.00%), are known, then it's (v= Variable Portion planned) is:

v= (S'[1+s][1-f-i]-D/{d[1-t]})/S
  = (48,851[1+0.0500][1-0.5800-0.0100]
    -2,370/ (0.3000[1-0.3000]})
    /51,294
  = 19.00%

Law-1970:
If (i= I/S= Interest Portion= 1.00%), (S= Sales or Revenues= 51,294 millions USD), (S'= Sales of Past Year= 48,851 millions USD), (D= Dividend Paid= 2,370 millions USD), (v= V/S= Variable Portion= 19.00%), (s= [S/S']-1= Sales Growth= 5.00%), (d= D/A= Dividend Portion or Payout= 30.00%), and (f= F/S=Fixed Portion= 58.00%), are known, then it's (t= Tax Rate planned) is:

t= 1-D/(d{S'[1+s]-Sv-$S'f$[1+s]-Si[1+s]})
  = 1-D/(d{S'[1+s][1-f-i]-Sv})
  = 1-2,370/(0.3000{48,851[1+0.0500]
    [1-0.5800-0.0100]
    -51,294*0.1900})
  = 30.00%

Law-1971:

If (i= I/S= Interest Portion= 1.00%), (S= Sales or Revenues= 51,294 millions USD), (S'= Sales of Past Year= 48,851 millions USD), (D= Dividend Paid= 2,370 millions USD), (v= V/S= Variable Portion= 19.00%), (s= [S/S']-1= Sales Growth= 5.00%), (t= T/B= Tax Rate= 30.00%), and (f= F/S= Fixed Portion= 58.00%), are known, then it's (d= Dividend Portion or Payout planned) is:

d = $D/([1-t]\{S'[1+s]-Sv-S'f[1+s]-Si[1+s]\})$
= $D/([1-t]\{S'[1+s][1-f-i]-Sv\})$
= 2,370/ ([1-0.3000]{48,851[1+0.0500]
 [1-0.5800-0.0100]
 -51,294*0.1900})
= 30.00%

Law-1972:

If (I= Interest Expense= 513 millions USD), (S= Sales or Revenues= 51,294 millions USD), (S'= Sales of Past Year= 48,851 millions USD), (v= V/S= Variable Portion= 19.00%), (s= [S/S']-1= Sales Growth= 5.00%), (t= T/B= Tax Rate= 30.00%), and (F= Fixed Cost= 29,750 millions USD), (d= D/A= Dividend Portion or Payout= 30.00%), are known, then it's (D= Dividend Paid planned) is:

D = $d[1-t]\{S'[1+s]-Sv[1+s]-F-I\}$
= $d[1-t]/(S'[1+s][1-v]-F-I\}$
= 0.3000[1-0.3000]{48,851[1+0.0500]
 [1-0.1900]-29,750-513}
= 2,370 millions USD

Law-1973:

If (I= Interest Expense= 513 millions USD), (S= Sales or Revenues= 51,294 millions USD), (v= V/S= Variable Portion= 19.00%), (s= [S/S']-1= Sales Growth= 5.00%), (t= T/B= Tax Rate= 30.00%), and (F= Fixed Cost= 29,750 millions USD), (d= D/A= Dividend Portion or Payout= 30.00%), are known, then it's (S'= Sales Past) must be:

$$S' = (F+I+D/\{d[1-t]\})/\{[1+s][1-v]\}$$
$$= (29,750+513+2,370/\{0.3000$$
$$[1-0.3000]\})/\{[1+0.0500]$$
$$[1-0.1900]\}$$
$$= 48,851 \text{ millions USD}$$

Law-1974:

If (I= Interest Expense= 513 millions USD), (S= Sales or Revenues= 51,294 millions USD), (v= V/S= Variable Portion= 19.00%), (D= Dividend Paid= 2,370 millions USD), (S'= Sales of Past Year= 48,851 millions USD), (t= T/B= Tax Rate= 30.00%), and (F= Fixed Cost= 29,750 millions USD), (d= D/A= Dividend Portion or Payout= 30.00%), are known, then it's (s= Sales Growth planned) is:

$$s = (F+I+D/\{d[1-t]\})/\{S'[1-v]\}-1$$
$$= (29,750+513+2,370/\{0.3000$$
$$[1-0.3000]\})/\{48,851$$
$$[1-0.1900]\}-1$$
$$= 5.00\%$$

Law-1975:

If (**I**= Interest Expense= 513 millions USD), (**$**= Sales or Revenues= 51,294 millions USD), (**s**= [S/S']-1= Sales Growth= 5.00%), (**D**= Dividend Paid= 2,370 millions USD), (**$'**= Sales of Past Year= 48,851 millions USD), (**t**= T/B= Tax Rate= 30.00%), and (**F**= Fixed Cost= 29,750 millions USD), (**d**= D/A= Dividend Portion or Payout= 30.00%), are known, then it's (**v**= Variable Portion planned) is:

$$v = 1-(F+I+D/\{d[1-t]\})/\{\$'[1+s]\}$$
$$= 1-(29,750+513+2,370/\{0.3000[1-0.3000]\})/\{48,851[1+0.0500]\}$$
$$= 19.00\%$$

Law-1976:

If (**I**= Interest Expense= 513 millions USD), (**$**= Sales or Revenues= 51,294 millions USD), (**s**= [S/S']-1= Sales Growth= 5.00%), (**v**= V/S= Variable Portion= 19.00%), (**D**= Dividend Paid= 2,370 millions USD), (**$'**= Sales of Past Year= 48,851 millions USD), (**t**= T/B= Tax Rate= 30.00%), and (**d**= D/A= Dividend Portion or Payout= 30.00%), are known, then it's (**F**= Fixed Cost planned) is:

$$F = (\$'[1+s][1-v]-I-D/\{d[1-t]\})$$
$$= (48,851[1+0.0500][1-0.1900]-513-2,370/\{0.3000[1-0.3000]\})$$
$$= 29,750 \text{ millions USD}$$

Law-1977:
If (**F**= Fixed Cost= 29,750 millions USD), (**$**= Sales or Revenues= 51,294 millions USD), (**s**= [S/S']-1= Sales Growth= 5.00%), (**v**= V/S= Variable Portion= 19.00%), (**D**= Dividend Paid= 2,370 millions USD), (**$'**= Sales of Past Year= 48,851 millions USD), (**t**= T/B= Tax Rate= 30.00%), and (**d**= D/A= Dividend Portion or Payout= 30.00%), are known, then it's (**I**= Interest Expense planned) is:

$I = (\$'[1+s][1-v] - F - D/\{d[1-t]\})$
  $= (48,851[1+0.0500][1-0.1900] - 29,750$
    $- 2,370/\{0.3000[1-0.3000]\})$
  $= \underline{513}$ millions USD

Law-1978:
If (**F**= Fixed Cost= 29,750 millions USD), (**$**= Sales or Revenues= 51,294 millions USD), (**s**= [S/S']-1= Sales Growth= 5.00%), (**v**= V/S= Variable Portion= 19.00%), (**D**= Dividend Paid= 2,370 millions USD), (**$'**= Sales of Past Year= 48,851 millions USD), (**I**= Interest Expense= 513 millions USD), and (**d**= D/A=Dividend Portion or Payout= 30.00%), are known, then it's (**t**= Tax Rate planned) is:

$t = 1 - D/(d\{\$'[1+s] - \$'v[1+s] - F - I\})$
  $= 1 - D/(d\{(\$'[1+s][1-v] - F - I\})$
  $= 1 - 2,370/(0.3000\{48,851[1+0.0500]$
    $[1-0.1900] - 29,750 - 513\})$
  $= \underline{30.00\%}$

Finance Construction-2, *Tim Asikin, Steve Asikin, Indra Senihardja*

Law-1979:

If (**F**= Fixed Cost= 29,750 millions USD), (**$**= Sales or Revenues= 51,294 millions USD), (**s**= [S/S']-1= Sales Growth= 5.00%), (**v**= V/S= Variable Portion= 19.00%), (**D**= Dividend Paid= 2,370 millions USD), (**$'**= Sales of Past Year= 48,851 millions USD), (**I**= Interest Expense= 513 millions USD), and (**t**= T/B= Tax Rate= 30.00%), are known, then it's (**d**= Dividend Portion or Payout planned) is:

$$\begin{aligned}\mathbf{d} &= \mathbf{D}/([1\text{-}\mathbf{t}]]/(\mathbf{\$'}[1+\mathbf{s}]\text{-}\mathbf{\$'v}[1+\mathbf{s}]\text{-}\mathbf{F}\text{-}\mathbf{I}\}) \\ &= \mathbf{D}/([1\text{-}\mathbf{t}]\{\mathbf{\$'}[1+\mathbf{s}][1\text{-}\mathbf{v}]\text{-}\mathbf{F}\text{-}\mathbf{I}\}) \\ &= 2{,}370/([1\text{-}0.3000]\{48{,}851[1+0.0500] \\ &\quad [1\text{-}0.1900]\text{-}29{,}750\text{-}513\}) \\ &= 30.00\%\end{aligned}$$

Law-1980:

If (**F**= Fixed Cost= 29,750 millions USD), (**$**= Sales or Revenues= 51,294 millions USD), (**s**= [S/S']-1= Sales Growth= 5.00%), (**v**= V/S= Variable Portion= 19.00%), (**d**= D/A=Dividend Portion or Payout= 30.00%), (**$'**= Sales of Past Year= 48,851 millions USD), (**i**= I/S= Interest Portion= 1.00%), and (**t**= T/B= Tax Rate= 30.00%), are known, then it's (**D**= Dividend Paid planned) is:

$$\begin{aligned}\mathbf{D} &= \mathbf{d}[1\text{-}\mathbf{t}]\{(\mathbf{\$'}[1+\mathbf{s}]\text{-}\mathbf{\$'v}[1+\mathbf{s}]\text{-}\mathbf{F}\text{-}\mathbf{\$i}\} \\ &= \mathbf{d}([1\text{-}\mathbf{t}]\{\mathbf{\$'}[1+\mathbf{s}][1\text{-}\mathbf{v}]\text{-}\mathbf{F}\text{-}\mathbf{\$i}\}) \\ &= 0.3000[1\text{-}0.3000]\{48{,}851[1+0.0500] \\ &\quad [1\text{-}0.1900]\text{-}29{,}750 \\ &\quad \text{-}51{,}294*0.0100\} \\ &= 2{,}370 \text{ millions USD}\end{aligned}$$

Law-1981:

If (**F**= Fixed Cost= 29,750 millions USD), (**$**= Sales or Revenues= 51,294 millions USD), (**s**= [S/S']-1= Sales Growth= 5.00%), (**v**= V/S= Variable Portion= 19.00%), (**d**= D/A=Dividend Portion or Payout= 30.00%), (**D**= Dividend Paid= 2,370 millions USD), (**i**= I/S= Interest Portion= 1.00%), and (**t**= T/B= Tax Rate= 30.00%), are known, then it's (**$'**= Sales Past) must be:

$'= (**F+$i+D/{d[1-t]}**)/{[1+**s**][1-**v**]}
   = 29,750+51,294*0.0100+2,370
       /{0.3000[1- 0.3000]})
       /{[1+0.0500][1-0.1900]}
   = 4,8851 millions USD

Law-1982:

If (**F**= Fixed Cost= 29,750 millions USD), (**$**= Sales or Revenues= 51,294 millions USD), (**$'**= Sales of Past Year= 48,851 millions USD), (**v**= V/S= Variable Portion= 19.00%), (**d**= D/A= Dividend Portion or Payout= 30.00%), (**D**= Dividend Paid= 2,370 millions USD), (**i**= I/S= Interest Portion= 1.00%), and (**t**= T/B= Tax Rate= 30.00%), are known, then it's (**s**= Sales Growth planned) is:

s= (**F+$i+D/{d[1-t]}**)/{**$'**[1-**v**]}-1
  = (29,750+51,294*0.0100+2,370
      /{0.3000[1- 0.3000]})
      /{ 48,851[1-0.1900]}-1
  = 5.00%

Law-1983:

If (F= Fixed Cost= 29,750 millions USD), (S= Sales or Revenues= 51,294 millions USD), (S'= Sales of Past Year= 48,851 millions USD), (s= [S/S']-1= Sales Growth= 5.00%), (d= D/A= Dividend Portion or Payout= 30.00%), (D= Dividend Paid= 2,370 millions USD), (i= I/S= Interest Portion= 1.00%), and (t= T/B= Tax Rate= 30.00%), are known, then it's (v= Variable Portion planned) is:

v= 1-(F+Si+D/{d[1-t]})/{S'[1+s]}
= 1- (29,750+51,294*0.0100+2,370
/{0.3000[1- 0.3000]})
/{48,851[1+0.0500]}
= 19.00%

Law-1984:

If (v= V/S= Variable Portion= 19.00%), (S= Sales or Revenues= 51,294 millions USD), (S'= Sales of Past Year= 48,851 millions USD), (s= [S/S']-1= Sales Growth= 5.00%), (d= D/A= Dividend Portion or Payout= 30.00%), (D= Dividend Paid= 2,370 millions USD), (i= I/S= Interest Portion= 1.00%), and (t= T/B= Tax Rate= 30.00%), are known, then it's (F= Fixed Cost planned) is:

F= S'[1+s][1-v]-Si-D/{d[1-t]}
= (48,851[1+0.0500][1-0.1900]
-51,294*0.0100-2,370
/{0.3000[1- 0.3000]})
= 29,750 millions USD

Law-1985:

If (v= V/S= Variable Portion= 19.00%), (F= Fixed Cost= 29,750 millions USD), (S'= Sales of Past Year= 48,851 millions USD), (s= [S/S']-1= Sales Growth= 5.00%), (d= D/A= Dividend Portion or Payout= 30.00%), (D= Dividend Paid= 2,370 millions USD), (i= I/S= Interest Portion= 1.00%), and (t= T/B= Tax Rate= 30.00%), are known, then it's (S= Sales or Revenues planned) is:

$$S = (S'[1+s][1-v]-F-D/\{d[1-t]\})/i$$
$$= (48,851[1+0.0500][1-0.1900]- 29,750$$
$$\quad -2,370/\{0.3000[1-0.3000]\})$$
$$\quad /0.0100$$
$$= 51,294 \text{ millions USD}$$

Law-1986:

If (v= V/S= Variable Portion= 19.00%), (F= Fixed Cost= 29,750 millions USD), (S'= Sales of Past Year= 48,851 millions USD), (s= [S/S']-1= Sales Growth= 5.00%), (d= D/A= Dividend Portion or Payout= 30.00%), (D= Dividend Paid= 2,370 millions USD), (S= Sales or Revenues= 51,294 millions USD), and (t= T/B= Tax Rate= 30.00%), are known, then it's (i= Interest Portion planned) is:

$$i = (S'[1+s][1-v]-F-D/\{d[1-t]\})/S$$
$$= (48,851[1+0.0500][1-0.1900]- 29,750$$
$$\quad -2,370/\{0.3000[1-0.3000]\})$$
$$\quad /51,294$$
$$= 1.00\%$$

Law-1987:

If (v= V/S= Variable Portion= 19.00%), (F= Fixed Cost= 29,750 millions USD), ($\$'$= Sales of Past Year= 48,851 millions USD), (s= [S/S']-1= Sales Growth= 5.00%), (d= D/A= Dividend Portion or Payout= 30.00%), (D= Dividend Paid= 2,370 millions USD), ($\$$= Sales or Revenues= 51,294 millions USD), and (i= I/S= Interest Portion= 1.00%), are known, then it's (t= Tax Rate planned) is:

t= 1-**D**/(**d**{$\$'$[1+$s$]-$\$'v$[1+$s$]-**F**-$\$i$}
   = 1-**D**/(**d**{$\$'$[1+$s$][1-$v$]-**F**-$\$i$})
   = 1-2,370/(0.3000[1- 0.3000]{48,851
      [1+0.0500][1-0.1900]- 29,750
      -51,294*0.0100})
   = 30.00%

Law-1988:

If (v= V/S= Variable Portion= 19.00%), (F= Fixed Cost= 29,750 millions USD), ($\$'$= Sales of Past Year= 48,851 millions USD), (s= [S/S']-1= Sales Growth= 5.00%), (t= T/B= Tax Rate= 30.00%), (D= Dividend Paid= 2,370 millions USD), ($\$$= Sales or Revenues= 51,294 millions USD), and (i= I/S= Interest Portion= 1.00%), are known, then it's (d= Dividend Portion or Payout planned) is:

d= **D**/([1-t]/{$\$'$[1+$s$]-$\$'v$[1+$s$]-**F**-$\$i$}
   = 1-**D**/(**d**{$\$'$[1+$s$][1-$v$]-**F**-$\$i$})
   = 2,370/([1- 0.3000]{48,851[1+0.0500]
      [1-0.1900]- 29,750
      - 51,294*0.0100})
   = 30.00%

Law-1989:

If (v= V/S= Variable Portion= 19.00%), (F= Fixed Cost= 29,750 millions USD), ($\$'$= Sales of Past Year= 48,851 millions USD), (s= [S/S']-1= Sales Growth= 5.00%), (t= T/B= Tax Rate= 30.00%), (d= D/A= Dividend Portion or Payout= 30.00%), ($\$$= Sales or Revenues= 51,294 millions USD), and (i= I/S= Interest Portion= 1.00%), are known, then it's (D= Dividend Paid planned) is:

$$D = d[1-t]\{\$'[1+s]-\$'v[1+s]-F-\$'i[1+s]\}$$
$$= d[1-t]\{\$'[1+s][1-v-i]-F\}$$
$$= 0.3000[1- 0.3000]\{48,851[1+0.0500][1-0.1900-0.0100]- 29,750\}$$
$$= 30.00\%$$

Law-1990:

If (v= V/S= Variable Portion= 19.00%), (F= Fixed Cost= 29,750 millions USD), (D= Dividend Paid= 2,370 millions USD), (s= [S/S']-1= Sales Growth= 5.00%), (t= T/B= Tax Rate= 30.00%), (d= D/A=Dividend Portion or Payout= 30.00%), ($\$$= Sales or Revenues= 51,294 millions USD), and (i= I/S= Interest Portion= 1.00%), are known, then it's ($\$'$= Sales Past) must be:

$$\$' = (F+D/\{d[1-t]\})/\{[1+s][1-v-i]\}$$
$$= (29,750+2,370/\{0.3000[1- 0.3000]\})/\{[1+0.0500][1-0.1900-0.0100]\}$$
$$= 48,851 \text{ millions USD}$$

Law-1991:

If (v= V/S= Variable Portion= 19.00%), (F= Fixed Cost= 29,750 millions USD), (D= Dividend Paid= 2,370 millions USD), (S'= Sales of Past Year= 48,851 millions USD), (t= T/B= Tax Rate= 30.00%), (d= D/A= Dividend Portion or Payout= 30.00%), (S= Sales or Revenues= 51,294 millions USD), and (i= I/S= Interest Portion= 1.00%), are known, then it's (s= Sales Growth planned) is:

$$s = (F+D/\{d[1-t]\})/\{S'[1-v-i]\} - 1$$
$$= (29{,}750 + 2{,}370/\{0.3000[1- 0.3000]\})$$
$$/\{48{,}851[1-0.1900-0.0100]\} - 1$$
$$= 5.00\%$$

Law-1992:

If (s= [S/S']-1= Sales Growth= 5.00%), (F= Fixed Cost= 29,750 millions USD), (D= Dividend Paid= 2,370 millions USD), (S'= Sales of Past Year= 48,851 millions USD), (t= T/B= Tax Rate= 30.00%), (d= D/A= Dividend Portion or Payout= 30.00%), (S= Sales or Revenues= 51,294 millions USD), and (i= I/S= Interest Portion= 1.00%), are known, then it's (v= Variable Portion planned) is:

$$v = 1 - i - (F+D/\{d[1-t]\})/\{S'[1+s]\}$$
$$= 1 - 0.0100 - (29{,}750 + 2{,}370$$
$$/\{0.3000[1- 0.3000]\})$$
$$/\{48{,}851[1+0.0500]\}$$
$$= 19.00\%$$

Law-1993:

If (s= [S/S']-1= Sales Growth= 5.00%), (F= Fixed Cost= 29,750 millions USD), (D= Dividend Paid= 2,370 millions USD), (S'= Sales of Past Year= 48,851 millions USD), (t= T/B= Tax Rate= 30.00%), (d= D/A= Dividend Portion or Payout= 30.00%), (S= Sales or Revenues= 51,294 millions USD), and (v= V/S= Variable Portion= 19.00%), are known, then it's (i= Interest Portion planned) is:

$$i = 1-v-(F+D/\{d[1-t]\})/\{S'[1+s]\}$$
$$= 1-0.1900- (29,750+2,370$$
$$/\{0.3000[1-0.3000]\})$$
$$/\{48,851[1+0.0500]\}$$
$$= \underline{1.00}\%$$

Law-1994:

If (s= [S/S']-1= Sales Growth= 5.00%), (i= I/S= Interest Portion= 1.00%), (D= Dividend Paid= 2,370 millions USD), (S'= Sales of Past Year= 48,851 millions USD), (t= T/B= Tax Rate= 30.00%), (d= D/A= Dividend Portion or Payout= 30.00%), (S= Sales or Revenues= 51,294 millions USD), and (v= V/S= Variable Portion= 19.00%), are known, then it's (F= Fixed Cost planned) is:

$$F = S'[1+s][1-v-i]-D/\{d[1-t]\}$$
$$= 48,851[1+0.0500][1-0.1900-0.0100]$$
$$-2,370/\{0.3000[1-0.3000]\}$$
$$= \underline{29,750} \text{ millions USD}$$

Law-1995:
If (s= [S/S']-1= Sales Growth= 5.00%), (i= I/S= Interest Portion= 1.00%), (D= Dividend Paid= 2,370 millions USD), (S'= Sales of Past Year= 48,851 millions USD), (F= Fixed Cost= 29,750 millions USD), (d= D/A= Dividend Portion or Payout= 30.00%), (S= Sales or Revenues= 51,294 millions USD), and (v= V/S= Variable Portion= 19.00%), are known, then it's (t= Tax Rate planned) is:

t = 1-D/(d{S'[1+s]-$S'v$[1+s]-F-$S'i$[1+s]})
= 1-D/(d{S'[1+s][1-v-i]-F})
= 1-2,370/(0.3000{48,851[1+0.0500]
[1-0.1900-0.0100]- 29,750})
= 30.00%

Law-1996:
If (s= [S/S']-1= Sales Growth= 5.00%), (i= I/S= Interest Portion= 1.00%), (D= Dividend Paid= 2,370 millions USD), (S'= Sales of Past Year= 48,851 millions USD), (F= Fixed Cost= 29,750 millions USD), (t= T/B= Tax Rate= 30.00%), (S= Sales or Revenues= 51,294 millions USD), and (v= V/S= Variable Portion= 19.00%), are known, then it's (d= Dividend Portion or Payout planned) is:

d = D/([1-t]{S'[1+s]-$S'v$[1+s]-F-$S'i$[1+s]})
= D/([1-t]{S'[1+s][1-v-i]-F})
= 2,370/([1-0.3000]{48,851[1+0.0500]
[1-0.1900-0.0100]-29,750})
= 30.00%

Law-1997:

If (s= [S/S']-1= Sales Growth= 5.00%), (i= I/S= Interest Portion= 1.00%), (d= Dividend Portion or Payout= 30.00% millions USD), (S'= Sales of Past Year= 48,851 millions USD), (F= Fixed Cost= 29,750 millions USD), (t= T/B= Tax Rate= 30.00%), (S= Sales or Revenues= 51,294 millions USD), and (v= V/S= Variable Portion= 19.00%), are known, then it's (D= Dividend Paid planned) is:

$$D = d[1-t]\{S'[1+s]-S'v[1+s]-F-S'i[1+s]\}$$
$$= d[1-t]\{S'[1+s][1-v-i]-F\})$$
$$= 0.3000([1-0.3000]\{48,851[1+0.0500][1-0.1900-0.0100]-29,750\})$$
$$= 2,370 \text{ millions USD}$$

Law-1998:

If (s= [S/S']-1= Sales Growth= 5.00%), (I= Interest Expense= 513 millions USD), (d= D/A= Dividend Portion or Payout= 30.00%), (D= Dividend Paid= 2,370 millions USD), (f= F/S= Fixed Portion= 58.00%), (t= T/B= Tax Rate= 30.00%), (S= Sales or Revenues= 51,294 millions USD), and (v= V/S= Variable Portion= 19.00%), are known, then it's (S'= Sales Past) must be:

$$S' = (Sf+I+D/\{d[1-t]\})/\{[1+s][1-v]\}$$
$$= (51,294*0.5800+513+2,370 /\{0.3000[1-0.3000]\})/\{[1+0.0500][1-0.1900]\}$$
$$= 48,851 \text{ millions USD}$$

Law-1999:

If ($'= $ Sales of Past Year= 48,851 millions USD), (I= Interest Expense= 513 millions USD), (d= D/A= Dividend Portion or Payout= 30.00%), (D= Dividend Paid= 2,370 millions USD), (f= F/S= Fixed Portion= 58.00%), (t= T/B= Tax Rate= 30.00%), ($= Sales or Revenues= 51,294 millions USD), and (v= V/S= Variable Portion= 19.00%), are known, then it's (s= Sales Growth planned) is:

$$s = (\$f + I + D/\{d[1-t]\})/\{\$'[1-v]\} - 1$$
$$= (51{,}294 * 0.5800 + 513 + 2{,}370$$
$$/\{0.3000[1 - 0.3000]\})$$
$$/\{48{,}851[1 - 0.1900]\} - 1$$
$$= 5.00\%$$

Law-2000:

If ($'= $ Sales of Past Year= 48,851 millions USD), (I= Interest Expense= 513 millions USD), (d= D/A= Dividend Portion or Payout= 30.00%), (D= Dividend Paid= 2,370 millions USD), (f= F/S= Fixed Portion= 58.00%), (t= T/B= Tax Rate= 30.00%), ($= Sales or Revenues= 51,294 millions USD), and (s= [S/S']-1= Sales Growth= 5.00%), are known, then it's (v= Variable Portion planned) is:

$$v = 1 - (\$f + I + D/\{d[1-t]\})/\{\$'[1+s]\}$$
$$= 1 - (51{,}294 * 0.5800 + 513 + 2{,}370$$
$$/\{03000[1 - 0.3000]\})$$
$$/\{48{,}851[1 + 0.0500]\}$$
$$= 19.00\%$$

Finance Construction-2, *Tim Asikin, Steve Asikin, Indra Senihardja*

The Rest will be Continued in
The Series of Book

Finance Cosntruction 3
Corporate IFRS-GAAP (B/S-I/S) Engineering
Technology No. 2,001-3,000 of 111,111 Laws

At
http: //www.Amazon.Com

Finance Construction-2, *Tim Asikin, Steve Asikin, Indra Senihardja*

REFERENCES:

A. Books:

1) **Amin, A.R. & Asikin, S.**(2011a- 110 pages), *CEO Time Matrix*: *Productivity Management Skill that Increase CEO Effectiveness by 12400%*, Seattle (US): Amazon Createspace. http://www.amazon.com/CEO-Time-Matrix-Productivity-Effectiveness/dp/1461066069/ref=sr_1_1?s=books&ie=UTF8&qid=1387091880&sr=1-1&keywords=ceo+time+matrix. ISBN-13: 978-1461066064, ISBN-10: 1461066069
2) **Amin, A.R. & Asikin, S.** (2014- 76 pages), *No Urgency CEO*: *A High-Tech Book in Modern Management*, Seattle (US): Amazon Createspace. http://www.amazon.com/No-Urgency-CEO-Hi-Tech-Management/dp/1489509356/ref=sr_1_1?s=books&ie=UTF8&qid=1396440688&sr=1-1&keywords=no+urgency. ISBN-13: 978-1489509352, ISBN-10: 1489509356
3} **Amin, A.R. & Asikin, S.** (2011b- 292 pages), *The Celestial Management*: *Spiritual Wisdom for All Human Beings*, Seattle (US): Amazon Createspace. http://www.amazon.com/Celestial-Management-Islamic-Wisdom-Beings/dp/1456450700/ref=sr_1_1?s=books&ie=UTF8&qid=1387092036&sr=1-1&keywords=celestial+management. ISBN-13: 978-1456450700, ISBN-10: 1456450700
4) **Asikin, S.** (2009- 548 pages), *Arigato, Obrigado... (Thanks)*: *Portuguese-Japan 1550-1647 Historic Novel*, Seattle (US): Amazon Createspace. http://www.amazon.com/Arigato-Obrigado-Thanks-Portuguese-Japan-1550-1647/dp/1453786368/ref=sr_1_1?s=books&ie=UTF8&qid=1396442162&sr=1-1&keywords=arigato. ISBN-13: 978-1453786369, ISBN-10: 1453786368
5) **Asikin, S.** (2013a- 824 pages), *Econometric Rex*: *888 Marketing Promo 6666 Six Sigma Parametric Theories (of 10 Data)*, Seattle (US): Amazon Createspace. http://www.amazon.co.uk/Econometric-Rex-Marketing-Parametric-Theories/dp/1482745259. ISBN-13: 978-1482745252, ISBN-10: 1482745259
6) **Asikin, S.** (2010b- 224 pages), *FINANCE Architecture*: *VISUAL art, for Corporate Finance's Beauty, Safety and Simplicity*, Seattle (US): Amazon Createspace. http://www.amazon.co.uk/FINANCE-Architecture-Corporate-Finances-Simplicity/dp/1453829547/ref=sr_1_14?s=books&ie=UTF8&qid=1385371842&sr=1-14&keywords=finance+architecture. ISBN-13: 978-1453829547, ISBN-10: 1453829547
7) **Asikin, S.** (2010c- 150 pages), *Geometric FINANCE*: *Useful concepts that work better than Economic Noble Laureates'* Seattle (US): Amazon Createspace. http://www.amazon.co.uk/Magic-Geometric-FINANCE-concepts-Laureates/dp/1453790284/ref=tmm_pap_title_0. ISBN-13: 978-1453790281, ISBN-10: 1453790284

8) **Asikin, S.** (2010a- 290 pages), *Investment ALGEBRA: Strategic Profitable Comprehensive Business Finance Engineering Technology,* Seattle (US): Amazon Createspace. http://www.amazon.co.uk/Investment-ALGEBRA-Profitable-Comprehensive-Engineering/dp/1453856811/ref=tmm_pap_title_0?ie=UTF8&qid=1385371309&sr=1-1, ISBN-13: 978-1453856819, ISBN-10: 1453856811
9) **Asikin, S.** (2011a- 352 pages), *State Economics: Comprehensive Macro-Micro Economics' Simple Fiscal-Monetary Export-Import Accouting, Integrated Supply-Demand Managerial ... Mathematical Engineering Visual* . Seattle (US): Amazon Createspace. http://www.amazon.co.uk/STATE-ECONOMICS-Steve-Asikin-ebook/dp/B00G89Q5SS, http:/www.amazon.co.uk State-Economics-Steve-Asikin-ebook/dp/B00BBPLQDI, ISBN-13: 978-1461066095, ISBN-10: 1461066093
10) **Asikin, S.** (2013b- 260 pages), *Technoeconomistat: Economic &Statistical Technology for Financial Planning,* Seattle (US): Amazon Createspace. http://www.amazon.co.uk/Technoeconomistat-Economic-Statistical-Technology-Financial/dp/1482304244, ISBN-13: 978-1482304244, ISBN-10: 1482304244
11) **Asikin, S.** (2011b- 540 pages), *Terima Kasih: Novel Internasional Sejarah Jepang-Portugis 1550-1647,* Seattle (US): Amazon Createspace. http://www.amazon.com/Terima-Kasih-Internasional-Jepang-Portugis-1550-1647/dp/146351106X/ref=sr_1_1?s=books&ie=UTF8&qid=1396442361&sr=1-1&keywords=terima, ISBN-13: 978-1463511067, ISBN-10: 146351106X
12) **Asikin, S.** (2010d- 294 pages), *TREASURY Planology: Strategic Profitable Comprehensive Business Finance Engineering Technology,* Seattle (US): Amazon Createspace. http://www.amazon.co.uk/TREASURY-PLANOLOGY-Steve-Asikin-ebook/dp/B00G88MTFW/ref=sr_1_1?s=books&ie=UTF8&qid=1385403690&sr=1-1&keywords=Treasury+Planology. ISBN-13: 978-1453891476, ISBN-10: 1453891471
13) **Asikin, S.** (2014a- 190 pages), *Calculus Accountancy: Leibnitz Newton Pacioli's Polynomial Quantitative Finance Risk Modelling Optimization,* Seattle (US): Amazon Createspace. http://www.amazon.com/Calculus-Accountancy-Polynomial-Quantitative-Optimization/dp/1495463028/ref=sr_1_1?s=books&ie=UTF8&qid=1396441338&sr=1-1&keywords=calculus+accountancy, ISBN-13: 978-1495463020, ISBN-10: 1495463028
14) **Asikin, S.**(2014b- 318 pages), *No Parametric:* No Parametric: Hyperbolic Octahedron Central 3-Dimensional Orthogonal Risk Distribution Topology, Seattle (US): Amazon Createspace. http://www.amazon.com/Parametric-Hyperbolic-Octahedron-3-Dimensional-Distribution/dp/1496089731/ref=sr_1_1?s=books&ie=UTF8&qid=1396441543&sr=1-1&keywords=No+Parametric, ISBN-13: 978-1496089731, ISBN-10: 1496089731

15) **Asikin, S.** (2014c- 138 pages), *Quick Help:* Religion vs Real-Legion Philosophy, Seattle (US): Amazon Createspace. http:/www.amazon.com/Quick-Help-Religion-Real-Legion-Philosophy/dp/1497466520/ref=sr_1_2?s=books&ie=UTF8&qid=1396441740&sr=1-2&keywords=quick+help, ISBN-13: 978-1497466524, ISBN-10: 1497466520

16) **Asikin, S.** (2014d- 824 pages), *Anne of Denmark:* King James I the Bible, Historic Plays Drama, Seattle (US): Amazon Createspace. http:/www.amazon.com/Anne-Denmark-James-Bible-Drama/dp/149299734X/ref=sr_1_1?s=books&ie=UTF8&qid=1396441857&sr=1-1&keywords=anne+asikin, ISBN-13: 978-1492997344, ISBN-10: 149299734X

17) **Asikin, S.** (2014e- 184 pages), *Constructive Spells:* Positive Spiritual Mentality Personal Development; Seattle (US): Amazon Createspace. http:/www.amazon.com/Constructive-Spells-Spiritual-Mentality-Development/dp/1497498899/ref=sr_1_1?s=books&ie=UTF8&qid=1397664512&sr=1-1&keywords=constructive+spells; ISBN-13: 978-1497498891, ISBN-10: 1497498899

18) **Asikin, S.** (2014f- 466 pages), *Elizabeth of Russia:* Another Virgin Queen Gloariana: A Romanov Plays Drama, Seattle (US): Amazon Createspace. http:/www.amazon.com/Elizabeth-RUSSIA-Another-Gloriana-Romanov/dp/1505542480/ref=sr_1_4?s=books&ie=UTF8&qid=1423888482&sr=1-4&keywords=elizabeth+russia; ISBN-13: 978-1505542486, ISBN-10: 1505542480

19) **Asikin, S.** (2014f- 708 pages), *Shinmen Munisai:Miyamoto Musashi's Father, A Sword Samurai Plays Drama,* Seattle (US): Amazon Createspace. http:/www.amazon.com/Shinmen-Munisai-Miyamoto-Musashis-Samurai/dp/1507857799/ref=sr_1_1?s=books&ie=UTF8&qid=1428094280&sr=1-1&keywords=munisai; ISBN-13: 978-1507857793, ISBN-10: 1507857799

20) **Asikin, S.** (2015a-767 pages), *Math Fin Law-1: Mathematical Financial Law for Public Listed Firms Rule 1-4441,* Seattle (US): Amazon Createspace. http://www.amazon.com/Math-Fin-Law-Mathematical-Financial-ebook/dp/B00WLNX3Y4/ref=asap_bc?ie=UTF8; ISBN-13: 978-1511792219, ISBN-10: 1511792213

21) **Asikin, S.** (2015b-813 pages), *Math Fin Law-2: Mathematical Financial Law for Public Listed Firms Rule 4442-8712,* Seattle (US): Amazon Createspace. http://www.amazon.com/Math-Fin-Law-Mathematical-No-4442-8712/dp/1512285080/ref=sr_1_1?s=books&ie=UTF8&qid=1432732199&sr=1-1&keywords=math+fin+law+2; ISBN-13: 978-1512285086, ISBN-10: 1512285080

22) **Asikin, S.** (2015c-813 pages), *Math Fin Law-3: Mathematical Financial Law for Public Listed Firms Rule 8713-12575,* Seattle (US): Amazon Createspace. http://www.amazon.com/Math-Fin-Law-Mathematical-8712-12575/dp/1514276763/ref=sr_1_1?s=books&ie=UTF8&qid=1433982149&sr=1-1&keywords=math+fin+law+3, ISBN-13: 978-1514276761. ISBN-10: 1514276763

Finance Construction-2, *Tim Asikin, Steve Asikin, Indra Senihardja*

23) **Asikin, S.** (2015d-813 pages), *Math Fin Law-4: Mathematical Financial Law for Public Listed Firms Rule 12576-16333,* Seattle (US): Amazon Createspace. http://www.amazon.com/Math-Fin-Law-Mathematical-12576-16333/dp/1514685132/ref=sr_1_1?s=books&ie=UTF8&qid=1435393628&sr=1-1&keywords=math+fin+law+4, ISBN-13: 978-1514685136, ISBN-10: 1514685132

24) **Asikin, S.** (2015d-816 pages), *Math Fin Law-5: Mathematical Financial Law for Public Listed Firms Rule 16334-19904,* Seattle (US): Amazon Createspace. http://www.amazon.com/Math-Fin-Law-Mathematical-No-16334-19904/dp/1515073009/ref=sr_1_1?s=books&ie=UTF8&qid=1437224693&sr=1-1&keywords=math+fin+law+5. ISBN-13: 978-1515073000, ISBN-10: 1515073009

25) **Asikin, S.** (2015d-728 pages), *Math Fin Law-6: Mathematical Financial Law for Public Listed Firms Rule 19905-23237,* Seattle (US): Amazon Createspace. https://www.amazon.com/Math-Fin-Law-Mathematical-No-19905-23238-ebook/dp/B0129ILVNK/ref=sr_1_2?ie=UTF8&qid=1478017624&sr=8-2&keywords=Math+Fin+Law+Asikin. ISBN-13: 978-1515158691, ISBN-10: 1515158691

26) **Asikin, S.** (2016a-728 pages), *Math Fin Law-7: Mathematical Financial Law for Public Listed Firms Rule 23238-27115,* Seattle (US): Amazon Createspace. https://www.amazon.com/Math-Fin-Law-Mathematical-No-23238-27115/dp/1508518521/ref=la_B00428V6JU_1_27?s=books, ISBN-13: 978-1508518525, ISBN-10: 1508518521

27) **Asikin, S.** (2016b-728 pages), *Math Fin Law-8: Mathematical Financial Law for Public Listed Firms Rule 27116-30330,* Seattle (US): Amazon Createspace. https://www.amazon.com/dp/1541091884, https://www.amazon.com/Math-Fin-Law-Mathematical-No-27116-30330/dp/1541091884/ref=la_B00428V6JU_1_27?s=books&ie=UTF8&qid=1479398538&sr=1-27&refinements=p_82%3AB00428V6JU, ISBN-13: 978-1508518525, ISBN-10: 1508518521

28) **Asikin, S.** (2017a-828 pages), *Math Fin Law-9: Mathematical Financial Law for Public Listed Firms Rule 30330-33373,* Seattle (US): Amazon Createspace. https://www.amazon.com/Math-Fin-Law-Mathematical-30331-33373-ebook/dp/B01MRA8HZQ/ref=la_B00428V6JU_1_4?s=books&ie=UTF8&qid=1485921040&sr=1-4, ISBN-13: 978-1541092433, ISBN-10: 1541092430

29) **Asikin, S.** (2017b-823 pages), *Math Fin Law-10: Mathematical Financial Law for Public Listed Firms Rule 33374-36242,* Seattle (US): Amazon Createspace. https://www.amazon.com/s/ref=nb_sb_noss?url=search-alias%3Dstripbooks&field-keywords=math+fin+law+10, https://www.amazon.com/s/ref=nb_sb_noss?url=search-alias%3Dstripbooks&field-keywords=math+fin+law+10, ISBN-13: 978-1541092761, ISBN-10: 1541092767

Finance Construction-2, *Tim Asikin, Steve Asikin, Indra Senihardja*

30) **Asikin, S**. (2017c-823 pages), *Math Finance Law-11: Mathematical Financial Law for Public Listed Firms Rule 36243-39158*, Seattle (US): Amazon Createspace. https://www.amazon.com/s/ref=nb_sb_noss?url=search-alias%3Dstripbooks&field-keywords=math+finance+law+11, https://www.amazon.com/dp/1541215265/ref=sr_1_1?s=books&ie=UTF8&qid=1488029978&sr=1-1&keywords=math+finance+law, ISBN-13: 978-1541215269, ISBN-10: 1541215265

31) **Asikin, S**. (2017d-823 pages), *Math Finance Law-12: Mathematical Financial Law for Public Listed Firms Rule 30150-42152*, Seattle (US): Amazon Createspace. https://www.amazon.com/s/ref=nb_sb_noss?url=search-alias%3Dstripbooks&field-keywords=math+finance+law+12, https://www.amazon.com/Math-Finance-Law-Mathematical-39159-42152/dp/1541215516/ref=sr_1_2?s=books&ie=UTF8&qid=1490990342&sr=1-2&keywords=math+finance+law+12, ISBN-13: 978-1541215511, ISBN-10: 1541215516

32) **Asikin, S**. (2017e-824 pages), *Math Finance Law-13: Mathematical Financial Law for Public Listed Firms Rule 42153-44938*, Seattle (US): Amazon Createspace. https://www.amazon.com/dp/1542675219/ref=sr_1_1?s=books&ie=UTF8&qid=1491146227&sr=1-1&keywords=math+finance+law+13, ISBN-13: 978-1542675215, ISBN-10: 1542675219

33) **Asikin, S**. (2017f-824 pages), *Math Finance Law-14: Mathematical Financial Law for Public Listed Firms Rule 44939-47745*, Seattle (US): Amazon Createspace. https://www.amazon.com/dp/1542675219/ref=sr_1_1?s=books&ie=UTF8&qid=1491146227&sr=1-1&keywords=math+finance+law+14, ISBN-13: 978-1544628967, ISBN-10: 154462896X

34) **Asikin, S**. (2017g-810 pages), *Math Finance Law-15: Mathematical Financial Law for Public Listed Firms Rule 47746-50465*, Seattle (US): Amazon Createspace. https://www.amazon.com/dp/1545118647/ref=sr_1_1?s=books&ie=UTF8&qid=1493626425&sr=1-1&keywords=math+finance+law+15, ISBN-13: 978-1545118641, ISBN-10: 1545118647

35) **Asikin, S**. (2017h-816 pages), *Math Finance Law 16: Mathematical Financial Law Public Listed Firm Rule No.50,466-53,320*, Seattle (US), Amazon CreateSpace, https://www.amazon.com/dp/1545320179/ref=sr_1_1?s=books&ie=UTF8&qid=1495692695&sr=1-1&keywords=math+finance+law+16, ISBN-13: 978-1545320174, ISBN-10: 1545320179

36) **Asikin, S**. (2017i-809 pages), *Math Finance Law 17: Mathematical Financial Law Public Listed Firm Rule No.53,231-56,124*, Seattle (US), Amazon CreateSpace, https://www.amazon.com/dp/1545320179/ref=sr_1_1?s=books&ie=UTF8&qid=1495692695&sr=1-1&keywords=math+finance+law+17, ISBN-13: 978-1545486474, ISBN-10: 1545486476

Finance Construction-2, *Tim Asikin, Steve Asikin, Indra Senihardja*

37) **Asikin, S.** (2017j-815 pages), <u>Math Finance Law 18</u>: *Mathematical Financial Law Public Listed Firm Rule No.56,125-58,942*, Seattle (US), Amazon CreateSpace,
https://www.amazon.com/dp/1545320179/ref=sr_1_1?s=books&ie=UTF8&qid=1495692695&sr=1-1&keywords=math+finance+law+18, ISBN-13: 978-1546912361, ISBN-10: 1546912363

38) **Asikin, S.** (2014-47 pages), <u>Doing Business in Comprehensive Interactive Interdisciplinary Micro-Macro Eco nomic International Accounting Architecture of Serial Well Known General Great Theories</u>, Proceedings in Finance and Risk Perspective 13, ACRN Oxford Publishing House, UK, p.197-243,
https://books.google.com/books?id=4e4KAwAAQBAJ&pg=PA219&lpg=PA219&dq=steve+asikin+oxford&source=bl&ots=qVq9KyAmem&sig=ABdBIiwuzUwRigIsuLimuJ5ZdW4&hl=en&sa=X&sqi=2&ved=0ahUKEwjZkMrF7KjTAhUDSiYKHREOB-qQ6AEILjAC#v=onepage&q=steve%20asikin%20oxford&f=false. ISBN 978-3-9503518-1-1.

39) **Asikin, T.& Asikin, S.** (2013- 828 pages), <u>448 Poems</u>: *KARAOKE Song Book: 24 Hours Story of 30 Languages*, Seattle (US): Amazon Createspace. http://www.amazon.com/448-Poems-KARAOKE-Song-Book/dp/1480034363/ref=sr_1_1?s=books&ie=UTF8&qid=1396442034&sr=1-1&keywords=448+poems, ISBN-13: 978-1480034365, ISBN-10: 1480034363

40) **Asikin, T, Asikin, S, Senihardja, I.** (2017- 508 pages), <u>Finance Construction 1</u>: *Corporate IFRS-GAAP (B/S-I/S) Engineering Technologies No. 1-1,000 of 111,111 Laws*, Seattle (US): Amazon Createspace. https://www.amazon.com/s/ref=nb_sb_noss?url=search-alias%3Dstripbooks&field-keywords=finance+construction+asikin. ISBN-13: 978-1544059525, ISBN-10: 1544059523

41) **Brigham, EF & Gapenski, LC,** . (1988- Textbook), *Financial Management: Theory and Practice (Study Guide)*, NY (US): Dryden Press.
http://www.amazon.com/Financial-Management-Theory-Practice-Study/dp/0030125391/ref=sr_1_1?s=books&ie=UTF8&qid=1429380835&sr=1-1 ;

42) **Chandonet, RE,** . (2011-32 pages), *Balancesheet Management*, Trends & Strategies, NY (US): Sandler O'Neil.
https://www.ncbankers.org/uploads/File/Meetings/Presentations/2011CONV-Chandonnet.pdf;

43) **Cornuejols G & Tutuncu, R,** . (2006-358 pages), <u>Optimization Methods in Finance</u>, Cambridge (UK): Cambridge University Press.
http://www.math.ku.dk/~rolf/CT_FinOpt.pdf;

44) **Crook, M.** (2015-6 pages), <u>Optimization Methods in Finance</u>, UBS FS (Swiss): CIO WM Research.
https://www.ubs.com/content/dam/WealthManagementAmericas/documents/investment-strategy-insights-2015-01-09-balance-sheet-optimization.pdf;

45) **Fee, Greg.** (2012-127 pages), *Catalan's Constant: (Ramanujan's Formula}: Catalan Constant to 300,000 digits,* Seattle (US): Amazon Kindle.
http://www.Amazon.com/gp/aw/d/B008ZZ5IEW/ref=mp_sa_1_4?ie=UTF8%Qqid=148 4367818&sr=8-4&pi=AC_SX236_SY340_QL65&keywords=ramanujan&dpPl=1&dpID=5lFqlwDZpRL&ref=plSrch

46) **Gardenfors, Peter.** (2014) *The Geometry of Meaning: Semantics Based on Conceptual Spaces.* Cambridge, Mass.: MIT Press. ISBN 0-262-02678-3.

47) **Gardenfors, Peter.** (2000), *Conceptual Spaces: The Geometry of Thought.* Cambridge, Mass.: MIT Press. ISBN 0-262-07199-1.

48) **Kanigel, Robert.** (2013-465 pages), *The Man Who Knew Infinity:A Life of the Genius Ramanujan,* Washington (US): Washington Square Press&Amazon Kindle.
http://www.Amazon.com/gp/aw/d/B008BW4VEGM/ref=pd_aw_sim35l_1?ie=UTF8&psc=1&refRID=VYN345GVDBJ6ATDQ35W2&dpID=9lZqpc4ETdL

49) **Limlingan, Victor S.** (1993-1 page), *The Limlingan Financial Model,* Manila: Asian Manager. http://limlingan.com/images/article.gif

50) **Newton, Isaac.** (2015-631 pages), *Principia:The Mathematical Principles of Natural Philosophy (Active Content),* London (UK): RSM&Amazon Kindle.
http://www.Amazon.com/gp/aw/sitb/B011SFJSJ0?ref=sib_dp__aw_kb_udp

51) **Nikisa, Evets.** (2014g- 828 pages), *Wonderful Live Spells: 144-Steps Recreative Nice Children Story, Succesful Development:* Seattle (US): Amazon Createspace. http://www.amazon.com/WONDERFUL-Live-Spells-Recreative-Development/dp/1499248423/ref=sr_1_1?s=books&ie=UTF8&qid=1418698379&sr=1-1&keywords=wonderful+asikin, ISBN-13: 978-1499248425, ISBN-10: 1499248423

52) **Nielson, SS.** (2007- 430 pages), *Practical Financial Optimization*: *Decision Making for Financial Engineers:* NY (US): Wiley.
https://books.google.com/books/about/Practical_Financial_Optimization.html?id=_XkwLAAACAAJ.

53) **Niswonger CR, Fess PE & Warren CE.** (1993-Textbook), *Accounting Principles:* Seattle (US): South Western Publishing Co.
http://www.amazon.com/Accounting-Principles-Chapters-C-Rollin-Niswonger/dp/0538818603

54) **Omerod, P.** (1993-Textbook), (1994), *The Death of Ecoomics,* St Martin's Press, London, . http://www.amazon.com/Accounting-Principles-Chapters-C-Rollin-Niswonger/dp/0538818603, ISBN-13: 978-0312134648, ISBN-10: 0312134649

55) **Paccioli, L.** (1989-Textbook), *Summa de Arithmetica Geometria Proportioni et Proportionalita*, Venice 1494. http://www.amazon.com/Arithmetica-Geometria-Proportioni-Proportionalita-Venice/dp/B00T4HXTIW/ref=sr_1_fkmr0_2?s=books&ie=UTF8&qid=1429597966&sr=1-2-fkmr0&keywords=summa+mhematica+arithmetica
56) **Persson, Torsten & Tabellini, Guido** (2000), *Political Economics – Explaining Economic Policy*, MIT Press. https://mitpress.mit.edu/authors/torsten-persson
57) **Persson, Torsten & Tabellini, Guido** (2003), *The Economic Effects of Constitutions*, MIT Press. https://mitpress.mit.edu/authors/torsten-persson
58) **Persson, Torsten & Tabellini, Guido** (2000), *Monetary and Fiscal Policy – 1 Credibility*, MIT Press. https://mitpress.mit.edu/authors/torsten-persson
59) **Persson, Torsten & Tabellini, Guido** (2003), *Monetary and Fiscal Policy – 2 Politics*, MIT Press. https://mitpress.mit.edu/authors/torsten-persson
60) **Puts, J** (2012-102 pages), *Balancesheet Optimization Under Basel III*, Amsterdam (Netherlads): Vrije Universiteit. http://www.math.vu.nl/~sbhulai/papers/thesis-puts.pdf
61) **Timmermans, K** (2012-30 pages), *Balancesheet Optimization Under Basel III*, Amsterdam (Netherlads): ING Investor Day. https://www.ing.com/web/file?uuid=09c803d9-79a1-47ec-bfe5...owner...
62) **Weston JF, Copeland TE & Shastri, K.** (2004-Textbook), *Financial Theory and Corporate Policy*: NY (US): Addison Wesley. http://www.amazon.com/Financial-Corporate-Copeland-Kuldeep-Paperback/dp/B008YSUE4M/ref=sr_1_1?s=books&ie=UTF8&qid=1429381700&sr=1-1&keywords=weston+copeland
63) **Zenios SA.** (2002-350 pages), *Financial Optimization:* Cambridge (UK): Cambridge University Press. https://books.google.com/books/about/Financial_Optimization.html?id=-FJdqjZOqSAC,.

Finance Construction-2, *Tim Asikin, Steve Asikin, Indra Senihardja*

B. Intellectual Properties:

1) **Indonesian Banker Certification 1993**, *Sertifikasi Bankir Indonesia 1993*;
2) **MATHFINPLAN** (Mathematical Financial Planning);
   Perencanaan Keuangan Matematis
3) **BANK MATHPLAN** (Bank's Mathematical Planning);
   Perencanaan Matematis Bank
4) **TECHNO-ECONOMISTAT** (Technology for Economical Statistics);
   Teknologi Statistik untuk Ekonomi
5) **FINANCIAL GEOMETRY** (IT Program); *Ilmu Geometri Keuangan*
6) **BAK DOBEL** (Car Gasoline Tank: Left and Right Side Fuel Fill-in).
   Tangki Mobil yang bisa diisi dari Kiri atau Kanan
7) **Mister GO-CHENG** (Cheap Food Franchising). *Waralaba Pangan Murah*

Finance Construction-2, *Tim Asikin, Steve Asikin, Indra Senihardja*

C. SCIENTIFIC PAPERS:

1) **Eichhornia Crassipes**, as Anti Polutant and Fertile Sediment, *Pemanfaatan Eceng Gondok untuk Menetralkan Polusi dan Pengurukan Subur pada Genangan Air*, LKIR, Lomba Karya Ilmiah Remaja, LIPI (1979)
2) **Archimedes Catesian Buoyancy Law**, *Penyimpangan Hukum Archimedes pada Model Pelampung Cartesian*, Lomba Karya Ilmiah Remaja, Depdikbud-RI (1980)
3) **SURAMADU**, Surabaya-Madura Inter Island Bridge, Design Supervisor, *Sumarsono Sumali Engineering Contractor*, Tambak-V, Tandes, Surabaya (1982),
4) **GAZOLINE Taxi-Pool Depot**, Jemur Handayani, Surabaya, *Sumarsono Sumali Engineering Contractor*, Tambak-V, Tandes, Surabaya (1982),
5) **SOEKARNO-HATTA**, Jakarta International Airport, Garuda Maintenance Facilities, *Junior Architect*, As Bulit Drawings, Coinstruction Management, Aerospatiale-Encona Engineering (1983-1985),
6) **ADISUCIPTO**, Yogya International Airport, *Junior Architect*, Preliminary Technical Drawings, Budiono Surasno-Encona Engineering (1986),
7) **BLOK-M PLAZA**, "New Garden Hall", Kebayoran Baru Supermall, *Preliminary Design Architect*, Danisworo-James Ferry- Jasa Ferry-Encona (1986),
8) **NORTH JAKARTA PUBLIC LIBRARY**, Semper, Tanjung Priok, *Chief Architect*, Agung Darma Sarana-Asep Hermawan (1987),
9) **The Banker's Trainer**: A Manual for The Training Head in Banking. *Profesi Bankir Pelatih, Tuntunan bagi Kepala Diklat Perbankan*, (1200 Pages, 1993),
10) **Indonesian Bankers Certification**, a Strategic Move for the National Banking Improvement. *Sertifikasi Bankir Indonesia, Langkah Strategis Pembenahan Perbankan Nasional*, (KOMPAS, 20 Juli 1993).
11) **Indonesian Rupiah's Exchange Rate**, on a Quiet but Steady Movement. *Nilai Tukar Rupiah, Diam-diam Meningkat Pesat*, (MANAJEMEN, Jan-Feb 1994).
12) **The Bank's Credit**, for the Best Practices. *Kredit Praktis* (200 pages, 1994),
13) **Down to Earth Credit Meeting Credo**, *Kiat Rapat Pemberian Kredit yang Realistik*. (MANAJEMEN, Sep-Oct 1994).
14) **SCHIPHOL**, Amsterdam International Airport, 1st Anthony Fokker Weg, Sloten, *Airport Engineer*, Samsung- Fokker-Beurlin, BV (1996),
15) **Ultrapeneurship: The Secret of Starting Business Without Capital**. *Ultrapreneur: Kiat Usaha Tanpa Modal*, (MANAJEMEN, Nov-Dec 1996),
16) **Recipe from the Corporate Raider's Kitchen**, *Menguak Dapur Corporate Raiders*, (MANAJEMEN, Jan-Feb 1997).
17) **The 10-Tips and Tricks of Engineered Conglomeration**. *Sepuluh Rekayasa Keuangan Konglomerat*, (MANAJEMEN, Mar-Apr 1997),
18) **The Management of Succession in Five Parts of the World**. *Manajemen Suksesi di Lima bagian Dunia*, (MANAJEMEN, Mei-Jun 1997),
19) **Auditing**: The Global Language for the 2003's Human Resources. *Audit : Bahasa Global untuk SDM 2003*, (MANAJEMEN, Jul-Aug 1997),
20) **The Elected Parliament 1997-2002**, and the Hope of Professionals. *DPR 1997–2002 dan Harapan Profesional*, (MANAJEMEN, Sep-Okt 1997),

Finance Construction-2, *Tim Asikin, Steve Asikin, Indra Senihardja*

21) **Disappearance of World's Soft Currencies**, and Proposition for Global Economic Thinking. *Lenyapnya Mata Uang Lunak Dunia dan Usulan Pemikiran Ekonomi Global,* (MANAJEMEN, Nov-Dec 1997),
22) **The Human Resources and Its Accountability**, in Critical Situation. *Akuntabilitas Sumberdaya Manusia di Masa Krisis,* (MANAJEMEN, Jan-Feb 1998), '
23) **The Investment Economic Model**, and the Failure of Growth Economic Theories. *Model Ekonomi Investasi dan Kegagalan Teori Ekonomi Pertumbuhan,* (MANAJEMEN, Nov 1998),
24) **The Exact Equilibrium of Capital Budgeting**, *Persamaan-persamaan akurat dalam Penganggaran Modal Perusahaan.* (MANAGEMENT EXPOSE, November 1998),
25) **The Corruptive Economic Model**, and Its Solution for Long Term Debt's Problem. *Model Ekonomi Korupsi dan Penyelesaian Utang Jangka Panjang,* (MANAJEMEN, Dec 1998),
26) **The Self Equity Budgeting**, and Operations Research in Investments. *Penganggaran Modal Sendiri dan Riset Operasi dalam Investasi,* (MANAJEMEN, Jan 1999),
27) **Actual Impact of Inflation**, and the Need of Geometric Approach in Statistic. *Pengaruh Riil Inflasi dan Perlunya Statistika Geometri,* (MANAJEMEN, Feb 1999),
28) **Multiplication of Money**, to Overcome the Crisis. *Melipat Gandakan Uang untuk Mengusir Krisis,* (MANAJEMEN, Apr 1999),
29) **The Interactive Mathematical Model**, for Multi Constraints Financial Budgeting, *Model Matematika Ekonomi Interaktif bagi Penganggaran Keuangan Multi Kendala.* (SCRIPTA ECONOMICA, Vol. 2, April 1999)
30) **Application of Ethics**, for the Management Accounting Practices. *Aplikasi Etika dalam Praktek Akuntansi Manajemen,* (MANAJEMEN, May 1999),
31) **The Curriculum of Magistrate of Management in Finance**: Its Needs and Practices. *Kurikulum Magister Manajemen Keuangan 1988-1998: Antara Kebutuhan dan Praktek,* (MANAJEMEN, Jun 1999),
32) **The Attractiveness**, Personal Power and Management. *Pesona Pribadi: "Personal Power" dan Manajemen,* (MANAJEMEN, Jul 1999),
33) **The Market Research**, for Magistrate Degree in Management. *Soal Jawab Riset Pasar untuk Pascasarjana Manajemen* (126 Pages, 1999),
34) **The Thesis Manual**, Question and Answer for Magistrate Degree in Management. *Soal Jawab Tugas Akhir untuk Pascasarjana Manajemen* (126 Pages, 1999),
35) **The Risk Management**, Question and Answers for Magistrate Degree in Management. *Soal Jawab Manajemen Resiko untuk Pascasarjana Manajemen* (126 Pages, 1999),
36) **The Credit Manual**, Question and Answers for Magistrate Degree in Management. *Soal Jawab Manajemen Kredit untuk Pascasarjana Manajemen* (126 Pages, 1999),

37) **Enhancing the Morality Structure**. *Solidkan Struktur Moralitas,* (MANAJEMEN, Oct 1999)
38) **Performance of PT. Astra Otoparts**, Plc, in its Press Releases and Public Financial Reports. *Kinerja PT. Astra Otoparts dalam Siaran Pers dan Laporan Keuangan Publik,* (MANAJEMEN, Oct 1999)
39) **The Management of Conflicts**, for the New Indonesian's "Rainbow Colored" Cabinet. *Manajemen Konflik dan Kabinet Pelangi,* (MANAJEMEN, Dec 1999),
40) **Technoeconometric**: An Engineering Elaboration to Matrix Multi Variate Econometrics, *Tekno Ekonometrik sebagai sumbangan teknologi untuk Ekonometrika Matrik Multi Variat.* (MANAGEMENT EXPOSE, November 1999),
41) **Technoeconometric** : A Mathematical Shortcut for Non Linear Bivariate Econometrics, *Tekno Ekonometrik sebagai Usulan Jalan Pintas Matematik untuk Ekonometrika non Linear Bivariat.* (SCRIPTA ECONOMICA, Vol. 2, No. 3, Desember 1999),
42) **Developing "New Image"**, for the New Indonesia. *Membangun Citra Indonesia Baru,* (MANAJEMEN, Jan 2000),
43) **Segregations of Duties**, the Needs of Both Indonesian Governments and Private Sectors. *Pemerintah dan Swasta Perlu Berbagi Tugas,* MANAJEMEN, Mar 2000),
44) **The 812.500 Financial Models**, for the Autonomous Country Governments. *Tentang 812.500 Model Keuangan Otonomi Daerah,* (MANAJEMEN, Mar 2000),
45) **The Forecast of 10 Most Attractive Share Values**, in the Jakarta Stock Exchange, April 2000. *Prakiraan Harga Sepuluh Saham Unggulan di Bursa Efek Jakarta, April 2000,* (MANAJEMEN, Mar 2000),
46) **Let Us Merchandise the Intelectual Properties!**. *Jual "Intelectual Property" Saja!* (MANAJEMEN, Apr 2000),
47) **The Forecast of ten Most Attractive Share Values**, in the Jakarta Stock Exchange May 2000. *Prakiraan Harga Sepuluh Saham Unggulan di Bursa Efek Jakarta Mei 2000,* (MANAJEMEN, Apr 2000),
48) **HEATHROW**, London International Airport, Second Runway, *Q-Basic Algorithmic Engineer,* Beasley-Krishnamoorthy, Imperial College, Cambridge, Post Doctoral Project (2000),
49) **Why the Banking Sector Should be Healthy**, Before Others?. *Mengapa Perbankan Harus Sehat?* (MANAJEMEN, Mei 2000),
50) **Recycling of Wastes as a Negative Costing**. *Memanfaatkan Sampah atau Negative Costing,* (MANAJEMEN, Agt 2000),
51) **The Guidance for Indonesia Corporate Restructuring**. *Kiat Restrukturisasi Perusahaan Indonesia,* (MANAJEMEN, Agt 2000),
52) **Redefinition of the Ethical Comprehension**. *Pemaknaan Kembali atas Pemahaman Etika,* (MANAJEMEN, Sep 2000),
53) **The Critical View the 1999's Indonesians Central Bank Financial Statement**. *Analisa Laporan Keuangan Bank Indonesia 1999,* (MANAJEMEN, Oct 2000),

54) **Becoming an "Achiever"**, not Just Common "Worker". *Bukan Sekadar "Kerja", Tapi Menghasilkan Karya!* (MANAJEMEN, Feb 2001),
55) **Ten Steps to Real Wealth!** *Sepuluh Langkah Menjadi Kaya,* (MANAJEMEN, Jun 2001),
56) **Creative Financial Breakthrough**, *Seminar Sehari tentang Terobosan Keuangan yang Kreatif,* (Gramedia-PPM dan MANAJEMEN, Jun 2001),
57) **The Marketing Practices of Financial Softwares**, for Publics Listed Corporation. *Praktek Pemasaran Program Perangkat Lunak untuk Keuangan Perusahaan Publik* (Univ Satyagama, Jakarta Juni 24, 2001).
58) **The Mathematical Financial Planning**, *Perancangan Keuangan Secara Eksak dengan Matematik.* (MANAJEMEN, Jul 2001),
59) **The Real Research and Development, not Rework and Destruction**. *Litbang yang Bukan "Sulit Berkembang",* (MANAJEMEN, Jul 2001),
60) **The Treatise on the Advantages and Disadvantages**, of Revolutionary Learning Methods. *Baik Buruknya: Revolusi Cara Belajar,* (MANAJEMEN, Jul 2001),
61) **Better Make an Indonesian "Corruption Prevention" Body**, not Just a "Corruption Watch". *Buat Indonesian Corruption Prevention (ICP), Bukan Indonesian Corruption Watch, ICW!* (MANAJEMEN, Okt 2001),
62) **Preparation of the Worlds Business Leaders**, for our Century and the Century After. *Mempersiapkan Manusia-manusia yang akan menjadi Khafilah Bisnis Dunia Abad ini dan Abad Mendatang.* (Univ Satyagama – Gregorio Araneta Univ, Jakarta, Oktober 27, 2001).
63) **The Good Corporate Governance**, is not just a Morale Condition. *Kepemerintahan Perusahaan yang Baik, Bukan Sekedar Moralitas,* (MANAJEMEN, Nov 2001),
64) **Corporate Healthiness and Cleanliness as a Challenge**. *Bersih, Sehat Perusahaan, sebuah Tantangan* (MANAJEMEN, Jan 2002),
65) **Developing the Cultural Base for Future Competitions**. *Budaya untuk Bersaing di Masa Depan* (MANAJEMEN, Jan 2002),
66) **"Catapult", "Bullet" or "Ballistic Missiles" Approach**, for Corporate Healthiness and Cleanliness. *Pola "Ketapel", "Peluru Biasa" atau "Rudal" untuk Keuangan Perusahaan yang Bersih Sehat.* (MANAJEMEN, Feb 2002),
67) **The Mathematical Invesment Solution Model**, for Indonesian Foreign Debt and Its Corruption, *Solusi Matematika Investasi bagi Penyelesaian Masalah Utang Luar Negeri dan Korupsi di Indonesia.* (MANAGEMENT EXPOSE, Mar 2002),
68) **The Attitude, Skills and Knowledge (ASK)**, in Educational System Approach. *Sikap, Keterampilan dan Pengetahuan Menurut Pendekatan Sistemik Pendidikan,* (MANAJEMEN, Apr 2002),

Finance Construction-2, *Tim Asikin, Steve Asikin, Indra Senihardja*

69) **Strategic Treatise on Regional Autonomy**, in Futurization, Digitalization, Securitization and Globalization Era in Accordance to the Awakening of the Mercantile Nations. *Kajian Strategik Otonomi Daerah : Era Futurisasi, Digitalisasi, Strukturisasi, dan Globalisasi dalam Kebangkitan Negara-Negara Dagang.* (Training for Provincial Government Regional Management, Indonesian Association of Provincial Governments, *Pelatihan Pengembangan Manajemen Pemerintah Daerah bagi Perangkat Pemerintah Propinsi, Angkatan II*, (APPSI), June 27, 2002.

70) **The Suggestion for the Doctoral Course**, for Islamic Financial Management System. *Usulan-usulan Perbaikan dalam Kurikulum Doktor Manajemen Keuangan Islam.* (December, 17-19, 2002, Canadian International Development Agency - CIDA, ADSGM, Universiti Utara Malaysia, Kedah, Da'arul Sintok).

71) **The Defensive Strategy for the Market Leader**, a Nokia Cellular Telephone's Case. *Strategi Bertahan Sang Pemimpin Pasar dengan Menggunakan Multi Merk, Kasus Telepon Seluler Nokia.* (March, 12, 2004, Univ Tarumanegara, Jakarta).

Finance Construction-2, *Tim Asikin, Steve Asikin, Indra Senihardja*

Finance Construction-2, *Tim Asikin, Steve Asikin, Indra Senihardja*

Finance Construction-2, *Tim Asikin, Steve Asikin, Indra Senihardja*

Finance Construction-2, *Tim Asikin, Steve Asikin, Indra Senihardja*

Finance Construction-2, *Tim Asikin, Steve Asikin, Indra Senihardja*

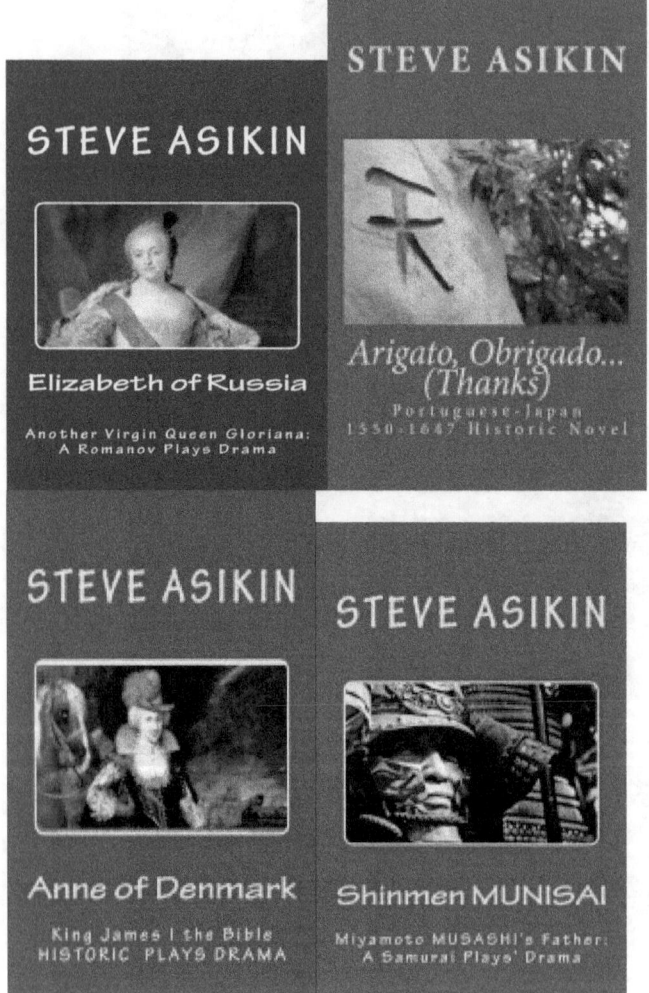

Finance Construction-2, *Tim Asikin, Steve Asikin, Indra Senihardja*

Finance Construction-2, *Tim Asikin, Steve Asikin, Indra Senihardja*

Finance Construction-2, *Tim Asikin, Steve Asikin, Indra Senihardja*

Finance Construction-2, *Tim Asikin, Steve Asikin, Indra Senihardja*

Finance Construction-2, *Tim Asikin, Steve Asikin, Indra Senihardja*

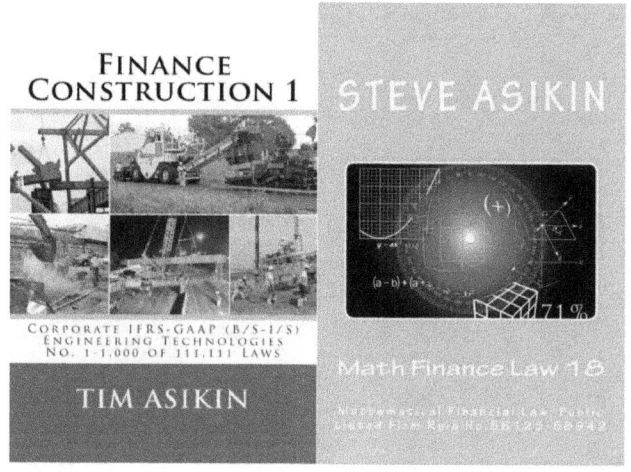

www.ingramcontent.com/pod-product-compliance
Lightning Source LLC
Chambersburg PA
CBHW071408180526
45170CB00001B/17